Structural Health Monitoring Using IoT

Structural Health Monitoring Using IoT

Dr. K. Aruna Bhaskar

ARCLER
PRESS

www.arclerpress.com

Structural Health Monitoring Using IoT

Dr. K. Aruna Bhaskar

Arcler Press
4164 Lakeshore Road
Burlington, ON L7L 1A4
Canada
www.arclerpress.com
Email: orders@arclereducation.com

ISBN: 978-1-77956-983-7 (Hardcover)

ABOUT THE AUTHOR

Dr. K. Aruna Bhaskar, currently working as an Associate Professor at KLEF, KL Deemed to be University, Vaddeswaram, Andhra Pradesh, in the Department of Computer Science & Engineering. He received Ph.D degree from Krishna University, Machilipatnam, and M. Tech in Computer Science from JNTUK, Kakinada.His academic background, which includes over fifteen years of teaching experience, demonstrates a commitment to both established and developing areas of technology.Dr. K. Aruna Bhaskar has advanced understanding in fields like biomedical image processing, computer networks, blockchain technology, Agile Technologies, artificial intelligence (AI), and machine learning (ML) while his long career. His multidisciplinary experience connects intelligence-driven approaches, secure network systems, decentralized computing (via blockchain), and imaging techniques with a healthcare focus.

TABLE OF CONTENTS

LIST OF FIGURES

PREFACE

In the modern era, the long life and integrity of the infrastructure have their counterparts in social safety, economic stability, and well-being. Bridges, roads, railways, dams, and tall buildings are the pillars of human progress but are vulnerable to degradation, overload, environmental loading, and unexpected events. Detection of vulnerabilities before they appear as catastrophic failures turned into an engineering issue and a social obligation. Conventional forms of inspection, as effective as they are, prove to be costly, time-consuming, and limited in providing continuous information. The entry of digital technologies has opened up a new era of possibilities for monitoring, measuring, and evaluating the structural health in real time. With the aid of advanced sensing and data communications technologies, designers can now design smart systems that continuously report on the condition of important assets and reduce risks and service life.

Among the many innovations revolutionizing this business, the Internet of Things has been a paradigm shift. IoT marries intelligent sensors, wireless connectivity, cloud computing, and analytics to enable an integrated environment where physical infrastructures are able to share details of their health themselves. Instead of relying on a scheduled manual check, infrastructure can now "speak" through embedded devices, generating copious streams of data around stress, vibration, cracks, and environmental conditions. This not only enables predictive and preventive maintenance but also guide policy and management choices. The intersection of IoT technology and structural engineering is building safer, smarter, and more secure infrastructure all over the world and reshaping the boundaries of how we are monitoring and maintaining what societies build.

Structural Health Monitoring Using IoT has been written to provide a comprehensive and practical textbook on this new discipline. It integrates the science of civil and structural engineering with the technology of IoT to present the subject matter in a manner that is easily comprehensible to students, researchers, and practitioners alike. The book is structured to move from basic concepts to advanced applications, balancing theoretical models with practical case studies. Combining engineering principles, sensor technologies, wireless communications, data analysis, and future directions, the book is both a guide to understanding the state of the art and a guide to exploring emerging trends. Chapters building on one another create an integrated learning path connecting principles to practice.

Chapter 1 is an introduction to an overview of structural health monitoring based on IoT. It presents the introduction, objectives, and importance of bringing smartness to infrastructure. The chapter explains how monitoring strategies result in enhanced safety, reduced maintenance costs, and extended structural life. The chapter also provides background information on how IoT is different from traditional monitoring techniques.

Chapter 2 revolves around the fundamentals of IoT in engineering applications. The reader gets familiar with the fundamental composition of IoT, including sensing devices, gateways, communication protocols, and cloud integration. The chapter highlights how these elements as a whole form an interdependent framework that can be tailored to structural monitoring needs.

Chapter 3 explains the diversity of sensors and data acquisition methods in structural health monitoring. Different sensing technologies such as strain gauges, accelerometers, piezoelectric sensors, fiber-optic sensors, and temperature sensors are explained elaborately. The chapter emphasizes precision, toughness, and issues related to calibration that must be addressed in order to achieve reliable data acquisition.

Chapter 4 discusses wireless sensor networks and their applications to structural health monitoring. It gives an overview of the design constraints, communication structures, energy conservation techniques, and scalability requirements. Real-life examples demonstrate how wireless networks reduce the complexity of wiring and enable the monitoring of large and intricate structures in real time.

Chapter 5 covers IoT platforms and integration techniques in the cloud. These include middleware, edge computing, data storage, and real-time processing frameworks. The chapter mentions how cloud services provide scalability, interoperability, and accessibility, while edge devices enable local filtering of data as well as reduction of latency.

Chapter 6 discusses data processing and analytics for structural health monitoring. Topics involve preprocessing and noise filtering up to machine learning models, anomaly detection, and predictive maintenance strategies. The chapter demonstrates how raw sensor data can be transformed into meaningful information using computational intelligence.

Chapter 7 reports case studies of the application of IoT-based structural health monitoring in real-world environments. Their applications to bridges, tunnels, heritage buildings, and industrial plants are presented. The practical challenges and benefits of applying IoT systems in real-world environments are illustrated by these case studies, providing lessons to researchers and practitioners.

Chapter 8 addresses structural health monitoring system security and reliability concerns. Data integrity, privacy, cyber-attacks, and communication network resilience are a few of the topics addressed. The chapter emphasizes the importance of good cybersecurity practices and fault-tolerant designs for ensuring trust in the monitoring systems.

Chapter 9 concentrates on the design and implementation of an IoT-based structural health monitoring system. It integrates ideas from earlier chapters to demonstrate the end-to-end development process. Realistic problems like system architecture, hardware selection, integration challenges, and testing approaches are explained with real-world applicability in view.

Chapter 10 showcases future areas of research and trends. It highlights future technologies such as analytics through AI, digital twins, blockchain for safe data storage, auto-powered sensors, and green IoT architectures. Chapter 10 concludes with the identification of research gaps and opportunities to further enhance this field.

This book has been written for a widespread variety of readers. Undergraduate and graduate students of civil engineering, structural engineering, and courses dealing with IoT will find it useful as a textbook. Researchers will be able to use it as a reference guide to current practices and directions for future research. Practicing engineers, infrastructure managers, and technology developers can be directed by its applied orientation and case studies, along with suggestions for implementing IoT solutions in real projects. Policy-makers and decision-makers with interests in the safety, sustainability, and cost-effectiveness of infrastructure will also benefit from the insights. Equipped with technical detail and practical relevance simultaneously, this book intends to be a bridge between governance, industry, and academia for the facilitation of safer and smarter societies.

CHAPTER 1

INTRODUCTION TO STRUCTURAL HEALTH MONITORING (SHM)

Contents

INTRODUCTION

Structural Health Monitoring (SHM) is a systematic approach to monitor and regulate the performance and integrity of structures throughout the course of their lifespan. This involves a combination of sensing technologies, data analysis procedures, and analysis techniques fused to assess structural integrity in real time. Continuously monitoring parameters such as stress, strain, and displacement, SHM has the potential to provide critical information about the condition of infrastructure. These observations enable engineers to detect potential weaknesses, minimize vulnerabilities, and thus prolong the life of critical structures from bridges to buildings. By doing so, SHM not only serves as protection but also as a proactive method in maintaining public safety and infrastructure dependability. SHM's definition inherently marries technological development with engineering theory to create a premise in real-time monitoring. Employing advanced sensor networks, including strain sensors and accelerometers, SHM systems can capture data that reflects the working state of a structure.

Figure 1.1. *The structural health monitoring (SHM).*

Source: https://media.springernature.com/lw685/springer-static/image/art%3A10.1007%2Fs11831-021-09665-9/MediaObjects/11831_2021_9665_Fig3_HTML.png.

This is analyzed and processed to identify degradation or anomalies that might weaken structural integrity. SHM also employs advanced algorithms, perhaps machine learning-based, to further enhance the predictive maintenance aspect of monitoring systems. This multilayered approach puts emphasis on SHM as a superior technology in modern civil engineering for ensuring that decisions can be made regarding maintenance and repair timetables (Wahab et al., 2017). Aside from such maintenance, the position of SHM is even beyond maintenance; it determines design processes and regulatory specifications. With engineers embracing the utilization of SHM, information obtained is worthwhile for use in subsequent improvement of design and standards to prevent failures. Intelligence from SHM systems assists in refining structural designs to maximize their performance and resist natural and man-made stresses. Ultimately, understanding and utilizing SHM not only enhances the reliability and safety of current structures but also guides the development of more sustainable infrastructure in response to evolving environmental conditions. This evolution illustrates the dynamic interaction between structure analysis, technology, and safety considerations in engineering practice (Hoell et al., 2014) (Figure 1.1).

Explanation of Structural Health Monitoring (SHM)

SHM is an important component in the preservation of the integrity and safety of various infrastructures, ranging from bridges to historic monuments. Essentially, SHM involves the use of a series of sensors and technologies to constantly monitor the physical state of such structures. By analyzing parameters such as strain, displacement, and acceleration, SHM can detect deviations from normal behavior, which are generally symptoms of expected decay or failure. Such a preventive measure is vital in the mitigation of structural failure risks, hence the safeguarding of human life and property. Moreover, the integration of sophisticated data analytics enhances interpretation capacity in large volumes of sensor data, allowing for efficient decision-making for maintenance and repair cycles. Application of SHM techniques is also extensive, involving monitoring present and historic structures. For instance, recent studies focus on the application of novel techniques, such as the stretching method, targeting evaluation of structural health in historic structures. In one example, the authors of a study on the Rethymnon Egyptian lighthouse in Crete employed three months of vibration data to show the effectiveness of the method. By comparing true structural conditions with a reference condition, the authors were able to detect frequency changes as small as 1%, which are relevant for early damage detection. Such precise measurements enable engineers to respond in time to emerging issues, thus ensuring the integrity of these valuable monuments and their longevity for future generations (Daskalakis et al., 2019). In addition, the functionality of SHM systems is enhanced by the development of numerical models that simulate structural response under varying conditions. The above study not only employed the stretching method but also substantiates a basic numerical model to predict the response of the lighthouse to stress. By correlating model output to measurement in seismic activities, researchers established the possibility of predicting structural responses in cases of potential damage scenarios. The models not only help enhance SHM systems but also facilitate enhanced planning for intervention maintenance. This puts into perspective the important role SHM plays in the grand picture of proactive infrastructure management, enabling stakeholders to implement sustainable practices in the maintenance and care of new and old buildings (de Lautour et al., 2014).

Importance of Structural Health Monitoring (SHM) in Engineering

SHM is at the heart of the engineering practice, particularly the durability and safety of the structures. Through continuous tracking of the health of structures such as bridges, buildings, and dams, SHM seeks to spot potential damages before they present themselves as catastrophic failures. Through proactive maintenance, SHM assists in containing maintenance costs and enhancing the consistency of structures bearing dynamic loads, an aspect vital in regions prone to natural disasters. As engineering projects become increasingly sophisticated in terms of nature and magnitude, the application of SHM technology is increasingly vital day by day. These technologies enable engineers to make informed decisions based on real-time information, ultimately ensuring the integrity of critical infrastructures and maximizing public safety. In SHM, the techniques in damage detection can be characterized as data-driven algorithms and model-based methods. Data-driven techniques use sensor measurements to improve the responsiveness of the monitoring system for real-time evaluations of the condition of a structure. The spatial and temporal resolution given by modern sensors allow engineers to detect flaws such as cracks or material fatigue prior to their evolving into catastrophic failures. Model-based techniques, however, depend heavily on the validity of analytical models to examine structural performance and interpret information. Although both of these methods are useful in structural assessment, there has been a growing tendency towards data-driven algorithms in recent times because they have the ability to deal with extensive amounts of data quickly, thus facilitating rapid decision-making in engineering applications (Bush et al., 2011). The use of sophisticated SHM systems does not only enhance safety but also provides the opportunity to develop smart structures. Smart structures involve the use of advanced materials and inbuilt technologies which respond in a timely manner to changing conditions. The interaction of real-time information aids to define structural response during unprecedented situations, thus facilitating adaptive action to mitigate resultant damage. The addition of data-driven algorithms has also enhanced the process of damage detection, setting a platform for preventive maintenance and repair strategies. These innovations facilitate a more sustainable approach to infrastructure management, where engineers can predict potential issues based on current data rather than waiting for failures to occur. By employing SHM, engineers can greatly enhance the resilience of critical infrastructures while also ensuring environmental and economic efficiencies (Cherta et al., 2020) (Figure 1.2).

Figure 1.2. Importance of structural health monitoring (SHM) in engineering.

Source: https://www.mekangroup.co.za/wp-content/uploads/2023/08/Structural-Health-Monitoring-Ensuring-Longevity-and-Safety.png.

Historical Context and Evolution of Structural Health Monitoring (SHM)

The historical background of SHM dates back to the early 20th century when engineers started realizing the significance of ongoing structural assessment towards maintaining infrastructure safety. Early monitoring was mainly dependent on manual inspections, whereby most of them were based on visual assessments for the identification of early indicators of distress or deterioration of structures. As more infrastructure development was taking place, particularly in the cities, the shortcomings of these rudimentary methods of doing things were starkly apparent. Hence, the evolution of SHM necessitated cutting-edge solutions that leveraged more sophisticated sensor technologies and data analysis methods for a transition from subjective opinion to objective, data-driven judgments. This ushered in an era where structural safety is supplemented with real-time information, ultimately leading to upgraded safety features and maintenance practices. In the latter half of the 20th century, advancements in computer technology had a major impact on the development of SHM systems. The application of sensors, such as accelerometers and strain gauges, facilitated constant observation of structures and allowed the recovery of vast amounts of data. This was also the time when computational power became accessible, and with that, advanced data analysis and modeling techniques. Specifically, damage-tolerant design theory began gaining momentum with strong connections to SHM concepts. The new methodology showed not only a move away from monitoring but a proactive approach in structural integrity as well. The integration of fault tolerance concepts with traditional structural design put great emphasis on responding not just to damage, but also on designing structures that would be able to endure and adapt with developing conditions, according to various interdisciplinary studies on control design approaches applied to mechanical structures (MECHBAL et al., 2012). Over the recent decades, the SHM field has continued to evolve, embracing cross-disciplinary methodologies that leverage advances in predictive computation, machine learning, and data fusion technologies. These new methods have transformed SHM from a repair-minded methodology to a predictive maintenance paradigm, significantly enhancing decision-making for structural management. Emerging studies focus on enhancing SHM systems' resilience to improve their robustness, not only to identify existing damage but also to predict likely degradation before it happens. This development is aided by the growing need for infrastructure to be resilient to natural catastrophes and climate change, resulting in greater emphasis being laid on maintenance practices focusing on long-term sustainability and safety. With increasing pressures of urbanization, the integration of SHM in smart city infrastructure has been a priority, and it has set the course for future infrastructure management (Dong et al., 2010).

1.1. OBJECTIVES OF STRUCTURAL HEALTH MONITORING (SHM)

The role of SHM is essential to ensuring the durability and safety of structures, particularly as they age or are subjected to environmental stresses. One of the central objectives is to identify any change in structural integrity that can happen as a result of internal or external factors, such as material fatigue, environmental corrosion, or earthquakes. Through the application of different sensor technologies, SHM enables constant monitoring and assessment of structural conditions and prompt maintenance and intervention initiation. This pre-emptive measure not only minimizes impending threats but also extends the lifespan of buildings, decreasing rates and costs of maintenance. Structural health monitoring is thus an essential civil engineering activity that marries technology and engineering know-how to ensure maintenance of infrastructure. Another fundamental objective of SHM is to facilitate maintenance and management of structures, especially in critical and high-value facilities like bridges, dams, and historical buildings. Through monitoring and analyzing data related to structural behavior, engineers have the ability to implement maintenance actions stratified on the basis of real-time estimates of performance of the structure. Data-driven decision-making enables the stakeholders to make effective decisions that enhance operational effectiveness as well as resource efficiency. For instance, the integration of damage tolerant active control methodologies, as highlighted in recent literature, directs towards the degree to which SHM can be complemented with fault detection and diagnosis systems, thereby offering a holistic approach to structural management. In establishing an organized framework for continuous evaluation, SHM not only ensures the safety of structures but also establishes a platform for sustainable infrastructure practices. Besides, the advent of open-source technologies has changed the objectives of SHM, making it not only more feasible but also less expensive. These new platforms enable monitoring of old structures, which are often underfinanced and vulnerable to deterioration. With new digital modeling and printing, as outlined in ongoing studies, researchers have proved it is possible to integrate low-cost elements into existing monitoring systems to deliver useful data on the state of the heritage structures. This democratization of technology promotes mass up-take of structural health monitoring practice, thereby maximizing the preservation of cultural heritage. These innovations symbolize the direction towards more flexible and reliable monitoring systems able to generate holistic data on the historic integrity of infrastructures, thereby fulfilling the broader ambitions of SHM in modern engineering (Figure 1.3).

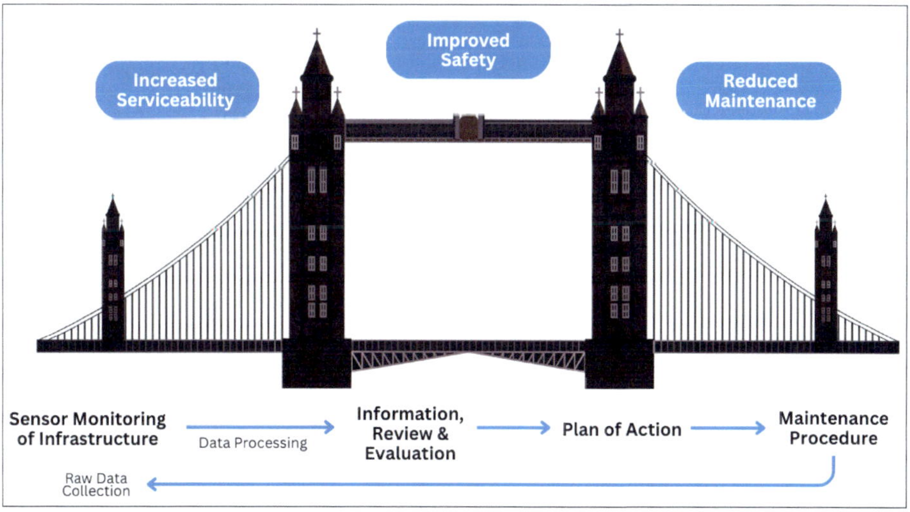

Figure 1.3. The objectives of structural health monitoring (SHM).

Source: https://aecbytes.com/illustrations/tipsandtricks/2024/issue109-Structural-HealthMonitoring-images/fig3.png.

1.1.1. Safety Assessment of Structures

The structural safety assessment of buildings is most important in their durability and trustworthiness, particularly for safety-critical applications such as civil engineering and aviation. Structural health monitoring (SHM) plays a central role in it, and various sensors and techniques are employed to detect potential defects and material degradation over time. Through the utilization of SHM systems, engineers can gain real-time data about the health of a structure and allow for timely interventions and maintenance. Continuous monitoring of structural integrity lessens the possibility of abrupt failures and optimizes resource allocation for rehabilitation, eventually leading to enhanced overall safety. As SHM technology evolves, it increasingly favors the utilization of advanced materials and methods that maximize the effectiveness of safety assessments to confirm that buildings can carry extreme loads and environmental conditions. One of the newer trends in recent years has been the use of light materials such as aluminum alloys and fiber-reinforced polymers (FRPs) for structural use. These materials have significant advantages, including their high mechanical stiffness in addition to lower weight, which is especially beneficial in aerospace engineering.

Figure 1.4. The safety assessment of structures.

Source: https://masterspmc.com/images/con1.jpg.

However, the unique properties of light-weight materials necessitate rigorous and continuous security screening in order to preserve their structural integrity properly. Notably, next-generation SHM systems are being planned to integrate lightweight sensors into the very structure itself. For example, a polymer paint founded on a multi-walled carbon nanotube (MWCNT) has been created to identify distributed damage in light structures at high speed. The technology is one instance of how material science innovation can enhance the structural safety evaluation capability by allowing issues to be identified early before they become failures that are critical in nature (Loh et al., 2017). Besides, technological advancement has enabled the application of sophisticated analytical methods in SHM plans. For instance, finite element modeling has been employed to characterize the electromechanical performance of thin coatings and model how real-life loads in the field can affect structure performance. Such models allow engineers to simulate various stress conditions and calculate how structures would respond under varying conditions. This prognostic capability is important to guarantee the level of safety, as it informs decision-making processes related to design modification and maintenance schedules. Ultimately, with more sophisticated SHM systems becoming integrated with smart materials, they are a promising avenue of research for enhancing the safety assessment of structures in different industries, thereby promoting safer environments and infrastructure (Limongelli et al., 2016) (Figure 1.4).

1.1.2. Maintenance and Lifecycle Management

Effective maintenance and lifecycle management are key components of Structural Health Monitoring (SHM), particularly in ensuring the longevity and integrity of structures such as wind turbines. As technology for renewable energy advances, the need for sound maintenance strategies becomes increasingly eminent. Through close monitoring of such installations' structural health, the stakeholders can spot imminent failures before they become significant, hence cutting downtime and maintenance costs. Custom-designed SHM systems enhance the ability to monitor live data from various structural components, allowing proactive rather than reactive maintenance regimes. This paradigm of data-centricity mirrors closely with asset management protocols, allowing knowledge-based decision-making on maintenance schedules and interventions. Finally, placing maintenance and lifecycle management at the forefront of SHM concepts ensures critical infrastructure safety and efficiency while maximizing its service life. Lifecycle management goes beyond conventional maintenance by covering the entire life cycle of a building, from design and construction right down to decommissioning. An integrated lifecycle strategy couples SHM data with predictive maintenance technologies, giving it a complete view of a structure's condition throughout its life. This consolidation is important because it makes it possible to identify patterns and anticipate potential structural interventions in advance when they reach a critical point. Furthermore, as with onshore wind turbines, randomness in external loads makes it proper to approach structural greatness with a systematic method of managing structural excellence. Through a customized structural health monitoring system, it is possible to continually assess and make an organization adapt to new challenges. This proactive lifecycle management not only optimizes operational efficiency but also promotes environmental responsibility through minimizing waste and maximizing resource utilization throughout a structure's life (Bush et al., 2015; Currie et al., 2012). Furthermore, the intersection of SHM with advanced lifecycle management methods attests to the importance of incorporating technology into maintenance activities. The creation of SHM technologies, including sensors and analytics, plays a critical role in the accuracy of maintenance procedures. The technologies help with continuous real-time monitoring, which allows operators to make decisions on the basis of accurate information. For instance, the application of embedded sensors within the foundations of wind turbines has the ability to reveal significant information about the structural integrity due to evolving load conditions. Failure mode identification at an early stage is essential in promoting effective maintenance cultures and, thus, extends the assets' life. As the world continues to rely on wind energy, ensuring optimal maintenance and lifecycle management for ensuring safety and reliability in renewable energy infrastructure (Bush et al., 2015; Currie et al., 2012) (Figure 1.5) becomes a priority.

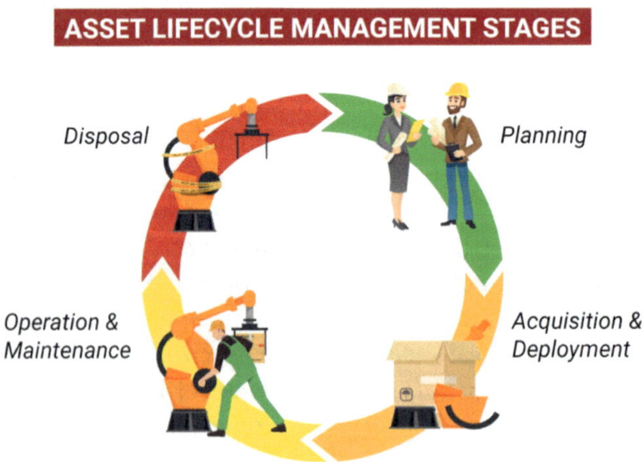

Figure 1.5. Maintenance and lifecycle management.

Source: https://limblecmms.com/wp-content/uploads/asset-lifecycle-management-stages.jpg.

1.1.3. Cost-Effectiveness and Resource Allocation

In the scenario of Structural Health Monitoring (SHM), cost-effectiveness and resource planning become seminal concerns, particularly in infrastructure management and maintenance. Effective SHM planning entails judicious outlay of funds to ensure optimal use of the monitoring technologies. This involves assessing the merits of various SHM technologies relative to their costs, such that there is a clear understanding of how resource deployment provides structural integrity and contributes to overall cost savings. As an example, it has been confirmed through research that efficient resource deployment directly impacts system performance, where the relationship between strategic investment in equipment and maintenance cultures and structural reliability and lifespan is positive. These strategic investments not only enhance performance but also optimize the cost of operations in the long term, necessitating more research into the optimal budgetary frameworks for SHM projects.

Comprehensive understanding of the economic effects of SHM necessitates examining the cost-effectiveness of various monitoring interventions. For instance, the use of next-generation sensor technologies can provide highly valuable information on structural health, but their upfront costs can be prohibitive. Therefore, a cost-benefit analysis must take into account not only the initial capital expenses but also the long-term benefits, such as reduced maintenance and increased service life. By comparing the lifecycle cost of traditional monitoring systems with SHM systems, stakeholders can make informed decisions towards more efficient resource utilization. Literature has shown that effective resource planning, much like strategies employed in healthcare facilities to improve patient satisfaction through optimal management of spending, can do much to maximize the outcomes of engineering activities, such as in McCrindle et al. (2010) and Ramírez Orellana et al. (2019). Ultimately, effective resource management in SHM relies on sound decision-making guided by an understanding of the cost-safety balancing act. The installation of SHM systems needs to be weighed not just by the upfront cost but by the role that they play in preventing catastrophic failure and enabling timely action to be taken. As infrastructure systems become increasingly complex and structures grow older, the necessity for systems with high quality monitoring increases, with the justification for taking out investments in SHM becoming stronger. Efficient programs can be seen across different industries, reflecting initiatives such as those of the Flying Start program that have shown dedication to ensuring that resources are maximized for greatest effect (McCrindle et al., 2010). Using knowledge from productive resource utilization, all stakeholders will be able to make certain that SHM projects are not just financially viable but are also consistent with larger infrastructural objectives.

1.2. KEY COMPONENTS OF STRUCTURAL HEALTH MONITORING (SHM) SYSTEMS

In comprehensive SHM systems, numerous key elements unite for optimal operation and reliability. Sensors are the backbone of all SHM systems, whose primary purpose is to acquire essential information on the structural integrity of buildings, bridges, and other infrastructures. Various sensors, including accelerometers, strain gauges, and displacement transducers, are employed to monitor different parameters, such as vibration responses and material deformations. The variety of sensor technologies available enables the SHM systems to receive a wide range of information, which enables engineers to identify issues before they manifest as large-scale failures. This pre-emptive approach not only adds safety but also sustains the lifespan of structures, echoing the need for proper selection of sensors suitable to specific structural applications. Data collected by sensors should be processed and analyzed effectively to yield useful information regarding structural condition. Central to this is the application of data acquisition systems, which consolidate sensor inputs and facilitate real-time monitoring. Advanced software packages and computation algorithms are needed to process the vast amounts of generated data since they are used for detecting anomalies and assessing structural performance under varying conditions. Predictive analytics is now an essential component of SHM systems, using past data and machine learning methods to forecast potential failures or maintenance actions required. This integration of technology not only speeds up decision-making but also maximizes the use of resources, allowing for timely repairs and improved safety interventions during the entire life span of a structure. Furthermore, the addition of an active control mechanism significantly increases the efficiency of SHM systems. By the incorporation of damage-tolerant active control methods, engineers are able to create adaptive systems capable of addressing structural vulnerabilities as and when they occur. This new approach enables real-time adjustments to reduce the effect of known issues, thereby lengthening a building's lifespan. With the exploitation of the synergy of fault detection, diagnosis, and active control design, modern SHM systems are capable of revolutionizing the manner structures are preserved and maintained. The research carried out to date in this area is crucial for enhancing the performance and reliability of SHM applications, particularly when used for ancient buildings, where cost-effective and efficient monitoring methods are mandatory. As recognized by recent research, the blending of new technologies has the potential to augment traditional monitoring approaches, paving the way for revolutionary breakthroughs in structural health monitoring (Basto et al., 2017; MECHBAL et al., 2012) (Figure 1.6).

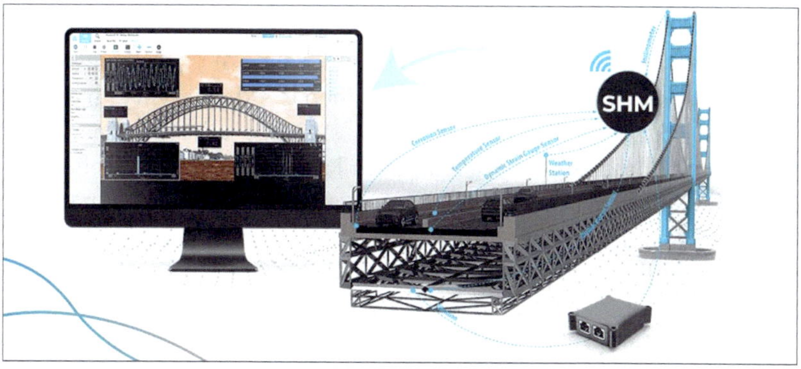

Figure 1.6. The comprehensive structural health monitoring (SHM) systems.

Source: https://aecbytes.com/illustrations/tipsandtricks/2024/issue109-Structural-HealthMonitoring-images/fig1.png.

1.2.1. Sensors and Data Acquisition

The function of sensors in structural health monitoring (SHM) is vital because they are the primary data acquisition devices for information regarding a structure's state. Various types of sensors, such as accelerometers, strain gauges, and displacement transducers, are employed to detect alterations in physical parameters that may indicate concealed damage. More particularly in vibration-based SHM, the sensors measure dynamic responses to external loads such that practitioners can evaluate structural integrity. Application of these traditional sensors, however, is hampered by mass-loading issues and elaborate installation processes, particularly when they are to be used on large structures. High spatial resolution upon deployment of such sensors is an intricate problem, with the tendency to lead to inefficiency in operations. Consequently, the transformation to non-contact measurement techniques, i.e., digital video cameras, is gaining momentum as it allows remote data collection with none of the compromises required by traditional instrumentation (Mao et al., 2018). Data acquisition systems (DAQ) play a key role in the integration of sensors into SHM systems. DAQ systems collect, store, and process data pulled from sensors, converting raw measurements into useful information to be analyzed. A well-designed DAQ system not only enhances the data accuracy but also makes the processed data available to analyze at real-time, offering timely decision-making information. Data-driven algorithms are typically paired with DAQ systems to efficiently process collected data. Algorithms pull out patterns and locate anomalies that indicate potential structural issues. While data-driven approaches can facilitate an ever-responsiveness of the SHM process, they necessitate high-performance computing capabilities due to the volume of data and the complexity involved. That this technology reliance underlines the relevance of ongoing advancement in sensor technology and data processing method in structural health monitoring (Cherta et al., 2020) is the implication. Continued advancement of sensor technology and data acquisition techniques has motivated researchers to find new ways of improving SHM outcomes. Advances in technology have led to the development of smart sensors with embedded processing capability to reduce dependence on bulky outside DAQ systems. The smart sensors may also coexist in a network to enhance data acquisition by sharing synchronized measurements. Machine learning and other sophisticated data-analysis techniques are being applied to improve the performance of such systems, allowing accurate damage identification, localization, and characterization. With the development of such technologies, increasingly autonomous and responsive SHM systems should become feasible, ultimately leading to safer and more resilient structures. Through the aid of smart sensors and innovative data acquisition methods, upcoming SHM projects can gain a lot in accuracy and efficiency (Cherta et al., 2020) (Figure 1.7).

Figure 1.7. Sensors and data acquisition.

Source: https://fprimec.com/wp-content/uploads/2020/01/Accelerometer.jpg.

1.2.2. Data Processing and Analysis Techniques

Sophisticated data processing and analysis methods in Structural Health Monitoring (SHM) play a central part in deciding the integrity and performance of the structure. These techniques include a large array of algorithms used to scan data from an array of sensors placed around a building. The efficiency of these techniques rests largely in their ability to accurately interpret the data, which can indicate damage or anomalies. They further enable real-time monitoring and decision-making, hence enhancing overall safety and reliability of structural systems. For instance, the implementation of damage tolerant active control systems reflects the dynamic relationship between detection, diagnosis, and monitoring capability, underscoring the requirement for resilient data processing structures in SHM (MECHBAL et al., 2012). Data-driven and model-based approaches are the two most common approaches towards damage identification in SHM. The data-driven approach employs real-time sensor readings, utilizing algorithms to identify possible damage without the need for heavy input models. This approach is particularly beneficial in that it can readily adapt to the particular conditions of the monitored structure and give real-time data to evoke immediate countermeasures. Model-based approaches, however, are supported by existing structural models such that the accuracy of the used models determines the quality and effectiveness of damage identification. Even though both methods are prevalent in application, data-driven methods are more favored due to their efficiency and applicability in various situations typical of SHM (Cherta et al., 2020). For further maximizing of structural health monitoring, researchers are always seeking innovations in the technology of data processing capable of handling the high level of computational load typical of such data-driven approaches. The technology in high-performance processors has been key to overcoming these challenges, allowing real-time examination of big data. Furthermore, the development of smart structures—material with embedded sensors and processing—again illustrates the potential of SHM's future. As data-based damage identification algorithms evolve, they incorporate significant functionality such as detection of damage, localization, and prognosis, which are essential to proactive maintenance. Therefore, at the intersection of technology and structural analysis, SHM practice is constantly being enhanced with an eye towards greater accuracy and efficiency in maintaining structural integrity (Cherta et al., 2020).

1.2.3. Communication and Reporting Systems

Communication and reporting systems are essential in Structural Health Monitoring (SHM) and exist to facilitate information sharing that relates to buildings, bridges, and other structures' structural integrity. These systems facilitate stakeholders like engineers, managers, and regulators to see real-time data from monitoring, thus enhancing decision-making. For instance, when structural health monitoring detects anomalies or degradation in the structure, automated notifications that notify relevant parties of any potential hazards are sent. Through this, communication ensures that interventions or maintenance can be programmed on schedule to prevent catastrophic failure. In addition, the use of user-friendly dashboards and visualization abilities allows interpreting advanced data intuitively, which simplifies the understanding of the implications of the outcomes on structural safety for non-specialists. Technology's influence on communication networks has also changed how information becomes processed and communicated in SHM systems. Advanced data analysis and machine learning methods can handle vast amounts of sensor readings to recognize patterns and predict potential failures, significantly improving reporting. Automated reporting systems can easily produce concise summaries of structures' health status, enabling speedy reviews from stakeholders. With increasing amounts of data being generated from Structural Health Monitoring technologies, handling and presenting the data have been of highest priority. Having robust data management procedures in place not only guarantees the retention of historical data and monitoring data, but also that these are searchable and readily retrievable for future use, critical in long-term asset management and evaluation. Additionally, some case studies verify that good communication and reporting systems are crucial in promoting safety levels in industries such as renewable energy. For onshore wind turbines, for example, a specially customized structural health monitoring system is needed to analyze foundation integrity because of the complex load dynamics to which such structures are exposed. According to Currie et al. (2012), a best-designed monitoring solution can potentially provide early indicators of structural failure, which is critical to maintain operational safety and efficiency. Similarly, large scientific projects like the DUSEL Project are aimed at ensuring that more sophisticated communication infrastructures are developed in order to promote multidisciplinary research and outreach activities. The development of such facilities goes beyond taking account of continuous monitoring by embracing provisions for communication frameworks which enable collaboration between different scientific disciplines, with an emphasis on how integrated reporting systems can optimize overall project success and sustainability in structural monitoring.

1.3. TYPES OF STRUCTURAL HEALTH MONITORING (SHM) TECHNIQUES

The field of SHM encompasses a suite of techniques that allow for the quantification of the structural performance and integrity over time. Of these techniques, one may broadly divide them into data-driven techniques and model-based techniques. Model-based techniques utilize analytical models of the structure, such that the observed behavior is compared to calculated responses. Data-driven techniques, on the other hand, use real-time measurements obtained from an array of sensors attached to or in close proximity to the structure. Both techniques are applied for particular applications in SHM, but data-driven approaches are primarily favored for their real-time performance and capacity to respond to prevailing conditions. This offers decision-making at the point of occurrence, which is critical for guaranteeing safety and effectiveness in structure utilization. The evolution of smart materials and the combining of these techniques is also revolutionizing how structures are monitored, stimulating the innovation of predictive maintenance and damage detection. Data-based algorithms have become a fundamental component in the technology of SHM, particularly in the area of damage detection and identification. Algorithms process sensor data to detect anomalies that might correspond to potential structural issues. Such methods' computational needs demand powerful processors for performing real-time analysis in an efficient manner. As suggested by new studies, the process of damage identification can be directed by a number of algorithms that evolve through continuous learning and updating to new data. Professionals could achieve enhanced accuracy in damage localization and classification and prognosis for probable failure through utilization of these algorithms. Advancements in smart structures also emphasize the merging of sensor technology and data-driven approaches to present a more integrated means of tracking a structure's condition over the life of the structure (Cherta et al., 2020). The second important aspect of structural health monitoring is active control techniques, particularly in the context of damage-tolerant practices. These strategies are pertinent to fault-tolerant control design, which focuses on the operational integrity of mechanical structures under failures. Structural integrity monitoring methods that provide fault detection and diagnosis enhance the robustness of a structure. The integration of control design and SHM introduces novel paradigms that go beyond conventional monitoring techniques. Although traditional SHM is predominantly about passive monitoring, the integration of active control further improves its performance and responsiveness to fault detection. New research in this area is seeking to lay the foundation for sophisticated controller design that not only monitors damage but actively adjusts operation to maintain structural integrity (MECHBAL et al., 2012). Such integration is considered to contribute potentially towards overall safety and durability of critical infrastructures.

1.3.1. Visual Inspection Methods

Visual inspection methods are central to the Structural Health Monitoring (SHM) practice, providing critical information about the structural health. The process is highly reliant on the experience of individuals who visually inspect the structural elements for signs of damage such as cracks, corrosion, and deformation. Despite advances in sensors and data-driven algorithms, the human eye remains a valuable asset in being able to spot issues which may not be apparent through instrumentation. Visual inspections can be performed during routine maintenance or due to specific problems, thereby ensuring long-term structural integrity. The effectiveness of such methods depends significantly on inspectors' training and experience, which must allow them to discern fine indications of distress that can be harbingers of more significant structural defects. While visual inspections are qualitative and therefore open to personal opinion, they can be aided by more quantitative methods for better decision-making in SHM. The integration of visual assessment with sensors' data analysis permits reaching a higher extent of knowledge about the condition of a structure. For instance, data-driven algorithms can read data from multiple sensors, such as accelerometers and strain gauges, to monitor structural behavior in real-time.

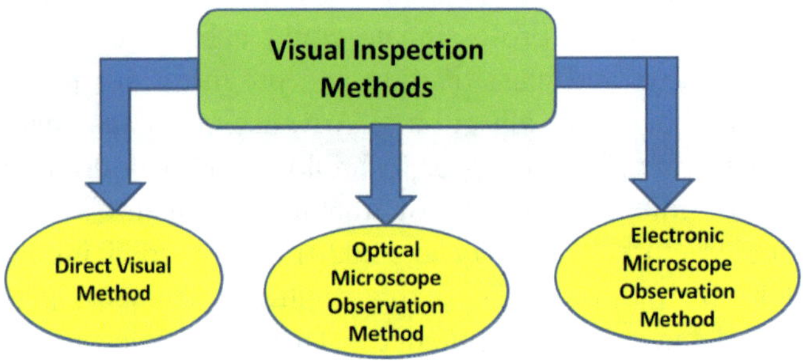

Figure 1.8. Visual inspection methods.

Source: https://www.researchgate.net/publication/377298536/figure/fig3/AS:11431281216777232@1704906045939/Classification-of-visual-inspection-methods.tif.

This holistic approach not only enhances the precision of damage detection but also aligns with the direction towards smart structures that are self-monitoring. These technological advancements in SHM methods are an indicator of greater emphasis on achieving structural safety while maintaining the minimum expenditure on resources to be utilized during inspection and maintenance operations, to achieve the utmost cost-effectiveness of infrastructures. The application of visual inspection methods in SHM also gives importance to developing standard operating procedures to be adopted when performing the inspection. A standardized process of visual inspections can prevent inconsistency and ensure thorough and systematic inspections. Standardization can promote more effective communication between stakeholders, such as inspection teams, civil engineers, and maintenance workers, to enhance a more integrated structure in assessing the damage. For the best identification of damage, it is important to link visual observations with recorded data from previous inspections and real-time monitoring systems. This greater alignment not only serves to improve the validity of inspections but ultimately contributes to a more secure support system for maintenance and policy for safety in civil engineering, illustrating the vital role of visual inspection methods in the larger picture of SHM practices (Figure 1.8).

1.3.2. Non-Destructive Testing (NDT)

NDT leads in the use of Structural Health Monitoring (SHM) since it provides crucial information on the health of infrastructure without changing its structural integrity. It is a technique essential in assessing structures that are aging, particularly those constructed decades earlier, and are now subjected to higher degradation under traffic loads and climatic conditions. Since highlighted by recent studies, the danger presented by deteriorating infrastructures is heightened, and what is needed are efficient monitoring techniques to anticipate probable failures proactively. By employing NDT methods such as ultrasonic testing, radiography, and acoustic emission monitoring, engineers can conveniently identify subsurface flaws and deterioration in material, which might not be visible using standard inspection methods. As such, NDT is a cornerstone in ensuring the sustainability and safety of aging structures, hence a key factor in overall urban sustainability and public safety. The geographical integration of NDT into SHM systems ensures geographical coverage of structure performance over time, accommodating evolving demands on infrastructure. With increasing emphasis on asset management, NDT enables decision-making in maintenance and rehabilitation strategies, ultimately optimizing resource utilization. Complexity of aging infrastructures is further compounded by the requirement to fulfill societal demands of durability, sustainability, and safety. As quoted in the literature, a majority of civil engineering infrastructures are coming towards the end of their design life, thereby underscoring the imperative to adopt innovative assessment approaches like NDT. By applying these methods under an SHM strategy, stakeholders have the capacity to enhance the resilience of structures, diminishing the likelihood of calamitous collapse with its consequential economic and social effects, as in the case of the Morandi bridge collapse. Furthermore, application of NDT does not only help in identifying existing issues but also aids in developing predictive maintenance models. By providing baseline data through repeated NDT inspections, the ability is present to monitor change over time and predict future patterns of decline. Forward-looking of this nature is especially critical in the public infrastructure arena where safety and operational efficiency are paramount. With mounting pressure on minimizing operating costs and lowering environmental footprints, NDT emerges as an essential tool, consistent with broader objectives of sustainability in infrastructure management. As explained in current literature, successful integration of NDT into an SHM strategy is crucial in response to the inherent challenges caused by aging infrastructure, hence facilitating the need for standard practice and protocols in the field (Cantini et al., 2019; Wan-Wendner et al., 2018) (Figure 1.9).

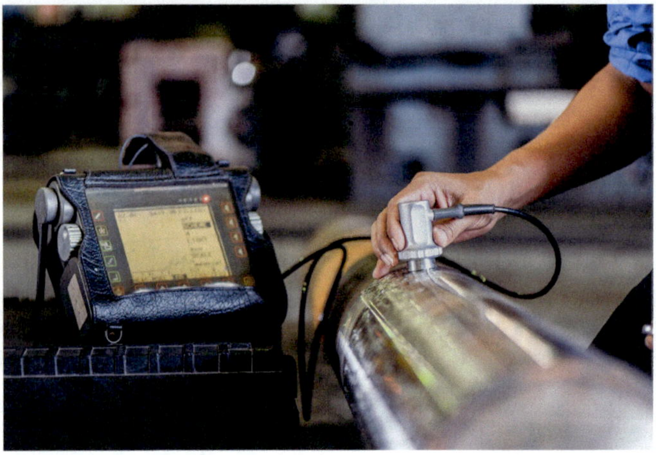

Figure 1.9. Non-destructive testing (NDT).

Source: https://aqcinspection.com/wp-content/uploads/2020/12/Ultrasonic-Testing-NDT.jpg.

1.3.3. Remote Sensing Technologies

The use of remote sensing techniques in Structural Health Monitoring (SHM) is a shift in paradigm in monitoring and assessment of infrastructure. Such advanced techniques allow for the non-invasive examination of structures and for the possible determination of integrity problems without massive human involvement. Remote sensing equipment, like air inspection and satellite photos, enable collection of information over large geographic regions, hence making them very convenient for tracking infrastructures such as bridges, dams, and windmills. The technologies can be utilized by engineers to track structural condition in real-time, significantly improving the accuracy and efficiency of health monitoring. In particular, remote sensing is particularly valuable in the renewable energy sector, where structural integrity of structures such as wind turbines is crucial for ensuring continued energy output and safety (Currie et al., 2012). The ability of remote sensing technologies to integrate SHM is supported by data-based algorithms, which are extremely crucial in processing the high amounts of data gathered. These algorithms work on information from an array of sensors mounted on or near buildings and enable damage location and identification. The data-driven nature implies that they can be used for real-time analysis and can provide a predictive maintenance feature that may preclude the risk of catastrophic failure. Conventional model-based structural analysis methods offer their benefits but often demand massive computational resources. Conversely, data-driven algorithms provide a more dynamic method of evaluating structural conditions because they react to new data and can provide real-time information for decision-making (Cherta et al., 2020).

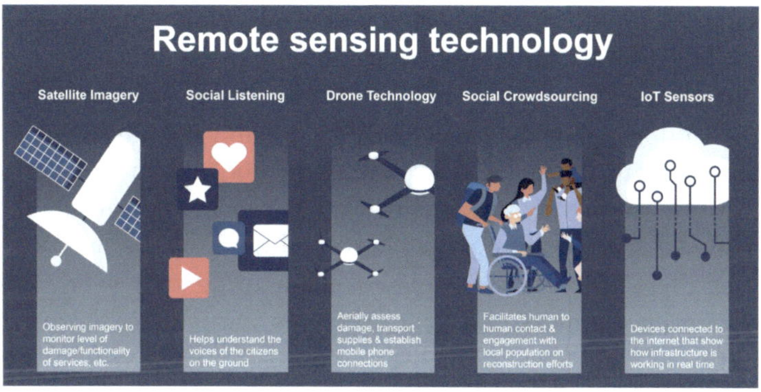

Figure 1.10. The integration of remote sensing technologies.

Source: https://www.ipsos.com/sites/default/files/inline-images/remote-sensing-ipsos.png.

Therefore, the combination of the two fields—remote sensing and data-driven analysis—provides a more proactive and intelligent means of structural health monitoring. Furthermore, application of remote sensing technologies in SHM has far-reaching implications for the extension of the entire lifespan and performance of structural systems. By providing ongoing monitoring and early detection of degradation, these technologies enable the implementation of maintenance strategies that extend the life of infrastructure, hence maximizing resource investment. In the context of critical infrastructures, for example, expanding wind farms, incorporation of effective monitoring systems can assist in reducing risks due to infrastructure failure. The implications of remote sensing therefore extend beyond immediate control measures; they instill a culture of preventative maintenance that leads to structures' resilience against environmental and operational stresses. The continuous development of remote sensing technologies is set to further shape the structural health monitoring scenario with a promise of increased safety and sustainability in engineering uses (Currie et al., 2012) (Figure 1.10).

1.4. APPLICATIONS OF STRUCTURAL HEALTH MONITORING (SHM)

Applications of SHM are extensive and substantial, varying across all phases of infrastructure, from buildings and bridges to marine structures. Through the use of advanced sensor technologies, SHM offers continuous monitoring of structural performance, enabling rapid assessment of potential issues. For instance, in bridge monitoring, SHM sensors are able to detect intrusive elements such as vibrations and strains that may indicate structural failure or fatigue. This forward-looking approach not only ensures public safety but also lengthens the lifespan of critical infrastructure via the provision of repair serviceability based on actual condition rather than mechanical timetables. With increasing urban cities grappling with issues of aging infrastructure, implementation of SHM is critical to ensuring structural integrity for the place we call home, thereby demonstrating its key position in modern civil engineering practice. Besides routine safety assessments, SHM plays an important role in identifying the optimal maintenance policies for most structures, particularly those susceptible to fatigue. Combining SHM with life-cycle management (LCM) techniques offers a robust framework for the assessment of the performance of structures vulnerable to fatigue under conditions of uncertainty.

The combination allows engineers to estimate the safe duration of structures' lifetimes while considering multiple factors that may compromise their integrity. For instance, the best inspection plans can be developed that are perfectly synchronized with performance prediction, avoiding unwanted disruptions and expenditure of traditional preventative maintenance strategies. By focusing on actual wear and tear, stakeholders can distribute resources more accurately towards repairs and upgrades where they are most required, making infrastructure systems more resilient and reliable in the long run (Frangopol et al., 2013). Also, the general climate of infrastructure resilience is enriched significantly through SHM implementation, particularly in disaster preparedness and response. Highly sophisticated monitoring systems can provide rich information before and during natural disasters, which enables rapid evaluation of structural conditions following an event. Such capability is particularly critical for critical infrastructure, such as emergency services and hospitals, where Operational Continuity is critical. Information derived from SHM can guide quick decision-making processes for evacuation, repair, and recovery measures. Furthermore, with growing urban environments and the possibility of climate change bringing forth new issues, SHM can be pivotal in determining adaptive strategies that enhance the sustainability and life of various structures, resulting in more resilient communities when confronted with environmental uncertainty (Bush et al., 2011).

1.4.1. Civil Engineering Structures (Bridges and Buildings)

Structural health monitoring (SHM) is a vital mechanism that ensures civil engineering infrastructure such as bridges and buildings are safe and enduring. The infrastructure is subject to various stressors such as environmental conditions, traffic loading, and materials fatigue. Lack of monitoring these factors may lead to catastrophic failures, calling for the integration of SHM methods. For instance, the application of advanced methods like carrier phase kinematic GPS has been reported to successfully track the three-dimensional translations and vibrations of structures. Such technologies are not only capable of detecting slight changes but can also give information about the dynamic responses caused by external forces like vehicular traffic, as seen in field trials conducted on the Forth Road Bridge (Atkins et al., 2012). With continuous monitoring, engineers can make informed decisions regarding maintenance and safety protocols, thereby lengthening the life of significant infrastructure. Methods applied in SHM can be categorized into model-based and data-driven approaches.

Data-driven algorithms rely primarily on current measurements supplied by an array of sensors placed strategically inside the structure. These sensors capture critical data on structural integrity, which are then filtered through advanced algorithms developed with the objective of damage detection and localization. This approach has found extensive application since it is capable of delivering immediate feedback on structural performance and hence facilitating real-time corrections. Model-based approaches, on the contrary, focus on available structural models in order to assess performance by giving importance to the validity of such models in damage detection. Each approach has strengths of its own, but there is a tendency to increasingly prefer data-oriented approaches, which are essential to developing smart structures that can self-monitor their own condition and health, thus providing more responsiveness to emerging issues (Cherta et al., 2020). The use of SHM extends beyond overt structural examination; it promotes structured understanding of infrastructure life cycle. Regular monitoring facilitates the identification of trends of wear and tear that would otherwise be unobserved, ultimately affecting maintenance timetables and investment strategies. Sophisticated technologies like fiber optics and wireless sensor networks facilitate long-term observation of structural behavior under varied conditions. The novel use of such technologies enables the collection of vast amounts of data, which can be analyzed to forecast potential issues before they can develop into significant safety concerns. As civil engineering continues to advance, the adoption of sophisticated SHM systems not only improves the safety and reliability of structures and bridges but also results in more effective practices in the management of infrastructure. The ongoing innovation in SHM technologies thus takes center stage in the future of civil engineering (Figure 1.11).

Figure 1.11. Civil engineering structures (Bridges and buildings).

Source: https://erusuconsultants.com/wp-content/uploads/2023/05/31e24b_6bd7e270981f45.webp.

1.4.2. Aerospace and Automotive Industries

The aerospace and automobile industries have progressively witnessed the importance of Structural Health Monitoring (SHM) as a central factor in guaranteeing safety and reliability. Such industries have the priority of structural integrity because it is a high-risk occupation where they are operating. SHM technologies facilitate real-time condition monitoring and predictive maintenance, significantly reducing the risk of the structure collapsing. With sensing technologies combined in structural components, companies can track parameters such as stress, strain, and damage continuously, thus making interventions proactive. Implementing SHM not only extends the lifespan of assets but also ensures greater confidence in the safety of passengers and cargo. Since the two sectors strive to make their operations more streamlined without facing risks, adhering to SHM is an instinctive and required progression from traditional forms of maintenance. NASAs commitment to implementing SHM technology is a testament to its use on the aerospace sector, particularly when it involves structural health assessment during crucial missions. The agency takes an active step towards the development of SHM systems that can scan launch vehicles and fuel tanks for structural damage. One such effort is the collaboration with Acellent Technologies Inc. with the objective of advancing the SMART Layer technology for aerospace structures. The hybrid technology has proven to be a valuable tool for monitoring and detecting damage, enabling timely assessments crucial for safety and performance in the aerospace sector. The success of these projects highlights the paradigm shift towards more advanced monitoring systems that can support better decision-making and operational effectiveness.

Figure 1.12. The aerospace and automotive industries.

Source: https://images.squarespace-cdn.com/content/v1/5e66303d2a35d86b999ddf82/1584455962810-QWB95GIIT9GP3IJ7ZRM40/image-asset.jpeg.

The general relevance of this technology goes beyond aerospace, with automotive and construction industries taking particular interest, where safety and reliability are equally high (N/A, 2010). The adaptability of SHM technology has enabled the benefits of SHM to spread into other industries outside aerospace, including automotive and utilities. As manufacturers increasingly use more advanced materials and designs in car production, the requests for efficient monitoring solutions have also grown. SHM systems provide critical information about the structural integrity of the automotive component, allowing pre-emptive alerting of wear and damage that may have safety relevance. Moreover, the application of SHM technology in utility infrastructure demonstrates its broader significance, with potential for performing cost-effective maintenance and management of vital systems. By tapping the benefits of the technologies that have been designed in the aerospace industry, sectors such as energy and transportation can achieve enhanced operational performance and reduced downtime. Such technology overlap indicates the extensive applicability of SHM as a key part of continuous improvement in safety and reliability in industries (N/A, 2025) (Figure 1.12).

1.4.3. Energy Sector (Wind Turbines and Dams)

The energy sector plays a pioneering function in the transition towards renewable energy systems, notably through technologies such as dams and windmills. These renewable sources noticeably constrain greenhouse gas emissions, thereby supporting global initiatives towards combating climate change. Structural health monitoring (SHM) is crucial to the long term life and effectiveness of such energy infrastructure such that both wind farms and hydroelectric dams are operated economically and securely. With the aid of high-level sensor devices, operators can receive real-time structural integrity determinations, which is particularly important in terms of the stress and environmental loading these plants are subjected to. Monitoring in real-time not only improves safety but also minimizes operational downtime, thus maximizing energy yield and efficiency in resources. Ideal SHM systems provide significant insights that are used to inform maintenance activities, ultimately guaranteeing the sustainability objectives concerning renewable energy installation, as explained in existing literature (Alessandra et al., 2006). The use of SHM in the operation of wind turbines and dams is more relevant with respect to sustainability. As renewable energy technologies gain prominence in electricity generation, their reliability has a direct relationship with energy security and availability. Structural issues may lead to catastrophic failure, demonstrating the need for stringent monitoring and maintenance protocols. Aside from their significance for the energy production industry, efficient SHM also carries far-reaching implications that are deeply connected with society development objectives favoring ecological equilibrium and resource efficiency. For instance, increased monitoring can guarantee the maintenance of sensitive ecosystems near dams, which aligns with the three-pillar sustainable development framework of economy, ecology, and society. This exchange suggests that investment in SHM technologies is not only technical advancement but also required to uphold broader sustainability objectives within the energy sector (Angerer et al., 2011). Apart from guaranteeing sustainability, SHM in the energy sector can help achieve economic benefits through the optimization of maintenance strategies. Through the extension of wind turbine and dam lifespan, SHM can help lower costs significantly in the long run. Predictive maintenance based on present structural inspections allows facility managers to address prospective issues in advance, reducing the likelihood of unexpected repairs that disrupt energy supply and add financial expense. Additionally, the cost of employing SHM technology is typically recovered by its ability to enhance operational efficiency and energy generation. As such, the economic implications of SHM are as important as its environmental benefits. For the context of growing global energy demands and the need for clean energy technologies, proper monitoring practices are critical to ensure that renewable energy plants like wind turbines and dams remain operational and productive in the long term to facilitate a transition to a more sustainable and resilient economy (Alessandra et al., 2006) (Figure 1.13).

Figure 1.13. Energy sector (Wind turbines and dams).

Source: https://images.stockcake.com/public/c/a/c/cac02c29-f4a1-474f-9ad8-7557d577dbce_large/hydropower-energy-production-stockcake.jpg.

1.5. CHALLENGES IN IMPLEMENTING STRUCTURAL HEALTH MONITORING (SHM)

Installation of SHM systems is accompanied by a multitude of challenges that can have a tremendous impact on their efficiency and performance. One of the most pertinent challenges is the selection and installation of appropriate sensors, which are pivotal in collecting reliable information pertaining to the structural integrity. The tremendous variety of sensors available, ranging from accelerometers to strain gauges, each has its own set of strengths and limitations. Choosing the correct type requires a thorough knowledge of the structure's characteristics along with the environment that it faces. Furthermore, the entire process is compounded by the need for data-based algorithms that can effectively interpret the readings from sensors. These algorithms would typically need high-performance processors depending on the vast computation to handle such huge datasets, indicative of the need for order and sophistication in hardware and software elements of SHM systems (Cherta et al., 2020). A further challenge in implementing SHM is to ensure reliability and accuracy of the data received from sensors. Outside influences, including temperature changes and electromagnetic interference, can also impact the sensors' performance, which may cause faulty readings. Also, the data interpretation algorithms need to be carefully developed and tested so as not to yield false positives or negatives in damage detection. The dependency on reliable data is very vital, given that failure to detect structural vulnerabilities could result in cataclysmic outcomes. Here, the intersection of control schemes and damage identification is key since new paradigms such as damage tolerant active control aim to enhance the precision of monitoring systems. Detection is merged with diagnosis that have been found to be essential for optimal structural management (MECHBAL et al., 2012). Furthermore, integrating SHM systems into traditional structures poses logistical and economic problems. Retrofitting existing older buildings with new monitoring technologies is highly capital-intensive and can interfere with ongoing operations. Every stakeholder, including city planners, engineers, and property owners, must coordinate and collaborate with each other to develop sustainable means of incorporating SHM in a budget-friendly manner. Maintenance and maintenance work on SHM systems over long periods also require a trained labor force familiar with sophisticated technologies and analysis techniques. This introduces added complexity to implementation since there is constant demand for professional training in a rapidly evolving discipline. Thus, while the potential benefits of SHM systems are immense, the impediments in their implementation need careful analysis and robust planning.

1.5.1. Technical Limitations of Current Technologies

One of the key technical limitations of available technologies in Structural Health Monitoring (SHM) is the cost and availability of advanced sensors. Despite the innovations that have led to the development of open-source and low-cost platforms, as noted by Basto et al. (2017), these alternatives may not provide the precision and reliability required for critical applications. The majority of traditional SHM systems are based on high-quality sensors, which are extremely expensive, thus limiting their utilization in large-scale monitoring efforts, especially for small structures or heritage buildings. The question remains, however, how to balance affordability and performance, because applying affordable sensors introduces data quality variability, which directly affects the effectiveness of health checks. In addition, complexity in processing data is another challenge when capitalizing on current technologies for SHM. Processing data in contemporary data-intensive algorithms requires an enormous amount of computational resources, as pointed out by Cherta et al. (2020).

Figure 1.14. *One significant technical limitation of current technologies.*

Source: https://www.sixense-group.co.uk/wp-content/uploads/sites/2/2020/09/eversense-shm-hero-image.jpg.

This kind of reliance on high-performance processors complicates the implementation of SHM systems, particularly in rural or remote locations where computing power may be scarce. This can similarly hinder processing time and cost and thus constrain real-time decision-making processes, even though instant availability of information may prove useful in identifying structural issues. Improvement in both sensor technology and processing is thus a necessity to enhance SHM techniques. Lastly, the use of multiple devices in SHM systems adds complexity that can limit overall system performance. Whereas data-driven algorithm methods have gained popularity on the basis of their promise for real-time analysis, numerous systems are still confined to old-school modeling techniques that are unable to dynamically adapt with shifting structural conditions. As further discussed in Cherta et al. (2020), the precision of the models has a major role in determining their applicability to damage detection and prognosis. The reliance can lead to challenges in accurate localization and identification of structural damage, particularly in heterogeneous materials or under complex loading conditions. It is therefore vital to address the disparities between model accuracy and real-world applications in order to develop SHM technologies and keep them efficient in preserving structural integrity (Figure 1.14).

1.5.2. Data Management and Interpretation Issues

For effective monitoring and decision-making in the case of Structural Health Monitoring (SHM), good data management is essential. It is concerned with the acquisition, storage, and analysis of large amounts of data received from multiple sensors placed on structures. These sensors give real-time readings that can be cumbersome to interpret and manage. Compatibility with advanced data handling systems enables processing of the data to make it reliable and usable. Additionally, advancements in the availability of cheap, open-source platforms offer alternatives to heritage buildings. By using these technologies, stakeholders can plan the implementation of monitoring systems that are efficient and cost-effective. These developments in data management can propel the evolution of structural monitoring practice, establishing a broader database of digital evidence to serve in favor of cultural heritage preservation (Basto et al., 2017). Also noteworthy is the differentiation between data-driven and model-driven methods of data interpretation in SHM. Data-driven approaches are rooted primarily on sensor data for damage detection, with the added advantage of real-time analysis of the utmost significance in terms of timely intervention. But with this dependence on voluminous data comes the need for high-performance processors because of the computational effort involved, with the need therefore to balance the capturing of data and processing capacity. Model-based methods, on the other hand, depend on structural models' accuracy and thus the need to incorporate quality information to get useful inferences. This bifurcation presents interpretation challenges as the success of the monitoring system relies on the approach taken and reliability of information it is based on. These interpretation puzzles are central to optimizing the effectiveness of SHM systems (Cherta et al., 2020). With SHM evolving, an optimally tuned method to address data management and interpretation issues is an urgent imperative. The use of open-source technology not only avoids the restrictive implementation costs but also brings SHM tools within reach of all, hence encouraging more participation in structural monitoring schemes. In this regard, the adaptability of such systems allows for their implementation in tailored forms that are effective for special structural conditions and monitoring requirements. Furthermore, researchers must continue working on data-driven algorithms for detecting damage in order to make them suitably applicable in various conditions. With the focus on continual optimization of data collection and analysis procedures, the SHM field will develop towards enhanced structure condition assessments, ultimately resulting in better safety protocols and preservation techniques for both modern and historic uses (Basto et al., 2017; Cherta et al., 2020).

1.5.3. Financial Constraints and Funding Sources

The integration of Structural Health Monitoring (SHM) technology becomes increasingly regarded as crucial to the sustenance of safety and integrity of infrastructure. Despite that, deployment and application of such technologies are consistently foiled by financial limitations at governance levels as well as by institutional administration. The funding issue with sufficient amounts is pervasive, in particular reference to locations where budgetary appropriations are limited. For example, in the majority of African nations, including Nigeria, under-funding is a core issue in the university system, something which reflects a recurring tale in public infrastructure development. Most of the funds allocated to learning institutions simply go towards human resource expenditure, with none remaining for essential innovation and technology, including SHM systems (Odebiyi et al., 2002). This shortage makes it crucial to embrace diversified financial strategies to make SHM projects worthwhile and productive. In the European context, local governments are subject to heavy economic pressures fueled by austerity measures and budget reductions. The economic reality of today necessitates an effective rethinking of how these funds are being invested and allocated towards SHM projects. Local authorities are countering these constraints through the creation of increased financial resilience, which includes innovative budgeting and strategic partnerships. Empirical investigations of a range of local authorities have quantified four financial patterns of resilience that include pro-active and adaptive resilience, which places emphasis on the importance of responses not only to existing financial challenges, but also the requirement for anticipatory planning in order to mitigate future vulnerabilities (Barbera et al., 2014). This with it emphasizes the significant role played by financial awareness and elasticity in ensuring effective deployment of SHM technology to public infrastructure, which requires a holistic approach to finance that utilizes available resources to the maximum. Ultimately, the quest for sustainable sources of financing SHM must enlist both traditional and contemporary channels. Public-private collaborations, grants, and government incentives are some of the possibilities that can assist in mitigating funding issues faced by authorities and institutions. Furthermore, increased appreciation of the long-term value proposition of SHM can assist in engaging stakeholder support through private sector companies and policy makers. Several sources of funding, if utilized to the full extent possible, enhance not only the implementation ability of SHM systems but also the culture of maintenance and infrastructure resilience. Thus, the funding strategy that embraces adaptation to innovation and collaboration is essential in overcoming financial barriers and advancing the structural health monitoring practice, enabling infrastructures to remain safe and operational against increased threats (Figure 1.15).

Figure 1.15. Financial constraints and funding sources.

*Source:*https://www.forbes.com/councils/theyec/2023/04/20/eight-common-funding-sources-that-often-come-with-disadvantages/

1.6. FUTURE TRENDS IN STRUCTURAL HEALTH MONITORING (SHM)

Future SHM is going to be vastly different, particularly with open-source technologies. With these now more widely accessible and less expensive, usage in heritage building monitoring is trending. The use of cheap, open-source digital modeling makes it possible to develop effective monitoring systems especially dedicated to the preservation of historical infrastructure. Research indicates that these types of technologies have the potential to facilitate large-scale deployment of SHM through promoting flexible and cost-effective methods that allow continuous assessment of structural condition over the long term. This shift also enhances the validity of data collection as well as encourages a culture where information regarding preserving cultural heritage is shareable and implementable across various disciplines and domains (Basto et al., 2017). Another very important trend within SHM is the progression toward data-driven algorithms in damage detection, which are increasingly being integrated into real-time assessment. These algorithms employ the large volumes of data collected by sensors to translate raw readings into useful information regarding a structure's condition. Data-driven approaches are increasingly being favored as they can provide timely and accurate evaluation, which decision-making is based upon. The efficiency of such approaches generally rests with the processor power, which necessitates ongoing design and advancement of high-performance technology. As researchers continue to develop these algorithms, the field can anticipate breakthroughs that will allow better detection, localization, classification, and prognosis of structural damage and push SHM technology further into the realm of smart structures and predictive maintenance (Cherta et al., 2020). Besides, future structural health monitoring will be affected by the introduction of artificial intelligence (AI) and machine learning within traditional approaches. They have the potential to revolutionize the analysis and interpretation of structural health data. Their combination with AI algorithms allows for the development of adaptive models that learn from history, with enhancing accuracy of prediction over time. Such an evolution can lead to the creation of independent SHM systems that, in addition to damage detection, can predict potential failures through patterns of change in stored data. With the incorporation of such advanced computational methods, the discipline of SHM will surely experience greater efficiency and dependability, enabling structures to be maintained proactively rather than repaired or replaced reactively, ultimately extending their lifespan and furthering public safety. The potential for a more intelligent, data-driven approach to monitoring infrastructure is a critical step forward in managing both new and heritage structures (Figure 1.16).

Figure 1.16. The future of structural health monitoring (SHM).

Source: https://pulseiot.tech/wp-content/uploads/2024/08/Pulse-Banner_1920x1080_SeptemberHD.jpg.

1.6.1. Advancements in Sensor Technology

The technology of sensors has advanced considerably, particularly in their use in Structural Health Monitoring (SHM). The development of micro-electromechanical systems (MEMS) has upended traditional ways of monitoring the integrity of structures such as bridges and pavements. The small sensors can be directly embedded in construction materials such that continuous information regarding the condition of a structure can be obtained. The method supports enhanced system reliability and performance and extended life for infrastructure. Besides, technology from MEMS enables real-time monitoring, with immediate feedback being critical to identify possible failures in an initial stage. Literature review and vendor surveys highlight the benefits with off-the-shelf MEMS devices, with their effectiveness proven in SHM applications. The advances not only indicate technological progress but also enable more cost-effective and sustainable infrastructure management (Buss et al., 2013). Wireless sensor networks (WSNs) are also an influential innovation in the field of sensor technology for SHM.

WSNs allow multiple sensors to be placed on various structures without the constraint of wiring systems and therefore make the installation and maintenance processes simpler. The integration of WSNs into structural monitoring systems can provide real-time data acquisition and analysis, which is essential in assessing the health of infrastructure in dynamic conditions. The comparatively low price of these technologies makes them a feasible option for long-term monitoring programs, particularly for projects involving green infrastructure. Moreover, the advancement of sensor technology over time has the potential to add better data analytics and machine learning, which may improve the interpretative role of SHM systems. These innovations ensure stakeholders of infrastructure are able to obtain valuable insights that ultimately translate to enhanced safety and operational efficiency (Buss et al., 2013).

The intersection of sensor technology innovation and emergency response capability cannot be overstated, particularly in the context of seismic activity. Real-time or near real-time information through earthquake early warning (EEW) systems is essential to averting loss due to structural collapse in earthquakes. As evolved from recent research, immersive virtual reality environments are used to model ground motions and how they impact structural integrity in order to assist emergency responders in understanding the dynamics of seismic activity and structural response better. The immersion experience, incorporating elements of performance-based earthquake engineering (PBEE), enhances situational awareness and decision-making during the critical event. With the integration of earthquake science, engineering, and the human factors of emergency response, sensor technology developments not only improve traditional SHM practice but also help extend community resilience when faced with natural disasters.

1.6.2. Integration of Artificial Intelligence (AI) and Machine Learning (ML)

The use of AI and ML for structural health monitoring (SHM) systems is a great advancement for determining the structure's integrity. Most classic SHM methods employ fixed sensors and manual interpretation of data, which can be time-consuming and prone to human errors. Machine learning algorithms, however, can process vast amounts of data swiftly and effectively, enabling damage detection and identification to be done more efficiently. For instance, through the application of sophisticated techniques such as principal component analysis combined with parametric t-distributed stochastic neighbor embedding (P-t-SNE), experts can correctly classify structural states. This intersection, aside from enhancing the precision of damage detection, also decreases the operational costs of traditional monitoring systems, hence enhancing the field of SHM (Mao et al., 2018). Also, the implementation of AI technologies, such as computer vision, allows non-contact measurement techniques that significantly increase the practicability of SHM. Traditional vibration-based monitoring typically requires the installation of numerous physical sensors, which can introduce mass-loading effects and are hard to apply on large structures. High-density spatial data can be recorded and real-time observation and analysis allowed by using remote digital video cameras with sophisticated algorithms. Techniques like Phase-based Motion Estimation (PME) process this visual data to detect subtle structural motions linked with damage. This new approach not only reduces the installation burden of sensors but also enhances spatial resolution and measurement quality as it does not have the drawbacks of traditional contact sensors (Mao et al., 2018). Apart from monitoring, AI and machine learning enable predictive maintenance through enhanced forecasting of structural health patterns. By studying historical data together with current performance indicators, machine learning models are able to identify patterns that suggest potential failures or interventions needed. For instance, predictive capabilities by machine learning can lead to preemptive maintenance practices, maximizing a structure's lifespan and safety without unjustified low-level shutdowns and costly overhauls. With models achieving high accuracy rates, i.e., 99.5%–99.75%, for damage classification tasks, the validity of such AI-based approaches is evident (Cherta et al., 2020). With the advance in SHM technology, the increasing interweaving of AI and ML will certainly redefine monitoring techniques' standards, ultimately leading to more resilient infrastructure systems.

1.6.3. Global Standards and Regulations for Structural Health Monitoring (SHM)

With the rapidly evolving nature of SHM, establishing global standards and regulations is essential in ensuring the effectiveness and reliability of infrastructure management. Standardization facilitates the global integration of SHM technologies into a common framework that bridges the gap between various national regulations. This integration is particularly necessary in Europe, where ageing structures such as bridges and viaducts designed over more than five decades ago are progressively being undermined by rising traffic loading and climatic stressors. As highlighted in the SHM discussions, these recent series of catastrophic events like the Italian Morandi bridge collapse highlight the need to embrace harmonized practices with enhanced resilience of such key infrastructure. Harmonized regulation guarantees not only improved maintenance measures but also compliance with public safety goals across nations (Cantini et al., 2019). The implementation of international standards in SHM is also meant to address operational problems encountered in various industries, including renewable energy. The tidal power industry, for instance, has witnessed notable commercialization achievements; yet, its advancement has been hindered by inconsistent operational availability due to routine maintenance.

By adopting standardized SHM practices, the industry can acquire more dependable infrastructure using uniform condition monitoring strategies. This response results in reduced downtime and higher overall efficiency, addressing the low operational availability reported for this industry. For this reason, setting up regulatory standards is likely to have significant impacts on investment and innovation in the tidal power industry, ultimately leading to overall energy sustainability goals (Avdelidis et al., 2016). Furthermore, international collaborative initiatives can further enhance the regulatory framework for SHM. Initiatives such as the European cooperation on standardization, which involves concerted efforts like COST Action TU1402 on valuation of the value of SHMs, are representative of the potential boost provided to performance monitoring and assessment in various infrastructures. With group research efforts and sharing of knowledge, stakeholders can highlight best practices that inform policymaking and regulatory frameworks. This double-pronged strategy not only addresses short-term requirements of aging infrastructures but also establishes an avenue for novel developments in SHM technology. With the convergence of public safety and environmental matters, a successful standardization process can lead to more resilient infrastructures supporting sustainable development initiatives (Cantini et al., 2019).

SUMMARY

SHM system implementation is a critical process in maintaining the integrity of most structures, ranging from bridges to cultural heritage sites. As described in the essay, the use of sophisticated technologies in SHM has revolutionized traditional practices, allowing real-time monitoring and analysis of structural performance. The potential for affordable cures, particularly via open-source mechanisms, is a significant opportunity for the widespread utilization of critical infrastructure. Notably, research into more mainstream digital modeling and fabrication technologies provides an avenue for functional restoration and repair of heritage buildings and hence promoting cultural preservation in addition to functional monitoring options (Basto et al., 2017). This unified approach not only facilitates timely interventions but also provides enhanced historic record concerning architectural monuments. The ability to detect, find, and quantify damage is pivotal in effective SHM procedures. Reduced finite element-based models and test approaches have emerged as breakthroughs that enable more accurate damage detection at minimal levels of impact. This is the way forward to achieve greater understanding of structural behavior using modal parameter modifications to assess potential weaknesses. Developing an All In One (AIO) package with modal analysis is a giant leap in the way forward to create tools easily usable by engineers as much as researchers. By correlating experimental measures with numerical calculations, this approach facilitates improved structural safety interpretation, open to increased decision-making in maintenance practices (Abrate et al., 2011). In the process, extending these techniques not only emphasizes the importance of SHM but also follows progressive safety practice for public infrastructure. Lastly, the career of Structural Health Monitoring is poised for evolutionary advancement through the fusion of new methodologies and technologies in maximizing efficiency and accuracy. As shown in this essay, the integration between digital innovation and traditional monitoring methods provides comprehensive understanding of structural integrity. The emphasis placed on low-cost solutions offered by open-source technologies holds the potential of being used by many, particularly in areas underfunded with conservation goals for heritage. Moreover, with advanced impact detection means, overall safety and longevity of buildings can be improved to a remarkable extent. Finally, the adoption of these innovations by SHM serves instant structural needs but also creates a forward-thinking attitude that appreciates environmental responsibility and historic respect.

In the field of SHM, critical factors are formulated that determine its significance in maintaining the integrity of structures. Fundamentally, SHM revolves around the ability to detect and diagnose defects in structures to avert catastrophic failure. The integration of advanced technologies makes it possible to conduct real-time monitoring of structural conditions using sensors that capture information on vibration, stress, and other indicators of health. This approach not only enhances safety but also directs maintenance planning and resource allocation, thereby extending the life of structures. Last but not least, proactive detection of likely damages transforms reactive maintenance practice into a more reasonable and efficient model, which is the vital contribution of SHM to modern infrastructure management. Damage identification process is one of the primary approaches employed in SHM and can be addressed from different perspectives. The first one is sensor-based algorithms that utilize sensor data to track structural performance in real time. These algorithms have become very popular since they enable immediate information, hence timely decisions as well as intervention. Conversely, model-based approaches focus on utilizing analytical models to extract information regarding the health of a structure. Although both methods are important, increasing reliance on data-driven algorithms is a sign of a trend towards the use of high-tech means to make more real-time and dynamic structural stability assessments. This new discipline is a sign of the requirement for high-level computational power to deal with the enormous amounts of data processing that define SHM applications (Cherta et al., 2020). In addition, the emerging paradigm of damage-tolerant active control is a new synthesis between fault-tolerant design and SHM paradigms. The study comprises techniques that not only detect but also permit active intervention techniques to suppress identified damages. Active control systems possess the ability to modulate structural responses in real time, which assists in modifying the way structures react to various stresses and environmental conditions. Through incorporating such sophisticated systems within a unified SHM plan, the resilience and longevity of buildings can be significantly enhanced. The shifting paradigms of SHM bring into focus the importance of interdisciplinarity in engineering fields towards laying down the foundations for new solutions that are centered on structural safety to counter increasing stresses on infrastructure systems (MECHBAL et al., 2012).

REFERENCES

1. Abdel Wahab, M., Hongyou, C., Zhen, N., et al. (2017). Review on structural damage assessment via transmissibility with vibration-based measurements. https://core.ac.uk/download/84044964.pdf (accessed on 2nd October 2025).
2. Abrate, S., Amir Shahdin, A., Bathe, K. L., Bendsoe, M. P., Bendsoe, M. P., Carden, E. P., Cook, R. D., et al. (2011). Correlating low energy impact damage with changes in modal parameters: Diagnosis tools and FE validation. https://core.ac.uk/download/12041677.pdf (accessed on 2nd October 2025).
3. Acheson, S., Alonso, J., Bauer, P., et al. (2011). *Deep Underground Science and Engineering Laboratory–Preliminary Design Report*. http://arxiv.org/abs/1108.0959 (accessed on 2nd October 2025).
4. Agis Cherta, D., & Pozo Montero, F. (2020). Vibration-based structural health monitoring using piezoelectric transducers and parametric t-SNE. https://core.ac.uk/download/294830446.pdf (accessed on 2nd October 2025).
5. Agis Cherta, D., Gomez Vargas, R. C., Pedraza, C., et al. (2020). Damage identification in structural health monitoring: A brief review from its implementation to the use of data-driven applications. https://core.ac.uk/download/288625955.pdf (accessed on 2nd October 2025).
6. Angerer, G., Bauer, C., Bazilian, M., et al. (2011). Renewable energy in the context of sustainable development. https://core.ac.uk/download/13750495.pdf (accessed on 2nd October 2025).
7. Atkins, C., Brown, C. J., Colford, B., et al. (2012). Deflection and frequency monitoring of the Forth Road Bridge, Scotland, by GPS. https://core.ac.uk/download/20089517.pdf (accessed on 2nd October 2025).
8. Avdelidis, N. P., Georgoulas, G., Kappatos, V., et al. (2016). Tidal stream generators, current state and potential opportunities for condition monitoring. https://core.ac.uk/download/74410681.pdf (accessed on 2nd October 2025).
9. Barbera, C., Jones, M., Saliterer, I., et al. (2014). European local authorities' financial resilience in the face of austerity: A comparison across Austria, Italy and England. https://core.ac.uk/download/30643410.pdf (accessed on 2nd October 2025).
10. Basto, C., Chacón Flores, R. A., & Pelà, L. (2017). Open-source digital technologies for low-cost monitoring of historical constructions. https://core.ac.uk/download/81578722.pdf (accessed on 2nd October 2025).

11. Bush, S., Henning, T., McCarten, P., et al. (2011). Risk-based bridge data collection and asset management and the role of structural health monitoring. https://core.ac.uk/download/29465888.pdf (accessed on 2nd October 2025).

12. Bush, S., McCarten, P., & Omenzetter, P. (2015). Guidelines for data collection and monitoring for asset management of New Zealand road bridges. https://core.ac.uk/download/29465830.pdf (accessed on 2nd October 2025).

13. Buss, A., Ceylan, H., Ceylan, H., et al. (2013). Highway infrastructure health monitoring using micro-electromechanical sensors and systems (MEMS). https://core.ac.uk/download/38934593.pdf (accessed on 2nd October 2025).

14. Cantini, L., Carosio, S., Isoird, J. M., et al. (2019). Lifecycle management, monitoring and assessment for safe large-scale infrastructures: Challenges and needs. https://core.ac.uk/download/199459255.pdf (accessed on 2nd October 2025).

15. Colli, A., Kirchsteiger, C., & Vetere Arellano, A. L. (2006). Development of a general scheme for fuel cycles and life cycles from all energy technologies as a basis for the European Energy Risk Monitor (ERMON). https://core.ac.uk/download/38613834.pdf (accessed on 2nd October 2025).

16. Currie, M., Quail, F., & Saafi, M. (2012). Development of a robust structural health monitoring system for wind turbine foundations. https://core.ac.uk/download/9558383.pdf (accessed on 2nd October 2025).

17. Daskalakis, E., Kalogeras, I., Melis, N. S., et al. (2019). Stretching method-based operational modal analysis of an old masonry lighthouse. https://core.ac.uk/download/323063841.pdf (accessed on 2nd October 2025).

18. de Lautour, O. R., & Omenzetter, P. (2014). Damage classification and estimation in experimental structures using time series analysis and pattern recognition. https://core.ac.uk/download/29465879.pdf (accessed on 2nd October 2025).

19. Dong, Y., Liu, H., & Song, R. (2010). *Bridges Structural Health Monitoring and Deterioration Detection: Synthesis of Knowledge and Technology*. https://core.ac.uk/download/286607584.pdf (accessed on 2nd October 2025).

20. Frangopol, D. M., & Soliman, M. (2013). *Life-Cycle of Fatigue Sensitive Structures Under Uncertainty*. https://core.ac.uk/download/294831424.pdf (accessed on 2nd October 2025).

21. Hoell, S., & Omenzetter, P. (2014). Detection and localization of

trailing edge disbond in a large wind turbine blade. https://core.ac.uk/download/29151287.pdf (accessed on 2nd October 2025).

22. Limongelli, M. P., Omenzetter, P., & Yazgan, U. (2016). A pre-posterior analysis framework for quantifying the value of seismic monitoring and inspections of buildings. https://core.ac.uk/download/131024284.pdf (accessed on 2nd October 2025).

23. Lin, T. (2017). *Virtual Reality of Earthquake Ground Motions for Emergency Response*. https://core.ac.uk/download/213074488.pdf (accessed on 2nd October 2025).

24. Mao, Z., Niezrecki, C., Poozesh, P., et al. (2018). Vibration-based damage detection in wind turbine blades using phase-based motion estimation and motion magnification. http://arxiv.org/abs/1804.00558 (accessed on 2nd October 2025).

25. McCrindle, L., & White, G. (2010). *Interim Evaluation of Flying Start*. https://core.ac.uk/download/4150624.pdf (accessed on 2nd October 2025).

26. Mechbal, N., & Nobrega, E. (2012). Damage tolerant active control: Concept and state of the art. https://core.ac.uk/download/29137540.pdf (accessed on 2nd October 2025).

27. N/A. (2010). Smart sensors assess structural health. https://core.ac.uk/download/pdf/10557490.pdf (accessed on 2nd October 2025).

28. N/A. (2025). *Space Benefits: The Secondary Application of Aerospace Technology in Other Sectors of The Economy*. https://core.ac.uk/download/pdf/42883414.pdf (accessed on 2nd October 2025).

29. Odebiyi, O. I. A. (2002). *Alternative Modes of Financing Higher Education in Nigeria and The Implications for University Governance*. https://core.ac.uk/download/75779358.pdf (accessed on 2nd October 2025).

30. Ramírez Orellana, A., & del Valls Martínez, M. C. (2019). Patient satisfaction in the Spanish National Health Service: Partial least squares structural equation modeling. https://core.ac.uk/download/286590550.pdf (accessed on 2nd October 2025).

31. Wan-Wendner, R. (2018). *Aging Concrete Structures: A Review of Mechanics and Concepts*. https://core.ac.uk/download/287939526.pdf (accessed on 2nd October 2025).

CHAPTER 2

FUNDAMENTALS OF IOT IN ENGINEERING APPLICATIONS

Contents

INTRODUCTION

Internet of Things (IoT) is a paradigm revolutionary that interconnects devices and collects data, transforming the way in which systems communicate and function in engineering applications. As technology constantly evolves, the footprint of IoT expands, encompassing a wide spectrum of devices from basic sensors to complex machines. It's not merely connectivity; it is self-contained data sharing that enables devices to act on information they ingest independently. As IoT innovations permeate industries, engineers get the ability to develop systems that make things more efficient, expand capabilities, and improve operating procedures. Thus, an understanding of these interrelated systems and the technologies involved becomes essential for engineering professionals as they are looking to bring out the full potential of IoT in various applications in addition to addressing problems like security, interoperability, and integrity of data. It is through education that students become well-versed with the principles of IoT, gaining a grip on theoretical principles as well as practical applications. An IoT program that puts students in hands-on learning experiences shows students how education can be mixed with real-world applications of technology. The program is designed to break down advanced topics and allow students to lead themselves through a set of IoT technologies, learning problem design and best practices in software engineering. Following Julien et al.'s (2019) suggestion, the course modules encourage students to step into the design process and let their imagination run wild to design their IoT applications. By learning through interactive projects and real-time data interactions, students not only learn technical skills but also gain confidence to step into engineering fields that intersect IoT, validating the hypothesis that the face of engineering education must change as per technological advancements. Moreover, the integration of IoT with philosophies such as Business Process Management (BPM) serves as a prime example of the varied consequences of the technology in actual uses. IoT facilitates real-time data collection, which leads to observations that can significantly enhance decision-making within industries. The integration of IoT and BPM facilitates the streamlining of operations and innovative solutions to intricate organizational problems of the day. As listed by Burattin et al. (2017), the overlap of these two domains points to the necessity of transforming BPM principles to adequately manage IoT application lifecycles. Both engineers and business executives will need to contend with the emerging IoT complexities, ensuring that systems are made to be robust, adaptable, and able to leverage data for strategic advantage. Consequently, understanding the way these technologies harmonize gives interdisciplinary insight into the field of engineering, with resulting pathways for emerging innovation and optimality in myriad situations.

Definition of the Internet of Things (IoT)

The IoT is a paradigm change that connects physical objects to the Internet, enabling them to share data, analyze, and act in real time. This expansive network consists of a massive number of devices from household appliances to industrial equipment, and all these are filled with sensors and software that track and transmit. IoT facilitates this connectivity on a massive scale, not just streamlining operations across various industries but also allowing users to make data-driven decision-making. For instance, smart home technology allows one to visually monitor energy usage and adjust living conditions, offering the double benefit of convenience and sustainability. With the power fueling many applications, IoT is an innovation enabler that highlights its inherent position in today's engineering landscape, stressing its worth in propelling individual and industrial spheres of endeavor.

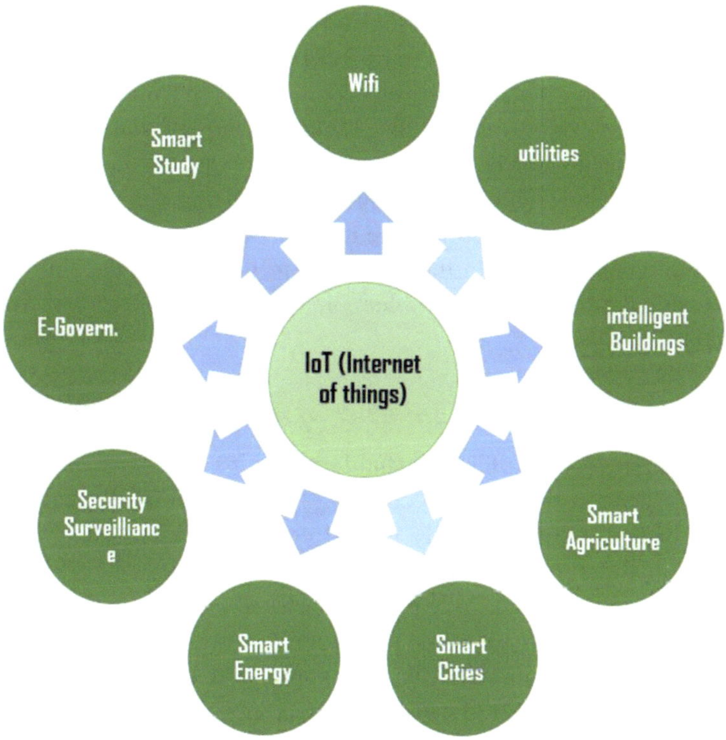

Figure 2.1. The Internet of Things (IoT).

Source: https://media.geeksforgeeks.org/wp-content/uploads/Internetof-thing.jpg.

Going deeper into the importance of Internet of Things, it is necessary to evaluate its impact on education, especially engineering and technology fields. IoT is a good platform for students to learn complicated software engineering ideas using practical projects. As described in recent studies, IoT courses include practical tutorials that allow learners to get familiar with crucial technologies in addition to developing key design thinking skills. This teaching approach is intended to create interest in computing, particularly among underrepresented populations, through demonstrating the practical application of theoretical knowledge in real-world contexts (Julien et al., 2019). Furthermore, the IOT project development process being cyclical in nature allows the students to learn software engineering methodology while fostering creativity in the domain of technological applications. Other than its pedagogical advancements, the Internet of Things also transcends boundaries with bigger trends in the information technology field. Initiatives like the IT2017 competency model place emphasis on integrating academic curricula with IoT competencies so that graduates are adequately prepared to address the future technology challenges of the global economy. Through the incorporation of IoT into learning models, organizations create opportunities for a future group of experts who are not only technically proficient but also strategically aware of their contributions' impact on the IoT environment (Alrumaih et al., 2016). With engineering fields increasingly utilizing IoT solutions, the need to have an overall understanding of the technology increases so that experts can tap into its capabilities while managing ethical and logistical problems inherent in an interconnected world (Figure 2.1).

Historical Development of Internet of Things (IoT) Technology

The development history of IoT technology begins from the initial ideas of networked devices in the late 20th century. The concept of devices talking to each other using networks initially gained popularity when the Internet took center stage in the 1960s and 1970s, with the possibility of various systems communicating information. Yet, it was not until the 1990s that the idea of networking everyday things became the focus of study, with the prime stimulus coming from advances in wireless communication technology. People like Mark Weiser advocated for the ubiquitous computing principle, laying the groundwork for IoT as we know it today. This period was marked by a vision above connectivity; it envisioned a way technology could be merged into everyday life in such a way that devices will gather and share information automatically to enhance human life. This base laid the grounds for IoT to set its base as an integral part of engineering applications. The start of the millennium witnessed enormous advances that further solidified IoT technology into mainstream. The introduction of IPv6 in 1998, which provided more space for addresses, made it easy to connect more and more devices. Additionally, the initial installment of objects receiving IP addresses, such as fridges and toasters, put the tone for the tangible progress being witnessed. With companies beginning to see the promise of operational efficiencies, business momentum gained. Various sectors, including manufacturing, medicine, and agriculture, began adopting IoT systems to enhance operations and drive innovation. This wider usage was characterized by a shift from theoretical explorations towards practical applications, making IoT relevant in every engineering use case possible. Thus, the convergence of IoT with Business Process Management (BPM) emerged, suggesting where earlier paradigms could converge to enhance operational outcomes (Burattin et al., 2017). By the 2010s, IoT's convergence with cloud computing and data analytics had become more pronounced, indicating a level of maturity in technology and application. Everywhere models increased, enhancing the capability of IoT with a multidimensional model of environmental information. This paradigm was built around a cycle of iterative modeling, continuously validated by real-time data, which has great potential for solving nuances across a broad spectrum of engineering disciplines (Ahmed et al., 2019). Important to this movement was the evolving sensor technology paired with machine learning algorithms that further enabled this movement, making possible progressively sophisticated IoT applications. With technology barriers reduced, the stakeholders embarked on examining the overall implications of IoT to transform industries, highlighting the imperative need for integrated research agendas that would be used to guide future innovations. The trajectory followed by IoT in its history emphasizes its pioneering role in driving efficiency and enhancing decision-making mechanisms in engineering applications (Figure 2.2).

Figure 2.2. The historical development of Internet of Things (IoT) technology.

Source: https://www.rishabhsoft.com/wp-content/uploads/2023/07/Banner-Image-IoT-in-Mobile-App-Development.jpg.

Importance of Internet of Things (IoT) in Modern Engineering

The IoT is an important driver of modern engineering in terms of optimizing efficiency and effectiveness of complex systems across various industries.

Furthermore, within the power industry, IoT technologies are transforming traditional power grids into Smart Grids (SGs) that can enable bi-directional energy transfer between end-users and service providers. This transition addresses most of the challenges associated with conventional grids, such as wastage of energy and unreliability.

Figure 2.3. The Internet of Things (IoT) plays a pivotal role in modern engineering.

Source: https://media.licdn.com/dms/image/v2/D4D12AQEc2GI_yt-2mQ/article-cover_image-shrink_720_1280/article-cover_image-shrink_720_1280/0/1691581216303?e=2147483647&v=beta&t=cauCcoLayktdessIiklL_1FsXYxD5vq7oaiSHQ2ofjs.

IoT devices like sensors, actuators, and smart meters are incorporated into SGs for real-time monitoring and control. The installation of these devices enables data-driven decision-making, optimizing energy consumption and enhancing system performance overall. In addition, remote monitoring and management of energy resources provide enhanced sustainability and robustness of energy infrastructure, and IoT thus becomes an unavoidable aspect of modern engineering solutions (Copeland et al., 2019). Apart from energy systems, IoT significance is observed in various fields of engineering, and its spread extends to logistics, manufacturing, and infrastructure management. The implementation of IoT technology allows for the creation of "smart" environments that can detect changes, adapt to new circumstances, and enhance performance. For instance, in manufacturing, IoT sensors can monitor the health of equipment, leading to more effective maintenance and less downtime. Similarly, in logistics, real-time tracking of assets allows for transparency and effectiveness in supply chains. As facilitators of more efficient operations, IoT solutions enable engineers to recommend innovative solutions that are commensurate with the increasing demand for automation and data analysis. Such developments indicate the transformative potential of IoT, positioning it as an imperative technology for engineering experts adopting a rapidly evolving technology landscape (Alrumaih et al., 2018). In addition, curricular blueprints guiding curricula development in engineering are increasingly centered on the role of IoT competencies. The ACM and IEEEs IT curricular report highlights the need to include contemporary technologies, including IoT, in information technology and engineering education. As the programs evolve, they must prepare students with an awareness and capabilities to manage the complications of IoT systems. This approach not only prepares entering engineers to apply IoT to real-world problems but also encourages innovation through imparting critical examination of technology's influence on society. By focusing on IoT within engineering courses, institutions can ensure that graduates possess a solid foundation to address persistent challenges and emerging opportunities in most branches of engineering, thus refining their skills in a technology-driven era (Alrumaih et al., 2018) (Figure 2.3).

2.1. KEY COMPONENTS OF INTERNET OF THINGS (IOT)

The application of IoT components in engineering use relies on a bundle of necessary elements that enable devices to communicate, work, and deliver tangible outcomes. At the heart of this framework stands the integration of sensors and actuators, which undertake essential roles in data gathering and physical execution, respectively. Sensors pick up information from the environment—such as temperature, humidity, or motion—whereas actuators turn signals into physical motion or action. Two-way flow is the foundation of IoT systems, enabling smart and responsive applications able to learn to adapt to changing conditions in real-time. Academic curricula, such as those that introduce high school students to IoT technology, underscore the importance of these parts by providing them with hands-on practice in programming and networking. This practical approach cultivates a deeper understanding of the technological intricacies involved in IoT solutions and therefore proves to be a new generation of engineers with in-depth knowledge of theory and practice (Julien et al., 2019). Besides sensors and actuators, connectivity is a pioneering component that characterizes the success of IoT systems. Establishing reliable communication among devices requires various networking standards and protocols, which facilitate data transfer among sprawling systems. Protocols such as MQTT (Message Queuing Telemetry Transport) and CoAP (Constrained

Application Protocol) are generally accepted for their efficiency in resource-constrained environments typical of IoT applications. Both protocols enable devices within an IoT ecosystem to communicate freely with each other, whether in home automation, industrial monitoring, or healthcare. Moreover, the advent of Smart Grids is a classic illustration of integration of connectivity in IoT, in which devices supply and receive power information in real time. By applying such technology, utilities optimize energy consumption and supply, addressing major challenges like energy efficiency as well as demand response (Copeland et al., 2019). Lastly, processing and analytics of data constitute another essential part of the IoT environment. The amounts of data collected through sensors are too large to withstand, and they need robust analysis frameworks that can filter actionable insights.

Artificial intelligence and machine learning methods are increasingly utilized to process such data, allowing predictive capabilities and enhanced decision-making processes. IoT devices are well able to manage data processing using the capabilities of cloud computing and edge computing, which ensure speedy responses to the underlying needs of various applications. This analytical dimension not only maximizes the potential of IoT applications but also empowers engineers and organizations to tap real-time insights towards operational improvement and innovation. As more intelligent solutions are in increasing demand, the synergy between data analytics and IoT will play a critical role in driving the engineering practice across sectors (Figure 2.4).

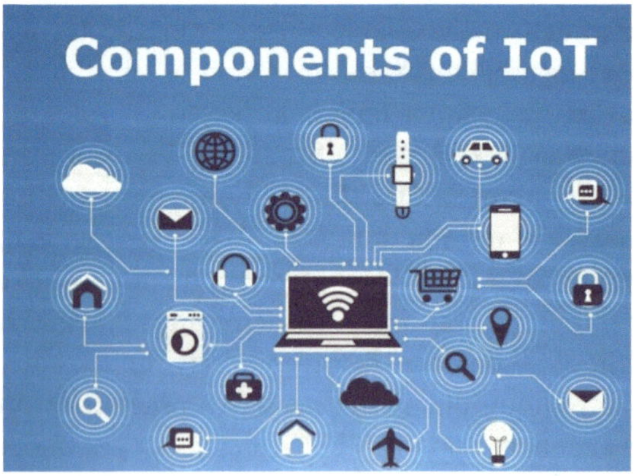

Figure 2.4. Key components of Internet of Things (IoT).

Source: https://cdn.educba.com/academy/wp-content/uploads/2020/01/Components-of-IoT.jpg.

2.1.1. Sensors and Actuators

The integration of actuators and sensors into Internet of Things (IoT) infrastructure is critical to the creation of intelligent spaces that adaptively respond to real-time data. Sensors carry out the vital role of collecting data from the physical world, be it temperature, humidity, movement, or other environmental parameters. The data are then relayed to a central processing unit or cloud platform, where sophisticated algorithms can analyze the data. Such a system makes it possible for conditions and actions to be automatically monitored without the involvement of human beings, thus enhancing efficiency in manufacturing, healthcare, and city management. IoT devices, based on sensor data, are capable of decision-making and action initiation autonomously, one illustration of how the interconnection of physical and virtual worlds gives rise to new engineering applications across various industries, all as guided by the guidelines discussed within debates on IoT (Burattin et al., 2017). Actuators bridge the gap between sensors by being the final device in the feedback loop, allowing systems to respond to information collected. After sensors have picked up on certain conditions, actuators execute physical tasks, like opening a valve, modulating a thermostat, or closing a door, based on pre-programmed commands or real-time data analysis. These actuators convert electrical signals into mechanical movement, presenting a physical response to the data-driven information they obtain from the sensors. It is this complementarity between actuators and sensors that underlies most IoT applications, particularly in automation procedures where timely responses are paramount. It is therefore important for engineering professionals to understand the operation and processes of these units of hardware to assure the successful deployment of IoT solutions.

Technological innovation in actuators will thus require continuous searching and innovating to enhance system responsiveness and reliability (Asensio et al., 2018). The mutual dependency between sensors and actuators not only enhances the functionality of IoT systems but also creates significant challenges in how to manage data and integrate systems. With the evolving IoT world, engineers are confronted today with the challenge of handling complexity related to the enormous amount of data from sensors that must be processed and analyzed via robust models. Second, actuators must be merged into existing infrastructure in a non-obtrusive way while being adaptable to different sensors, protocols, and communication standards. Resolving such concerns demands inter-disciplinary approaches that adapt concepts from disciplines such as business process management and data science. Through the use of these strategies, engineers can better design IoT systems that are efficient and responsive to the changing nature of modern environments and therefore leverage the entire potential of actuator and sensor technology to engineering ends (Asensio et al., 2018; Burattin et al., 2017) (Figure 2.5).

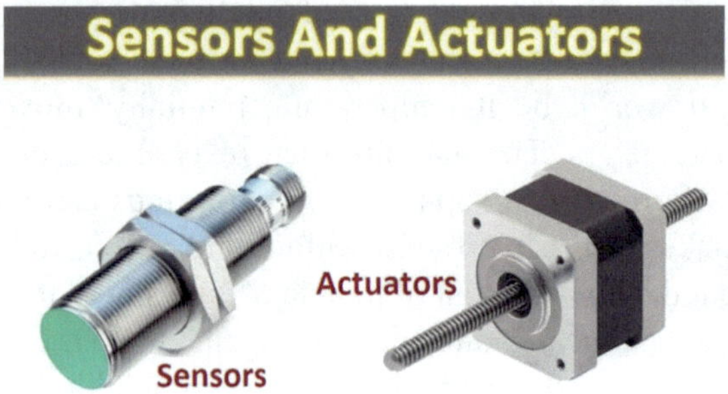

Figure 2.5. Sensors and actuators.

Source: https://cranesvarsity.com/wp-content/uploads/2023/07/csq1.jpg.

2.1.2. Connectivity Protocols

Connectivity protocols are critical to the functionality and scalability of Internet of Things (IoT) applications, facilitating straightforward interaction between different devices. The protocols are the foundation for data exchange in various industries, from industrial automation to smart cities. Some of the key protocols include the Constrained Application Protocol (CoAP) and Message Queuing Telemetry Transport (MQTT), which are specifically designed for low-bandwidth and high-latency networks. They address the specific challenges posed by resource-limited devices and sparse network conditions. The efficacy of these protocols is evident through their ability to optimize energy efficiency and minimize data transfer overhead, thus making them suitable for different IoT devices, particularly wireless sensor networks where energy efficiency is of significant importance (Leung et al., 2014). With the rise in IoT deployments, the significance of robust and efficient connectivity protocols cannot be stressed enough, as this decides the fate of such technological endeavors. In addition, the development of traditional power grids into Smart Grids (SGs) illustrates the pivotal role of connectivity protocols in modern-day engineering applications. SGs require two-way communication for monitoring, management, and automation to facilitate effective energy management. Connectivity protocols allow Smart Grids to integrate various IoT devices, such as sensors and smart meters, that send real-time energy consumption and grid status data. Such integration becomes critical to enhance reliability and security and react to issues of wastage of energy and rising demand. With the emergence of IoT-supported Smart Grids, the performance capacity of such systems to improve diverse network functions throughout the energy lifecycle from production to end-use can be attributed to the effective deployment of robust communication protocols (Copeland et al., 2019). In response, optimization of such connectivity protocols is fundamental in facilitating innovation in resilient and sustainable energy systems. In addition to technical competence, connectivity protocols also assist in guaranteeing security and privacy of IoT applications, which are increasingly exposed to cyber-attacks. Assurance that data is shared securely across various devices is essential, particularly in sensitive information-exchange applications such as healthcare and financial institutions. Advanced security features integrated into connectivity protocols such as encryption and authentication can significantly reduce exposure to unauthorized access. As the topography of the IoT evolves, there is an increasing need for adaptive protocols that not only enhance connectivity but also reinforce the security envelope surrounding IoT deployments. The simultaneous focus on connectivity and security will be critical to establishing user confidence and attainment of mainstream acceptance of IoT technologies. An integrated strategy, with seamless data exchange and strong security provisions, will be essential for the long-term sustainability of IoT in engineering applications (Figure 2.6).

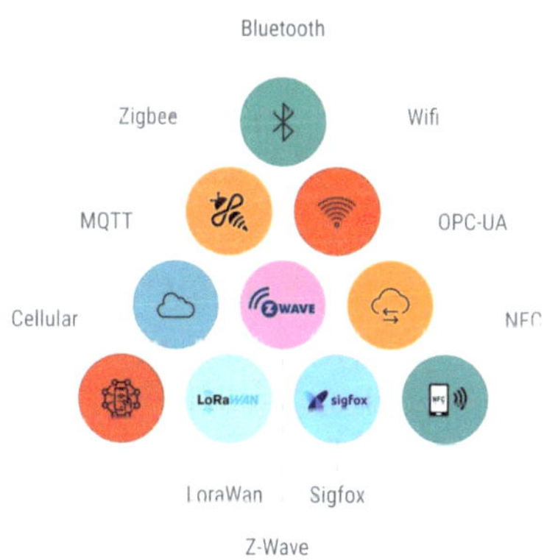

Figure 2.6. Connectivity protocols.

Source: https://images.ctfassets.net/8cjpn0bwx327/2ujUbTUYbDtLqdTKiRm1sw/ca9ec9d7d81b8397a1fa7cac268828d0/Exploring_Secure_Communication_Protocols_for_IoT_Security.webp.

2.1.3. Data Processing and Analytics

Incorporation of advanced data processing and analytics in Internet of Things (IoT) systems is essential for their performance and functional efficiency in a variety of engineering applications. Data generated by IoT sensors tends to be of high velocity, volume, and variety; thus, robust analytics frameworks are the driving force behind drawing useful insights from this deluge of data. For real-time use cases such as autonomous driving or physiological monitoring, the ability to analyze data on-the-fly not only increases responsiveness but also taps the maximum possible potential for predictive insights. For instance, application of machine learning algorithms can enable real-time decision-making using real-time sensor data and enhance operating efficiencies. Consequently, organizations employing such IoT architectures can expect problems in advance before they become actual issues, thereby ensuring smooth functioning and improved outcomes across various fields of engineering. Although processing of data enhances the functionality of IoT systems, it also raises issues related to security of data and associated risks of unauthorized access.

With evolving analytics processes, particularly where there is edge computing, solutions need to focus on data integrity as they handle large volumes of data. The introduction of near-sensor data analytics, as noted in the literature, minimizes energy consumption on communication and also restricts network traffic. However, with this new paradigm comes its own challenges, particularly ensuring safe passage of sensitive data when being transmitted. Forthcoming directions like Fulmine, which is a System-on-Chip integrating encryption with data processing capability, indicate the future direction through its high energy efficiency in secure analytics applications. Through the judicious use of encryption methods across the analytics stack, organizations can eliminate risk while taking advantage of the robust aspects of IoT technology to offer secure, scalable solutions to practical use cases (Benini et al., 2017).

The promise of real-time analytics of data to transform complex human activity tracking is beautifully illustrated by the development of wearable technologies. It is essential to develop systems to process and analyze IoT sensor streaming data, particularly in the healthcare sector. An excellent example is using deep neural networks to detect unusual human activity patterns, say falls, from accelerometer data. Achieving an impressive accuracy rate of 98.75% validates the effectiveness of machine learning in real-time application. Moreover, the system architecture ensures that one can monitor the systems extensively in real-time, enabling timely interventions and notifications. All these advancements not only enhance the quality of care provided in facilities like rehabilitation clinics but also empower users with the capability to proactively assume control over their health management. As the processing and analysis capabilities expand, the usage of IoT in engineering applications will most likely rise significantly, encouraging innovation and improving overall efficiency (Isah et al., 2018) (Figure 2.7).

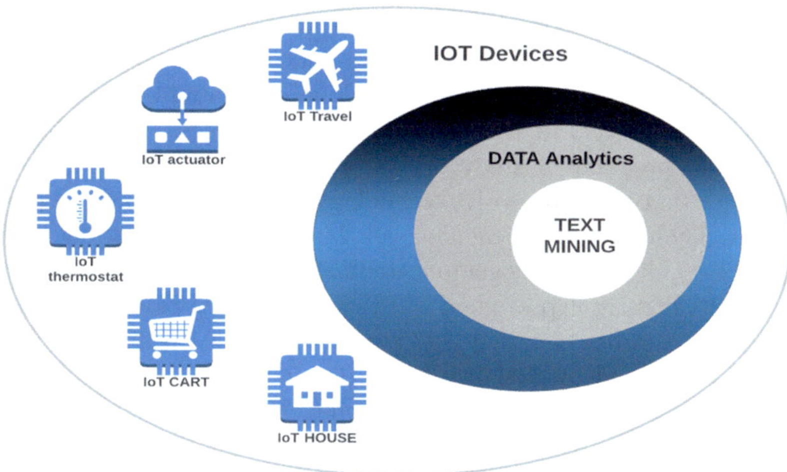

Figure 2.7. The data processing and analytics in Internet of Things (IoT) systems.

Source: https://www.researchgate.net/publication/354252613/figure/fig1/AS:1063096412028929@1630473344842/IoT-Data-Analytics-and-Text-Mining.png.

2.2. INTERNET OF THINGS (IOT) ARCHITECTURE

The IoT architecture is the infrastructure behind which devices can be easily connected and can communicate smartly with one another. Essentially, IoT architecture has several integral elements, including sensors, connectivity protocols, processing layers, and user interfaces. Sensors glean data from the environment, which is transmitted over several networking protocols. This data is analyzed, often in the cloud or on-prem edges, to facilitate real-time decision-making and analytics. The IoT architecture's modular and open design supports a range of applications, from the smart home to industrial automation, to allow engineers to design solutions that not only work efficiently but also scale and are secure. One of the emerging trends in IoT architecture is support for RESTful web services, particularly in energy-constrained wireless sensor networks (WSNs). RESTful architecture accommodates efficient data transfer with light-weight protocols, which is appropriate for devices with limited resources. Support for IPv6 in WSNs introduces connectivity and supports a large number of devices, increasing the applications of IoT systems. Using powerful gateway solutions, systems can render any internet client to be controllable to communicate with the sensors, enabling greater interoperability and functionality. With such IoT architecture advancements, overall data traffic is greatly enhanced without raising energy consumption, which is still of prime concern in both industrial and consumer applications (Leung et al., 2014). Educational programs have a significant role to play in introducing the future generation of engineers to IoT architecture and its uses. As an illustration, a high school program of study that aims to instruct students in software engineering using IoT can involve technology in practical, hands-on projects. Through technology experimentation, allowing students to try out different IoT technology, they not only get to learn the technology involved but also the design factors that are part of good software engineering practice. The course is designed to nurture creativity and problem-solving capabilities, allowing students to design and develop their own IoT solutions. The practical approach not only makes the complexity of IoT architecture easy to comprehend but also inspires students to envision its greater potential in upcoming engineering innovations (Julien et al., 2019) (Figure 2.8).

Figure 2.8. The Internet of Things (IoT) architecture.

Source: https://www.digi.com/getattachment/Blog/post/IoT-Architecture-Topology-and-Edge-Compute/GettyImages-968289874-1280x720.jpg?lang=en-US.

2.2.1. Layered Architecture of Internet of Things (IoT) Systems

In the case of IoT systems, a layered method provides a systematic process of addressing complex interactions among devices, networks, and applications. A layered system tends to be composed of many layers that are responsible for specific functionalities, which collaborate to enhance the overall system efficiency and effectiveness. Embedded devices and sensor networks at the base layer acquire real-time data from the environment. These sources of information are the primary input to the next layers, i.e., data processing, storage, and application deployment. By setting responsibilities between layers, this architecture not only promotes scalability but also facilitates interoperability across heterogeneous devices, thereby making it easier to integrate into current engineering applications based on IoT technologies. As IoT technologies continue to advance, increasing computational power and data handling capability are required, which results in the inclusion of fog and cloud computing in their layer architecture. Fog computing goes one step beyond the cloud service model by enabling processing at or near the data source, thus reducing latency as well as bandwidth usage. As noted in recent research, several flavors of fog computing offer assistance with decentralized processing of data between edge devices, which can perform data decision-making and data storage without being solely reliant on the cloud.

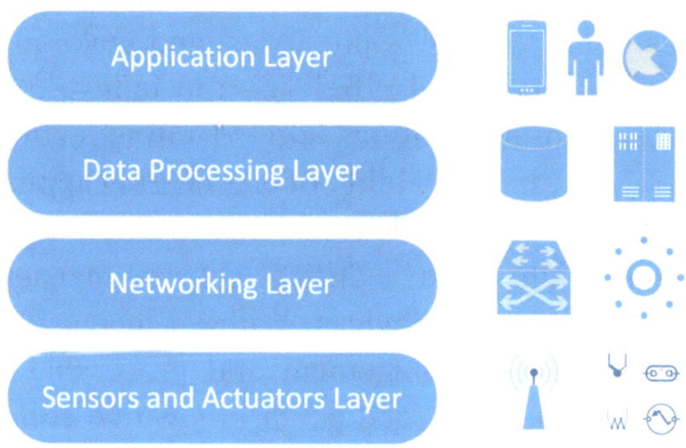

Figure 2.9. Layered architecture of Internet of Things (IoT) systems.

Source: https://www.hiotron.com/wp-content/uploads/2019/07/The-architectural-layers-of-IoT-systems.png.

This arrangement of fog services in the layered architecture is a manifestation of evolving towards intelligent and smart systems as they can react in real-time to the subtleties of smart manufacturing and smart city applications, their operational resilience and sustainability significantly enhanced. In actual use, the IoT systems' layered structure enables the development of smart scenarios consisting of applications like smart transportation systems and smart grids. For example, traditional power grids are being upgraded into Smart Grids (SGs) that utilize IoT to facilitate bi-directional communication among the consumers and suppliers of energy. This is aimed at highlighting how IoT devices such as smart sensors and smart meters have a critical contribution in monitoring and control throughout the energy cycle, and affirming the need for a well-designed architecture for integrating such high-end features together. By adequately addressing the issues of communication flow and energy management, layered architecture enables innovations that aim at enhancing reliability, security, and efficiency, placing IoT in the bracket of a force of change in engineering applications (Copeland et al., 2019; Jukan et al., 2016) (Figure 2.9).

2.2.2. Role of Cloud Computing in Internet of Things (IoT)

The combination of the IoT and cloud computing offers a strong platform for existing engineering applications, with enhanced data analysis, processing, and storage. The cloud offers a central location where vast amounts of data gathered from IoT devices are consolidated, processed, and stored securely. This unified approach not only makes data more accessible, but also enables advanced computational techniques such as machine learning and artificial intelligence, which can extract valuable insights from raw data. Engineers thus utilize these insights to visualize trends, optimize processes, and make informed decisions in real-time, enhancing operational efficiencies in industries. The frictionless integration of cloud computing solutions into IoT infrastructures plays a critical role for the responsiveness and scalability of engineering applications, making it possible for organizations to respond proactively to emerging market demands. Secondly, cloud computing plays a critical role in managing the compute load generated by numerous IoT devices. With an increase in interdependent devices, so does data traffic management and processing requirements. In response, fog computing has emerged as an innovative solution that extends cloud capabilities close to the network edge, thereby reducing latency and improving response times for mission-critical applications. Fog computing acts as a middleman by putting resources closer to IoT devices, thereby ensuring latency-critical tasks can be executed in real time. For instance, applications that need real-time analytics in sectors such as transportation or healthcare significantly exploit this architecture. Not only does this distributed system improve efficiency by offloading some work from centralized cloud resources but also optimizes resource utilization on the network and effectively monitors the state of attached devices. Communication protocols employed within IoT systems significantly contribute to effective data exchange between devices and cloud infrastructures. The analysis of request-reply and publish-subscribe protocols illustrates the application of diverse approaches to meet some of the needs of IoT communications. According to recent studies, the choice of protocol tremendously impacts performance characteristics such as latency, energy, and total throughput in fog and cloud environments (Carpio et al., 2019). By survey findings, the engineers can learn about the implication of each protocol in an IoT-to-fog-to-cloud converged system to help inform decision-making on system architecture. This is crucial in designing robust communication structures that complement interoperability, responsiveness, and scalability of IoT applications towards the ultimate goal of improved service delivery across all engineering disciplines (Figure 2.10).

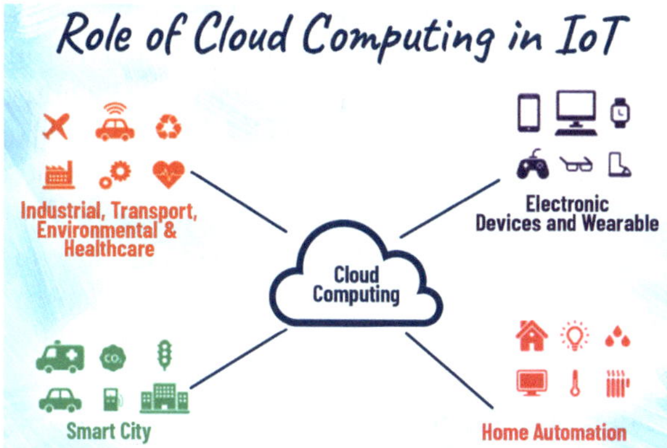

Figure 2.10. Role of cloud computing in Internet of Things (IoT).

Source: https://images.theengineeringprojects.com/image/main/2022/03/1-14.jpg.

2.2.3. Edge Computing and Its Significance

The term edge computing emerged as a change driver to the world of the Internet of Things (IoT) by fixing a significant issue related to data analysis and processing. Traditionally, enormous quantities of data gathered by IoT devices are forwarded to core cloud infrastructure, where they get analyzed. But with the rise in connected devices, this approach is marred by bandwidth limitations, latency issues, and concerns regarding privacy. Local processing of data is facilitated through edge computing by bringing computation closer to sources of data. With this paradigm shift, responsiveness of applications is not merely enhanced—critical for real-time processing—but also bandwidth is saved, hence making network resources more effectively utilized. With edge solutions, engineering applications can tap into real-time understanding, bringing faster efficiency in critical segments such as manufacturing, healthcare, and transportation, where quicker decision-making is of paramount importance. Also, integrating edge computing into IoT systems significantly enhances data privacy. Processing of data in the center has problems concerning data vulnerabilities and unauthorized access, particularly with sensitive data. By processing data at the edge in local devices, organizations reduce exposure of sensitive information but gain insightful results. Local processing is compliant with the data transmission minimization policy and hence sustains compliance with privacy regulations such as GDPR. Furthermore, data fusion across various sources in the edge allows systems to provide context to information and enhance results. This ability to process data at the source can potentially lead to faster response times and better user experiences, which are key in industries that rely on timely and accurate information for success. Finally, the significance of edge computing in engineering applications extends to innovation in the design of user experiences. With more and more professionals seeking to create intuitive interfaces that can easily interface with IoT devices, the technical barrier for those without expertise can be overwhelming. Present advancements, as described, mean that streamlined coding platforms can provide the ability to create sophisticated applications without extensive programming knowledge. Using block-based programming languages, such as the ones adapted to incorporate new stream processing features, enables one to develop applications that can ingest, process, and represent streams of data from different sensors efficiently. This democratization of technology encourages enhanced levels of innovation in smart application design and, generally, encourages greater interaction with IoT solutions across disciplines and therefore propels the future of engineering applications forward (Alturki et al., 2017; David et al., 2019).

2.3. APPLICATIONS OF INTERNET OF THINGS (IOT) IN CIVIL ENGINEERING

Application of IoT in civil engineering is made possible through an extensive range of applications that enhance infrastructure reliability and efficiency. One of the most critical applications is in structural health monitoring (SHM), where sensors are embedded into bridges, buildings, and other infrastructure to track stress, vibration, and environmental conditions in real-time. This information informs engineers on when to perform maintenance and add years to critical infrastructure, as well as ensure public safety. As the latest research has demonstrated, the transition from traditional monitoring systems to IoT-based solutions allows real-time assessment of structural integrity that is not only less costly but also poses fewer threats to possible breakdowns. All these innovations are characteristic of the forward-thinking nature facilitated by IoT, departing from the earlier practices of reactive maintenance to a predictive approach in civil engineering. IoT applications extend beyond health and safety tracking; they are also crucial in improving construction and resource management processes. For instance, IoT devices assist in ensuring efficient project management by monitoring materials and equipment in real time to guarantee that resources are best allocated during the construction period. With the help of systems that use connected devices, project managers can avoid delays caused by shortages or misplaced equipment. Furthermore, these systems, which are IoT-based, enable stakeholders to communicate with one another seamlessly, thereby improving coordination between various parties involved, such as contractors and suppliers. Networked by design, IoT technology fosters better decision-making and resource management, which is pivotal in an industry that is normally marred by time and cost concerns. Besides, the integration of IoT technologies in civil engineering is complemented by advancements in concepts such as the Internet of Robotic Things (IoRT), which enhances operational capabilities of construction sites further. Robotics, if paired with IoT systems, can automate work such as heavy lifting or assembly which involves putting together minute details, thus enhancing efficiency and reducing cost elements. As specified in the literature on the subject, the convergence of IoT and robotics addresses urgent issues like productivity and security. Autonomous devices can be programmed to carry out synchronized functions with enhanced coordination of efforts in usually perilous conditions. These advancements reflect not only the technological progress of civil engineering but also how applications of IoT can improve construction practices and make them safer and more efficient, eventually transforming the industry into a more productive and technology-driven sector (Bacciu et al., 2017; Johnson et al., 2016) (Figure 2.11).

Figure 2.11. Applications of Internet of Things (IoT) in civil engineering.

Source: https://www.planradar.com/wp-content/uploads/2019/04/proptech-construction-industry.jpg.

2.3.1. Smart Infrastructure Monitoring

The arrival of smart infrastructure monitoring through the Internet of Things (IoT) is a paradigm shift in the manner engineering applications monitor systems in real time. This solution uses an array of various sensors and devices with IoT technology integrated for continuous physical infrastructure collection and monitoring. Physical infrastructure such as bridges, roads, and buildings can now be fitted with sensors that track stress, temperature, vibration, and material condition parameters. These sophisticated monitoring systems enable engineers to assess the health of crucial infrastructure in a trustworthy way, which promotes safety and lifespan while keeping maintenance costs to a minimum. Moreover, by providing real-time data, smart infrastructure monitoring supports proactive decision-making to allow timely intervention of minor issues before extreme failure. The IoT contribution in transforming traditional operations to wiser grids highlighted in smart infrastructure monitoring reinforces its relevance in engineering application. As per literature, traditional power grids suffer from one-way information transfer and energy wastage when the demand is high. On the other hand, Smart Grids make use of bidirectional communication through IoT devices such as sensors and smart meters. The technologies allow for monitoring, analysis, and control of energy systems at the generation, transmission, and distribution levels.

Figure 2.12. The emergence of smart infrastructure monitoring through the Internet of Things (IoT).

Source: https://cf-images.eu-west-1.prod.boltdns.net/v1/static/1813624294001/f6981a14-a7ea-4b5a-a4ad-0cf8a074534f/b782da08-aa5e-4efe-89ae-1c7e77d12063/1280x720/match/image.jpg.

By combining a set of networked devices, Smart Grids are capable of streamlining resource consumption and system reliability, which is essential with an emphasis on sustainable energy systems. Deployment and management of such IoT systems are reflective of the shift in engineering culture, marrying infrastructure robustness with technology innovation (Copeland et al., 2019). Also, the integration of smart infrastructure monitoring with Business Process Management (BPM) emphasizes a significant experimentation in optimizing engineering processes. While IoT and BPM have been primarily dealt with as separate fields, the recognition of the synergies that arise from the integration of these models is becoming increasingly important. IoT applications generate big data, which is processable and can be utilized to optimize business processes in engineering sectors. This relationship is difficult since current BPM practices need to adapt to manage the complexity of the IoT system in an effective way. The potential for optimized processes through IoT technology-BPM methodology convergence not only enhances operating efficiency but also generates innovation in engineering processes. Such an integration can lead to wiser infrastructure, opening the door to increased adaptability and a wiser built environment (Burattin et al., 2017) (Figure 2.12).

2.3.2. Traffic Management Systems

Traffic management systems have emerged as key components in the intelligent transport sector, making use of the Internet of Things (IoT) to optimize urban mobility. These systems benefit from networked devices and sensors to facilitate the real-time monitoring of traffic, which allows for effective data collection and analysis. With the application of technologies such as mobile edge computing, traffic management systems can process data close to the source. This eliminates latency and facilitates an easy quick decision-making process, which is crucial in responding to evolving traffic conditions. With MEIoT architecture that adopts principles of cloudlet and software-defined networking, these systems not only reduce traffic flow but also contribute to the reduction of on-grid energy consumption, emphasizing the importance of sustainable approaches in engineering applications (Ansari et al., 2017). The adoption of machine-to-machine (M2M) communication within traffic management systems offers a paradigm shift towards cooperative applications that encompass broad industry verticals. This paradigm shift not only enhances data collection but also allows for interoperability between different traffic management systems. With cloud-based IoT architectures, these systems are able to offload the processing of massive amounts of sensor data from the devices to the cloud. Use of Fog Computing adds to this paradigm even more by supporting distributed computing and control between embedded devices and central processing units (Akrivopoulos et al., 2017). Such a distributed system is crucial in processing the increasing volume of data generated within cities since it prevents congestion in the central data centers and enables localized processing that is sensitive to real-time traffic variation. Moreover, the efficiencies achieved as a result of advanced traffic management systems create not only improved traffic flow but also improved road safety. Through real-time data analytics and predictive algorithms, they can predict traffic congestion, accidents, and other potential hazards, and take action in a timely manner. For instance, real-time data-based traffic light control can significantly reduce waiting time at intersections and ultimately lead to reduced levels of pollution. Emergency response teams can similarly leverage real-time access to traffic conditions in enhancing operational efficiency during the progression of critical incidents. By the integration of such advanced techno-solutions in urban planning and urban development, traffic management systems assist in bringing about the creation of wiser, eco-friendly cities in line with modern engineering concepts and practices.

2.3.3. Environmental Monitoring and Management

Environmental monitoring and management have been revolutionized by the integration of Internet of Things (IoT) technologies. Historically, environmental monitoring involved tedious manual data collection methods that were prone to errors and took a long time. With the advent of IoT, however, an interconnected network of working devices can now provide live data from numerous sensors spread all over various environments. These sensors record critical parameters such as air and water quality, temperature, and soil. By utilizing cloud platforms for data aggregation and analysis, stakeholders are in a position to better understand environmental change more efficiently, which results in timely decision-making for management. With this innovation, accuracy in data is not only enhanced but also helps organizations to act ahead of time in solving environmental challenges, thereby facilitating sustainable development and conservation in a dynamic environment (Akrivopoulos et al., 2017). In addition, the use of Fog Computing for environmental monitoring greatly mitigates the impediments of massive data volumes created by IoT devices. Fog Computing provides localized processing near the origin, which decreases latency and bandwidth consumption related to the transfer of massive datasets to cloud platforms. This shift in paradigm guarantees that actionable information can be derived in real-time, allowing for instant reactions to environmental events like outbreaks of pollution or natural calamities. For instance, real-time monitoring of air quality through the use of IoT devices can invoke immediate responses from the authorities, thereby protecting public health. Besides, Fog Computing facilitates cooperative applications that enable multiple stakeholders, including governments, researchers, and non-profits, to efficiently access and use environmental information, ensuring an overall strategic perspective on environmental management (Akrivopoulos et al., 2017). Meanwhile, Smart Grids (SGs) development is a prime example of the significant contribution IoT makes in sustainable management of energy resources, with a direct impact on the environment. By enabling two-way energy transfer and time-domain monitoring of energy consumption, SGs optimize resource utilization and loss reduction. It is achieved by the application of sensors and smart meters that track the energy consumption pattern in real-time, which allows dynamic re-allocation and greater consumer involvement. The integration of IoT in SGs offers a more effective and secure power delivery, addressing the shortcomings of traditional energy systems. As environmental sustainability is increasingly becoming a mounting issue, the development of IoT-based Smart Grids acts as evidence of the extent to which engineering solutions contribute to enhancing environmental monitoring and management activities as well as energy efficiency (Figure 2.13).

Figure 2.13. Applications of Internet of Things (IoT) in mechanical engineering.

Source: https://www.digi.com/getattachment/Blog/post/What-Is-Environmental-Monitoring/GettyImages-1130498419-1280x720.jpg?lang=en-US.

2.4. APPLICATIONS OF INTERNET OF THINGS (IOT) IN MECHANICAL ENGINEERING

The use of IoT in mechanical engineering has changed the dynamics of how industries design, produce, and service equipment. Through connecting mechanical components and systems with smart sensors and data processing, engineers can achieve increased efficiency and reliability. IoT technology enables the monitoring of machinery in real-time, facilitating predictive maintenance strategies, reducing downtime and expense. For instance, sensors may constantly track machinery performance and predict potential failures prior to their occurrence, thereby preventing costly shutdowns. Such forward-looking practice falls under the Industry 4.0 umbrella, which focuses on intelligent manufacturing and process implementation in production as per accurate data monitoring to effectively serve customer demands, as in most case studies illustrating its effective application in car customization and beyond (Chen et al., 2016). And neither should the role of IoT contribution be downplayed in facilitating collaborative designs since it enhances an environment where innovation is permissible. In mechanical engineering, designers rely more and more on networked systems in order to facilitate communication between stakeholders in design. Through the implementation of cloud-based solutions employing big data analytics, groups can work simultaneously on product designs while considering customer preferences and tailor features accordingly. Such targeted strategy is particularly beneficial in mass customization, presenting the greatest of both worlds when it comes to mass-producing personalized products.

Figure 2.14. The integration of the Internet of Things (IoT) into mechanical engineering.

Source: https://alpha-numero.com/wp-content/uploads/2022/05/in-Mechanical-Engineering.webp.

As a platform, amassing and dissecting user data fuels a cycle of ongoing improvement, enabling product development to occur as feedback and market trends are traced in real time (Chen et al., 2016). Besides, the concept of Federated Embedded Systems (FES) shows the manner in which IoT solutions are incorporated in mechanical engineering environments. Such systems are characterized by their interconnectedness and common functionalities and enable distinct mechanical components to function as integral units in networks. This system not only amplifies the operation efficiencies but also elevates the capabilities of the existing mechanical systems through incorporating them into larger cyber-physical systems. As mechanical engineers follow the prospects of FES, they operate in a world of interdisciplinary innovation through fields such as ubiquitous computing and multi-agent systems. The intersection of these diverse technologies advances intelligent devices that address current engineering problems, leading to an era of enhanced cooperation and intelligent designing (Figure 2.14).

2.4.1. Predictive Maintenance of Machinery

The advent of the Internet of Things (IoT) has revolutionized the practice of predictive maintenance in machines, revolutionizing the approach industries take with their machines and operation efficacies. Predictive maintenance employs data analysis, machine learning, and connectivity to make future predictions of equipment failure before it happens. This forward-thinking approach reduces downtime and maintenance costs significantly, creating a more effective manufacturing environment. By integrating IoT technologies, business enterprises can monitor real-time data from sensors on machines and thus monitor health metrics such as temperature, vibration, and fluid levels. Consequently, this data-driven decision-making not only enhances maintenance schedules but also extends equipment life, directly affecting productivity levels and resource planning in the organization. As industries shift towards this strategy, it is more pertinent to align predictive maintenance with Industry 4.0's strategy, resonating with the dominant interaction of strategy and technology (Holtkamp et al., 2017). Furthermore, integrating predictive maintenance into the framework of Industry 4.0 plays a central role in enhancing energy efficiency in production systems. Where businesses are grappling with higher energy prices and energy consumption limits from the regulatory agencies, optimization of the working capacity of equipment is paramount. Predictive maintenance can identify inefficiency and wastage of energy and hence allow organizations to rectify issues before catastrophic failures shut down operations. Research has proven that the application of these forecasting mechanisms not only saves energy but also meets sustainable production schemes through optimization of resources (Casta Dñeda et al., 2019).

Through efficiency in energy, companies are not only improving their bottom line but also achieving corporate social responsibility goals, showing how predictive maintenance innovation can have stakes beyond efficiency. Finally, the success of predictive maintenance through IoT is also contingent on the collaboration between different industries, blending the know-how of manufacturing, IT, and universities. For predictive maintenance to be used to its maximum capacity, stakeholders must develop collaborations that have an eye for the share of knowledge, technology, and resources. Collaborations can yield the creation of customized solutions that can address the specific challenges of a certain industry, in this case, scalability and flexibility. Through the fostering of such collaborations, organizations can speed up the deployment of efficient predictive maintenance practices that take advantage of IoT capabilities to yield innovations leading to operational excellence. As more and more industries embrace the principles of Industry 4.0, emphasis on predictive maintenance will definitely enhance machine uptime and reliability, further solidifying its place in the shifting paradigm of engineering applications (Holtkamp et al., 2017) (Figure 2.15).

Figure 2.15. Banner image Internet of Things (IoT) in predictive maintenance.

Source: https://www.rishabhsoft.com/wp-content/uploads/2023/07/Banner-image-IoT-In-Predictive-Maintenance.jpg.

2.4.2. Automation in Manufacturing Processes

The integration of automation into manufacturing processes is a revolutionary transformation, driven mainly by advancements in the Internet of Things (IoT). The technology revolution allows the synchronization of machines with digital platforms, allowing real-time monitoring and control. Automation not only reduces the requirement for human interference but also significantly eradicates the occurrence of human errors, resulting in better quality and consistency of products. Through the use of sensors and connected devices, creators can harvest gargantuan volumes of information to inform decision-making. This data-driven strategy forms a progressive management methodology, with troubles of any kind being located and resolved with promptness, optimizing the production line. In this case, automation serves as an innovation driver that makes manufacturers more responsive to market demands and changes in consumer behavior and thus maintain their competitiveness in changing markets. As manufacturing environments continue to change, the Industry 4.0 idea puts emphasis on the requirement to interlink various production domains through smart automation solutions. This concept focuses on predictive analytics that will assist in understanding consumer requirements since they are important in ensuring customized mass production is effective. The most challenging aspect here is quality analysis of big data for forecasting future trends in consumer preferences. This is where emerging technologies such as cloud computing and big data analysis come in handy. Using computational intelligence methods, such as cluster k-means methods, producers can segment and analyze customer information in order to obtain actionable insights. This not only guides product development but also enables the matching of production capacity with the particular preferences of the intended target market segments, improving overall customer satisfaction and propelling sales increase (Chen et al., 2016). Additionally, the level of automation extends far beyond the substitution of simple tasks; it is an entire transformation of paradigms of production. The convergence of automated systems allows for enhanced management of resources and reduced wastage, advancing sustainable manufacturing. The convergence of IoT with automation also opens avenues for infinite learning and modification as machines can reset their action in accordance with analytical feedback. This advancement not only improves the productivity of operations but creates a responsive manufacturing environment that adapts quickly to changes in supply and demand as well as the complexities of the supply chain. Through investments in smart manufacturing initiatives, IoT integration not only improves workflow but sets the stage for continuous innovation and technology enhancement in the manufacturing sector, with assurances of sustainability and applicability in an increasingly more connected industrial world (Li et al., 2015).

2.4.3. Supply Chain Optimization

Use of the Internet of Things (IoT) in supply chain operations has revolutionized the traditional operating models, allowing far more optimization potential across several dimensions. With the use of networked devices and real-time analytics, companies now have unprecedented visibility into supply chain operations. The visibility allows decision-making on time and enhances resource utilization, allowing companies to respond swiftly to market fluctuation and customers' needs. Use of IoT technology is one of the major factors to predict supply chain disruption, hence enabling businesses to anticipate and intervene in risks efficiently. Basically, strategic application of IoT not only streamlines operations but also forms an Agile supply chain infrastructure, thus improving overall efficiency and sustainability. In addition, energy efficiency of supply chains matters more in today's manufacturing climate. With companies migrating towards greener processes, supply chain optimization includes waste and energy reduction strategies. IoT technology can assist in tracking energy consumption in real-time across various phases of the supply chain, making it possible to identify inefficiencies that would have otherwise remained hidden. Research has confirmed that IoT-powered energy management technologies can be instrumental in reducing operational costs and carbon footprint. This synchronism with overall sustainability goals is essential, as depletion of fossil fuel deposits and rising energy prices call for a more sustainable model for logistics and manufacturing operations (Casta Dñeda et al., 2019). Thus, the supply chain optimization through energy efficiency by IoT has emerged as a central theme in modern supply chain optimization. Furthermore, supply chains' ever-changing environment requires a continuous reassessment of management and control measures to keep pace with developing technology. IoT facilitates easy development of high-level algorithms that can analyze large volumes of data and identify patterns and trends. The results bring better forecasting accuracy and inventory management and enable companies to maximize inventory levels and minimize surplus inventories and carrying costs. In addition, the application of IoT-based predictive analytics equips businesses with the ability to adapt their operations to accommodate evolving consumer demands without compromising delivery times. Consequently, the construction of supply chain operations founded on IoT variations necessitates the pursuit of continuous innovation and responsiveness when meeting market requirements. This adaptability is necessary to stay competitive in an increasingly complex and globalized economy, further solidifying the role of IoT in supply chain optimization (Chiew et al., 2017).

2.5. APPLICATIONS OF INTERNET OF THINGS (IOT) IN ELECTRICAL ENGINEERING

The integration of the IoT with electrical engineering has revolutionized how engineers create many different applications across the discipline. One of the key applications is the development of Smart Grids, which technically address the built-in issue of traditional power systems. Instead of data following a one-way flow, Smart Grids provide a two-way flow of information and power between the user and provider. This transformation provides higher reliability and efficiency and eventually, more sustainable energy management. IoT plays a critical role in this revolution through the deployment of tools such as sensors and smart meters that allow for real-time monitoring and control of the processes of energy distribution. With IoT, engineers can optimize the operation of the grid, improve the use of energy, and dynamically respond to changes in demand, marking a shift towards intelligent infrastructure. Educational courses in electrical engineering based on IoT have also been initiated, focusing on inspiring the next generation of engineers. An example of such a pioneering course teaches software engineering to high school students through project-based education with IoT technologies. Such a pedagogy contributes significantly towards enhancing students' knowledge of programming and networking and engaging them in hands-on technology applications. With the completion of some tasks and creation of a "do-it-yourself" project, the students gain not just technical knowledge but learn the spirit of a design mindset too. The process educates the value of software engineering beyond coding, focusing on creativity and problem-solving. These models of education assist in enrolling underrepresented populations into the field and create a workforce able to address the intricacies of modern engineering, particularly for IoT applications that engage interdisciplinarity in skill and knowledge (Julien et al., 2019). Furthermore, IoTs applications also make inroads into new systems like Smart Buildings that prioritize energy efficiency and automation. In these settings, various IoT devices interact to drive and regulate lighting, cooling, and heating systems. Through the use of smart sensors and actuators, engineers can create environments that react to occupancy and energy demands, reducing waste as well as promoting sustainability. Additionally, benefits of IoT-based Smart Buildings also complement rising urbanization trends, in which energy usage needs to be carefully managed. The integration of IoT devices into such infrastructures is an indicator of the peak of innovation in electrical engineering because professionals are challenged to use and implement innovative connectivity solutions. The innovation marks a landmark for engineering disciplines because it offers evidence of how the use of IoTs can drastically change the manner in which engineers can contribute towards energy optimization and environmental stewardship (Copeland et al., 2019).

2.5.1. Smart Grid (SG) Technology

The shift of traditional power grids to SGs is a revolutionary move towards the effective regulation of energy, which is inevitable due to the need to address inefficiencies in existing systems. Traditional grids encompass primarily one-way flow of information, thus resulting in wastage of energy, unreliability, and failure to keep up with rising energy demand. On the other hand, Smart Grids facilitate peer-to-peer communication between consumers and providers in a reversible direction, enhancing the regulation of power generation, transmission, and supply in general. This development is essential for modern energy networks to manage uncertain demand, integrate renewable resources, and provide users with real-time information on their energy consumption. In addition to that, SGs deploy IoT technologies to fit many devices such as sensors and smart meters with advanced monitoring and control functionality that aims to improve system performance and reliability (Copeland et al., 2019). The incorporation of IoT in Smart Grid technology is the ultimate solution to achieve higher levels of connectivity and automation within the energy sector. IoT devices are the building blocks of SG systems and enable intercommunication between all the components of the grid, from power generation plants to residential homes. Not only does this unification simplify operations, but it also empowers consumers to actively engage with their energy usage. With IoT devices, energy distributors can collect massive amounts of data, enabling them to analyze and predict energy usage more effectively. These analytics can potentially lead to optimized energy distribution, waste removal, and better resource management. Besides, edge computing technology, as discussed in recent studies, offers the potential for real-time in-source processing, minimizing latency and decision-making capability enhancement within SG systems (Ansari et al., 2017).

Figure 2.16. Smart grid (SG) technology.

Source: https://iotdunia.com/wp-content/uploads/2019/04/what-is-smart-grid-technology-featured.webp.

As Smart Grid technology continues to advance, various challenges and possibilities emerge as it pertains to implementation and incorporation into existing infrastructures. The first of the key concerns is the maintaining of public privacy and security, given the enormity of sensitive information that is amassed by IoT sensors. Protection of this data against hacking is essential to the maintenance of consumer trust and grid stability. Interoperability between various IoT devices and platforms is also a key issue, as the integration of various technologies has the potential to cause operational inefficiencies. Resolving these challenges requires ongoing research and creating standardized practices that enable devices to communicate with each other seamlessly. Smart

Grid technology, in turn, fueled by future advancements in IoT, promises to create a greener, more resilient energy world and the basis for smart cities and distributed energy systems (Copeland et al., 2019) (Figure 2.16).

2.5.2. Home Automation Systems

Home automation systems are a significant application of the Internet of Things (IoT) in engineering that provides security and convenience to the users. Through the incorporation of various devices like smart thermostats, lighting, and security cameras, home automation systems create an automated networked space that can be remotely controlled. Integrating the devices not only creates a comfortable user environment, but it also maximizes energy efficiency by minimizing the consumption of resources. While individuals grow more environmentally aware, the possibility of monitoring and adjusting energy consumption in real-time aligns with the growing demand for sustainable lifestyle choices. With IoT technology, home automation systems reflect the ways in which engineering innovation is turning traditional home settings into smart homes with the capability to make smart decisions and react. The functioning of home automation devices is sustained by a sophisticated network of actuators and sensors in communication through the IoT platform.

The devices collect and exchange data, enabling homeowners to efficiently control and manage their surroundings. Establishing this two-way exchange of information is significant; it enables real-time monitoring and adjustments that may significantly enhance security and energy management protocols. For instance, smart home security systems can detect unusual activity and notify home owners in real-time, while automatic lighting can be controlled based on occupancy to prevent wastage of energy. The performance of such systems relies on how reliable the supporting technology is, as well as on the careful embedding of intuitive interfaces. This intelligent orchestration highlights the critical role engineering needs to play in designing and deploying home automation systems. As home automation advances, engineering disciplines must contend with several issues, particularly security, interoperability, and consumer privacy. Since these systems rely on constant data transfer and connectivity, they are exposed to intruders and hackers, and engineers must devise robust security measures. Moreover, heterogeneous devices that are developed by numerous companies must co-exist and communicate effectively, adding another dimension of complexity in making them compatible and function in one framework. The implementation of Business Process Management (BPM) principles in IoT-based solutions can provide a process-oriented solution to manage such issues effectively (Burattin et al., 2017). Though research is underway, it is essential to explore innovative solutions that not only enhance the technical aspect of home automation systems but also ensure a secure, effective, and user-friendly experience for the occupants.

2.5.3. Energy Management Solutions

The integration of energy management solutions with Internet of Things (IoT) technology is a significant advancement in engineering applications. IoT makes it possible to design systems with the ability to remotely monitor and regulate energy use, leading to enhanced efficiency and little wastage. For example, energy management systems utilize wireless sensor networks (WSNs) for acquiring current information on energy use in various facilities. Such information can be analyzed to identify patterns in consumption, and companies can utilize energy in an efficient manner accordingly. Most importantly, the use of a RESTful web service architecture makes it possible to have free access to information accumulated in sensor nodes, thus making a more efficient means of energy management. This architecture, in addition to supporting the integration of energy management solutions, also emphasizes the critical role of IoT towards speeding up information and communication technologies in engineering applications (Leung et al., 2014). The shift away from traditional power grids to Smart Grids (SGs) is commendable evidence of the paradigm-shifting capability of IoT in managing energy assets. Unlike the conventional grids, which use the one-way approach of energy transfer, Smart Grids use the bi-directional approach of energy transmission, allowing enhanced communication between consumers and suppliers. This new process necessitates the deployment of numerous IoT devices, such as smart meters and sensors, at different levels in the energy distribution network. These devices are tasked with the job of monitoring, analysis, and control necessary in the building of grid reliability and avoiding wastage of energy. In addition, the inclusion of IoT in SG systems stimulates automation and connectivity, making possible real-time tracking and management of data, which are highly crucial in responding to the growing demand for energy and the complexities of modern power systems (Copeland et al., 2019). In the light of energy management solutions, IoT has the potential to catalyze tremendous growths in sustainability and efficiency. With the ability to automate and optimize energy processes comes not only cost savings in operational costs but also sustainable behavior among industries and consumers. IoT products generate actionable information that allows users to make informed decisions on the energy-using behaviors. As more organizations adopt these technologies, the generated data is used to create new innovations in energy management practices that aid in promoting sustainability. This convergence is critical because global energy demands continue to rise along with the imperative to mitigate climatic change impacts.

Thus, IoT and energy management integration is crucial to building smarter systems that can react to the future challenges of energy consumption and environmental protection.

2.6. CHALLENGES AND LIMITATIONS OF INTERNET OF THINGS (IOT)

IoT is a revolutionary way of technology integration across various applications in engineering. Yet, with such innovation comes unprecedented challenges which can impede its effects. The scalability and interoperability challenge between multiple IoT ecosystems is one of the concerns. With numerous devices and sensors communication over varying platforms, the lack of standard protocols poses the risk of slowing down intercommunication and data exchange. This fragmentation creates increasing complexity, where different systems do not cooperate, thus preventing potential gains in efficiency and data utilization. Problems with network traffic also become evident as more devices are connected, which endangers the offering of stable service and real-time processing of data. As IoT technology increases, data security and privacy issues become more critical. The pervasive interconnectivity and data-sharing inclinations of IoT devices pose a gigantic number of sensitive and personal data that are an attractive target for cybercriminals. Moreover, the devices lack adequate security tools, and they become vulnerable to being compromised that would harm not only private privacy but also societal infrastructure as a whole. The rapid expansion of IoT devices in tandem with the delayed action by regulatory and protective measures is a dangerous bet. With weak security infrastructure, the potential for misuse is actualized through unauthorized access and data manipulation, slowing down IoT systems' trust building with users and stakeholders alike. In specific applications, such as healthcare monitoring, integrating IoT technology is faced with unique challenges that require careful consideration. For instance, the fall prediction system shows how multi-component health complexities render IoT deployments challenging. While existing detection techniques have mostly emphasized physiological factors, they frequently ignore vital environmental and behavioral factors that precipitate fall hazards. This shortcoming demonstrates the current technology's inability to respond to the more expansive process of caring for a patient. Without revolutionizing user interfaces to allow greater interaction and offer holistic feedback, current systems may prove useless or inefficient. Thus, IoT development involves not just technological advancements but an overarching plan that includes dealing with realities of actual use, as indicated in the need for enhanced integration of contextual data into predictive models (Jung et al., 2017).

2.6.1. Security and Privacy Concerns

The rapid expansion of the Internet of Things (IoT) in engineering applications has ushered in a multitude of security and privacy concerns that warrant serious attention. As IoT devices proliferate, they are increasingly likely to function within complex networks known as Cyber-Physical Systems (CPS), which themselves dynamically change in response to user needs and the environment. This connectivity introduces severe control challenges across system elements, as the usual engineering assumptions regarding design integrity no longer apply. Network vulnerabilities are liabilities, which can mean sensitive user data becoming available to unauthorized entities or being targeted for attacks by malicious users. Thus, engineers must develop sound security models to safeguard the infrastructure of IoT devices as well as individual users' privacy whose data may get commodified in this changing digital world, as highlighted in the context of context-aware data trading by Azorin-Lopez et al. (2018). The interaction between mechatronics and cloud technologies also underscores the complexity of ensuring user privacy and security in IoT applications. Increased reliance on cloud computing technologies for storing and analyzing data raises questions regarding legal and ethical principles of data ownership, retention, and utilization.

Figure 2.17. Security and privacy concerns.

Source: https://www.smwllc.com/wp-content/uploads/2016/09/Mark-Petersons-Viewpoint-Security-and-Privacy-Concerns.jpg.

Current and future legislation must be considered, as legislation adapts to address such pressing concerns, guiding how mechatronics systems are designed and deployed. The privacy standards regime and legislation environment are changing, and the engineers must not only be technically competent but also cognizant of these regulatory rules in order to implement the fine line between privacy rights of the users and innovation, as suggested in Azorin-Lopez et al. (2016). Privacy's user-centered nature in engineering design is more than technical controls and requires an integrated approach where the user interest takes precedence. With IoT devices collecting huge volumes of data, the threat of accidental surveillance and profiling increases, and hence, there should be an emphasis on ethical design principles. Data processing and collection procedures must be transparent, with end-users understanding how their data is being used and revealed. There should also be the development of a culture of privacy awareness among end-users to facilitate a mitigation of risks associated with accidental data release. Engineers will have to contend with both the hard aspects of security and the soft factors of user perception and trust, so that privacy becomes an inherent part of IoT systems, consistent with the conclusions regarding privacy challenges and users' perceptions in Azorin-Lopez et al. (2016) (Figure 2.17).

2.6.2. Interoperability Issues

Internet of Things (IoT) interoperability issues are a significant barrier to the effective deployment of various engineering applications. With IoT systems increasingly incorporating heterogeneous devices and platforms, interoperability between such systems in being able to communicate and work together becomes essential. Interoperability issues arise due to the lack of standardized protocols and device specification heterogeneity, rendering various technologies incompatible with one another. This heterogeneity results in disjointed systems that fail to exchange data successfully, finally negating the promise of IoT in improving process efficiency and decision-making. In the absence of interoperability, organizations might not be able to extract useful insights from the huge amount of data being produced across connected systems, negating the overall promise of IoT in revolutionizing engineering practice. Smart Grids evolution is a great example of the importance of interoperability in IoT implementation. These networks need two-way energy transfer to enable un-interrupted communication amongst a series of devices—ranging from control systems to smart meters—dispersed throughout different environments. Each device will have to send information from and receive commands from different parts of the grid, which by its nature involves challenging interoperability concerns. As per the established literature, it is important to bring together and harmonize these complex elements with a view to having effective energy use and management, with good standards and protocols highlighted as being critical (Copeland et al., 2019). The quest for interoperability, thus, not only facilitates improved operational effectiveness but also increases energy infrastructures' reliability and security, which form the foundation of modern engineering application. Moreover, interoperability is a concern in IoT complicated by the need for security and privacy, particularly in intelligent spaces like buildings. With more interconnectivity among systems, the risk posed by data exposure and unauthorized access increases the stakes for engineers and stakeholders in such a project. For instance, the deployment of ambient intelligence systems for personalized living environments not only demands technical interoperability of the system but also security and privacy warranties (Alexandris et al., 2018). A SPDI approach emphasizing Security, Privacy, Dependability, and Interoperability can offer the requisite confidence for these concerns, ultimately leading to safer and more resilient smart environments. By focusing attention on these interrelated factors, engineering applications are in a position to leverage the full potential of IoT, without risking user safety or data integrity through creative solutions.

2.6.3. Scalability and Infrastructure Challenges

The emergence of the Internet of Things (IoT) revolutionized the landscape of engineering applications, but in the process, has also introduced significant connotations of scalability and infrastructure challenges. As soon as companies adopt IoT systems, they must be able to accommodate variable volumes of data and additional devices in tow. This surge of demands calls for an elastic and scalable infrastructure that can efficiently accommodate diverse sensors, actuators, and other components. Without sufficient infrastructure, IoT's potential can lose out on delays, downtime, or suboptimal performance. To deal with these problems, the use of next-generation architecture solutions such as Network Functions Virtualization (NFV) has grown in popularity. NFV supports the homogenization of diverse devices in favor of one ecosystem that can dynamically allocate resources based on existing needs, thus rendering continuity and reliability in IoT networks, as emphasized by principles discussed in Abu-Lebdeh et al. (2016). In addition to technological architectures, fighting scalability challenges also entails understanding diversified contexts in which such systems operate. For instance, urban environments, universally referred to as Smart Cities, have unique infrastructure demands that necessitate integrating various information and communication technologies (ICT). As urban populations expand and try to make themselves smart, underlying software platforms need to handle and sustain a variety of applications that have distinct needs such as responsiveness, throughput, and latency. Hence, reference architecture development is of prime importance for city planners and engineers to provide scalable, adaptable applications that leverage the maximum capability of IoT. The subtleties of such a project highlight the necessity of strategic investment and planning in scalable infrastructure to facilitate the ease of transition to intelligent urban spaces, again referring to the observations made by Chaves et al. (2017). All in all, scalability and infrastructure challenges that accompany IoT applications call for an integrated response that balances technological development with strategic planning and stakeholder action. Engineering teams need to collaborate with urban planners, policymakers, and technology suppliers to ensure that infrastructure built is ready for future requirements and that it is flexible enough to be capable of supporting evolving technologies and use. A successful IoT model must establish interconnectivity and interoperability between various elements of infrastructure and must ensure data streams are reliable and secure. Investments in state-of-the-art solutions like NFV, with focus on addressing problems in real life in Smart City applications, will be the driver in overcoming scalability barriers. Identifying on these pillars, stakeholders will be able to capitalize on the potential revolution that IoT has to offer in engineering applications and create the basis for smart and adaptive environments to the good of society as a whole.

SUMMARY

The adoption of the IoT into engineering applications is bringing about a revolutionary era of heightened efficiency, broader access to information, and rich interactivity. With ongoing advancements in the technology, it is also crucial to value its expansive implications in various fields, including industrial automation and smart infrastructure. The integration of IoT devices and advanced analytics can possibly automate processes and optimize resource utilization, thereby leading to significant operating savings. However, the deployment of these sophisticated systems is a challenge to be tackled if the complete potential of the new technologies has to be realized. Effective strategies that consider the strengths of the new technologies and resolve interoperability and data security problems are essential for sustainable IoT applications development. Understanding the evolution of IoT architectures, one can appreciate developments in such frameworks as the Mobile Edge Internet of Things (MEIoT). This forward-looking approach, which optimizes data processing by being proximate to devices, showcases unparalleled improvements in real-time processing and handling of data. Employing a cloudlet framework, MEIoT minimizes latency as well as simplifies access control by employing proxy Virtual Machines that may be adapted for IoT devices. These advances, as explained, identify the absolute priority of effective resource management for the purposes of delay and energy reduction, indicating how a well-designed IoT system can produce spectacular performance outcomes in various branches of engineering, as reflected in extensive simulations conducted in comparative studies (Ansari et al., 2017). Furthermore, the evolution of IoT is also significantly tied to progress toward Industry 4.0, aimed at mass customization in manufacturing. This vision is made possible through the application of smart analytics that can meaningfully interpret huge amounts of data. Creation of predictive analytics platforms that integrate cloud computing, big data analytics, and intelligent production systems is a prime example of how engineering is leveraging IoT capabilities to become more responsive and efficient in meeting consumer demand. Industries can use computational intelligence methods like k-means clustering to adjust dynamically to evolving consumer tastes, thus making them more agile and responsive. This shows the convergence of IoT technologies with new manufacturing paradigms, which ultimately facilitates the demand for constant research and innovation in the development of smart and sustainable engineering solutions (Chen et al., 2016).

The IoT has revolutionized the environment of engineering by enhancing the efficiency and effectiveness of many designs and systems. With the convergence of networked technologies and intelligent devices, engineers are now capable of obtaining real-time information and intelligence that make knowledge-based decision-making possible. This paradigm shift allows for the development of solutions previously unimaginable, in industries ranging from manufacturing, logistics, and transportation to urban planning. With emerging IoT technologies, engineers can now design systems that not only function independently but also communicate with other devices, leading to even more advanced and holistic engineering applications. Similarly, the role of engineers is shifting towards problem-solving and innovation, with engineers taking a more complete approach that includes technical as well as design considerations in their work. Today's education programs emphasize the relevance of IoT in engineering classes, but another manner in which it is making its impact felt in the discipline. A case in point is an introduction program that familiarizes high school students with software engineering through experience-based learning using IoT devices. By engaging students in real-world uses of IoT technology, the course aims to recruit underrepresented populations into computer science while instilling a design mentality. This experiential learning process allows students to understand the theoretical frameworks of IoT and apply them in practicality. As highlighted by Julien et al. (2019), not only do such initiatives promote technical skills, but they also motivate future engineers to consider their activity as a component within a broader design context, which ultimately contributes to an inclusive and innovative world of engineering. Furthermore, evolution of IoT also involves the shift towards Cloud and Fog Computing, revolutionizing the processing and utilization of data in engineering applications. This is changing the development paradigm from independent, single-function systems to interconnected, multi-function applications that can be deployed across several industries. The Fog Computing paradigm enhances IoT functionality by offering data processing in different levels that allow edge devices to exchange computational and control resources with cloud infrastructure. As illustrated in Akrivopoulos et al. (2017), the new strategy signals potential for the creation of more responsive and effective engineering solutions to address challenges arising from the exponential nature of data generated in IoT environments. Therefore, the employment of these technologies encourages engineers to re-think traditional methods and include cooperative and flexible practices in their working processes.

REFERENCES

1. Abu-Lebdeh, M., Glitho, R., Morrow, M., et al. (2016). Network functions virtualization architecture for gateways for virtualized wireless sensor and actuator networks. http://arxiv.org/abs/1601.03390 (accessed on 2nd October 2025).
2. Adelantado, F., Martinez, B., Melia, J., et al. (2017). Understanding the limits of LoRaWAN. http://arxiv.org/abs/1607.08011 (accessed on 2nd October 2025).
3. Ahmed, A., Atzori, A., Hankin, B., Basco-Carrera, L., Bastin, L., Beven, K., et al. (2019). Models of everywhere revisited: A technological perspective. https://core.ac.uk/download/322489447.pdf (accessed on 2nd October 2025).
4. Akrivopoulos, O., Amaxilatis, D., Chatzigiannakis, I., et al. (2017). Enabling stream processing for people-centric IoT based on the fog computing paradigm. https://core.ac.uk/download/158328004.pdf (accessed on 2nd October 2025).
5. Alexandris, G., Askoxylakis, I., Fysarakis, K., et al. (2018). Towards a security, privacy, dependability, interoperability framework for the Internet of Things. https://core.ac.uk/download/323057627.pdf (accessed on 2nd October 2025).
6. Alrumaih, H., Impagliazzo, J., & Sabin, M. C. (2016). An information technology competency model and curriculum. https://core.ac.uk/download/86427251.pdf (accessed on 2nd October 2025).
7. Alrumaih, H., Impagliazzo, J., & Sabin, M. C. (2018). An overview of the new ACM/IEEE information technology curricular framework. https://core.ac.uk/download/215538141.pdf (accessed on 2nd October 2025).
8. Alturki, B., Perera, C., & Reiff-Marganiec, S. (2017). A hybrid approach for data analytics for Internet of Things. http://arxiv.org/abs/1708.06441 (accessed on 2nd October 2025).
9. Ansari, N., & Sun, X. (2017). Mobile edge computing empowers Internet of Things. http://arxiv.org/abs/1709.00462 (accessed on 2nd October 2025).
10. Asensio, E. Serral, Brouns, N., Grefen, P., et al. (2018). Modeling IoT-aware business processes–A state of the art report. https://core.ac.uk/download/577055769.pdf (accessed on 2nd October 2025).
11. Axelsson, J., & Kobetski, A. (2012). Federated embedded systems–A review of the literature in related fields. https://core.ac.uk/download/11434985.pdf (accessed on 2nd October 2025).

12. Azorin-Lopez, J., Bradley, D., Hehenberger, P., et al. (2016). Privacy matters: Issues within mechatronics. https://core.ac.uk/download/228177687.pdf (accessed on 2nd October 2025).
13. Azorin-Lopez, J., Bradley, D., Hehenberger, P., et al. (2018). Mechatronics & the cloud. https://core.ac.uk/download/228178466.pdf (accessed on 2nd October 2025).
14. Bacciu, D., Bahr, R., Bröring, A., et al. (2017). Internet of robotic things: Converging sensing/actuating, hypoconnectivity, artificial intelligence and IoT platforms. https://core.ac.uk/download/84045908.pdf (accessed on 2nd October 2025).
15. Benini, L., Conti, F., Gautschi, M., et al. (2017). An IoT endpoint system-on-chip for secure and energy-efficient near-sensor analytics. http://arxiv.org/abs/1612.05974 (accessed on 2nd October 2025).
16. Burattin, A., Di Ciccio, C., Gal, A., et al. (2017). The Internet-of-Things meets business process management: Mutual benefits and challenges. https://core.ac.uk/download/299377534.pdf (accessed on 2nd October 2025).
17. Carpio, F., Dizdarevic, J., Jukan, A., et al. (2019). A survey of communication protocols for Internet of Things and related challenges of fog and cloud computing integration. https://core.ac.uk/download/187266039.pdf (accessed on 2nd October 2025).
18. Chaves, A. P., Gerosa, M. A., Kon, F., et al. (2017). Software platforms for smart cities: Concepts, requirements, challenges, and a unified reference architecture. http://arxiv.org/abs/1609.08089 (accessed on 2nd October 2025).
19. Chen, Y., Flores Saldivar, A., Goh, C. S. F., et al. (2016). Identifying smart design attributes for Industry 4.0 customization using a clustering genetic algorithm. https://core.ac.uk/download/74230589.pdf (accessed on 2nd October 2025).
20. Chiew, K. L., Ong, C. T., Tay, K. G., et al. (2017). Simulation of undular bores evolution with damping. https://core.ac.uk/download/17297499.pdf (accessed on 2nd October 2025).
21. Copeland, R., Crespi, N., Rehmani, M. H., et al. (2019). Internet of Things-aided smart grid: Technologies, architectures, applications, prototypes, and future research directions. http://arxiv.org/abs/1704.08977 (accessed on 2nd October 2025).
22. David, H., Ruiz-Rube, I., Rodríguez Corral, J. M., Mota, J. M., Dodero, J. M., et al. (2019). Block-based development of mobile learning experiences

for the Internet of Things. https://core.ac.uk/download/322903564.pdf (accessed on 2nd October 2025).

23. Díaz Castañeda, J. L., & Ocampo-Martínez, C. (2019). Energy efficiency in discrete-manufacturing systems: Insights, trends, and control strategies. https://core.ac.uk/download/288625543.pdf (accessed on 2nd October 2025).
24. Holtkamp, B., & Iyer, A. (2017). *Industry 4.0: The Future of Indo-German Industrial Collaboration.* https://core.ac.uk/download/322353876.pdf (accessed on 2nd October 2025).
25. Isah, H., Mahfuz, S., Nicholls, P., et al. (2018). Detecting irregular patterns in IoT streaming data for fall detection. http://arxiv.org/abs/1811.06672 (accessed on 2nd October 2025).
26. Johnson, C. (2016). Securing the participation of safety-critical SCADA systems in the Industrial Internet of Things. https://core.ac.uk/download/46664188.pdf (accessed on 2nd October 2025).
27. Jukan, A., Marín Tordera, E., Masip Bruin, X., et al. (2016). Foggy clouds and cloudy fogs: A real need for coordinated management of fog-to-cloud computing systems. https://core.ac.uk/download/81582274.pdf (accessed on 2nd October 2025).
28. Julien, C. (2019). Using the Internet of Things to teach good software engineering practice to high school students. https://core.ac.uk/download/287648521.pdf (accessed on 2nd October 2025).
29. Jung, T. P., Litvan, I., & Rajagopalan, R. (2017). Fall prediction and prevention systems: Recent trends, challenges, and future research directions. https://core.ac.uk/download/323065612.pdf (accessed on 2nd October 2025).
30. Leung, V. C. M., Sheng, Z., & Zhu, C. (2014). Surfing the Internet-of-Things: Lightweight access and control of wireless sensor networks using industrial low power protocols. https://core.ac.uk/download/30610557.pdf (accessed on 2nd October 2025).
31. Li, Y., & Yu, H. (2015). *Special Session on Industry 4.0.* https://core.ac.uk/download/42366589.pdf (accessed on 2nd October 2025).

CHAPTER 3

SENSORS AND DATA ACQUISITION IN SHM

Contents

INTRODUCTION

Structural Health Monitoring (SHM) is a vital field that deals with the assessment and maintenance of the structural integrity. The general aim of SHM is the detection and prevention of potential structural failure through real-time monitoring of various parameters, thus ensuring safety and durability. The technique utilizes a combination of sensors to acquire data on the health of structures, which are employed to track the changes over time. The use of advanced data acquisition systems is particularly useful for real-time monitoring, a key factor in early damage detection as well as preventive maintenance. By using modal analysis as well as other methods, SHM can detect anomalies in the natural frequency of a structure, normally pointing to issues. By evolving with advancements in SHM, it further sophisticates through the implementation of non-contact measuring devices, again elevating its efficiency and effectiveness in safeguarding structures from unexpected failure. Among the most well-proven techniques in SHM is vibration-based monitoring, founded mainly on the information given by sensors like accelerometers. The method relies on the premise that structural damage manifests through changes in the dynamic response of the structure under external loading. Data collected by sensors are then attacked with processing, usually experimental modal analysis (EMA) or operational modal analysis (OMA), leading to a clearer picture of the state of the structure. However, traditional contact sensors have limitations in that they induce mass-loading effects and instrument installation involving manual labor, particularly in massive structures. To address such limitation, researchers are increasingly interested in non-contact sensing techniques like digital video cameras, which can measure high-density spatial information. Using techniques like Phase-based Motion Estimation (PME), these technologies provide stronger and less intrusive methods of structural monitoring, thereby allowing stronger determinations of structural condition over time (Mao et al., 2018). Damage detection and structural performance measurement rely on data-driven algorithms that are critical to the processing and interpretation of data collected by sensors. This data-centric paradigm allows decisions to be made in real time and enhances the effectiveness of SHM practices. The effectiveness of such algorithms relies significantly on sensor readings quality and accuracy, as well as models' complexity applied for the analysis. Data-centric strategies are more versatile and more adaptable than model-based methods because they are directly applicable to any given structure and conditions of the surrounding environment. This cooperative relationship between data collection and algorithmic processing reflects the current advancements in SHM, which continue to advance in tandem with technological developments. Current research in the field is, in turn, venturing into new algorithms and new sensor technologies to enhance the precision and reliability of damage detection systems (Cherta et al., 2020).

Definition and Importance of Structural Health Monitoring (SHM)

SHM is a fundamental process that involves continuous or routine monitoring of the condition of a structure through different types of technologies. Multi-dimensional system is designed to gather information about the structural health of infrastructures such as bridges, buildings, and dams, hence enabling decision-making by operators and stakeholders on maintenance and safety. The most important aspect of SHM is its capability to sense changes in the material properties or the overall condition of a structure caused by externally applied loads or environmental conditions. Through the ability to identify potential defects prior to their growth into significant issues, SHM plays a critical role in the lengthening of the life of structures and the safety of the public, ultimately resulting in cost savings and enhanced reliability of civil infrastructures. The worth of SHM may be defined in terms of its usefulness in damage identification and prevention.

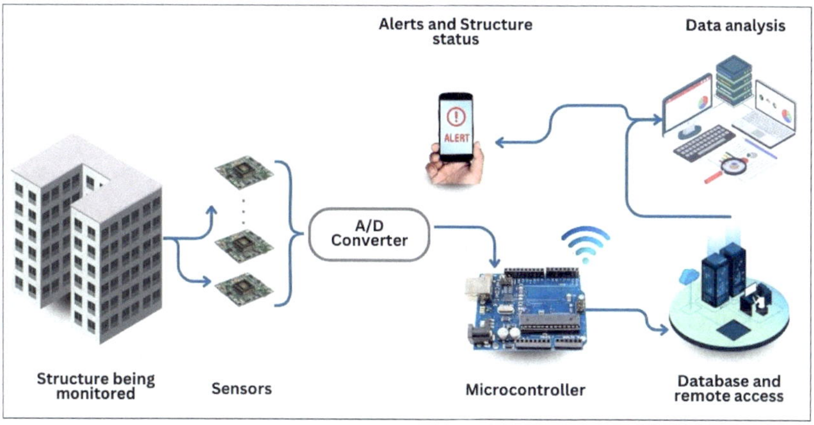

Figure 3.1. The importance of structural health monitoring (SHM).

Source: https://aecbytes.com/illustrations/tipsandtricks/2024/issue109-Structural-HealthMonitoring-images/fig2.png.

As discussed, SHM applies a combination of data-driven algorithms and model-based analysis to monitor the structural performance over time. This combination is flexible in approach in that operators can select methods suitable for their own circumstances. Increasingly, data-driven programs are being favored since they enable real-time decision-making and analysis from sensor-acquired data, maximizing the responsiveness of maintenance groups. Furthermore, the advance in SHM technologies, including integrating complex sensors, has maximized the robustness of the data acquisition system, thereby fostering a proactive maintenance culture. This shift from reactive to proactive methods not only optimizes safety but also reduces the cost of unexpected structural failure or prolonged downtime for inspection. In the last 10 years, reasons for adopting SHM have grown increasingly complex, since the requirement for modern infrastructures to be safe and functionally sustainable in the long term has been of especial urgency. Some civil engineering applications have witnessed significant development in SHM technologies, with greater insight into load and response mechanisms. This has placed greater emphasis on real-time information regarding the structural integrity, especially in the case of aging structures that have higher probabilities of unforeseen risks. Moreover, as implied by recent case studies, the state-of-the-art advancement in instrumentation and data communication systems has boosted data mining and presentation procedures for more effective diagnosis and evaluation of structural health. Here, the importance of SHM extends beyond immediate damage detection; it is a strategic method of ensuring the long-term economic viability and security of critical infrastructure, as debated in the literature (Brownjohn et al., 2016) (Figure 3.1).

Overview of Structural Health Monitoring (SHM) Applications in Various Industries

In a wide range of industries, SHM has become a key technology in ensuring structure integrity and safety. In civil engineering, SHM is an important technology for real-time monitoring of bridges, buildings, and dams. With the help of advanced sensor technologies, engineers are able to monitor vibrations, strain, and temperature variation in real-time, allowing early intervention whenever structural abnormalities are detected. The merging of Radio Frequency Identification (RFID) and sensor networks systems, with Building Information Modeling (BIM), enhances data collection and analysis effectiveness. The technology not only allows real-time tracking of structural elements but also provides detailed performance information on them during key construction phases, hence significantly improving project delivery and mitigating risks of structural failure (Liu et al., 2019). The aerospace industry also has a major reliance on SHM for guaranteeing the safety and reliability of aircraft. With the use of sensor technologies, engineers can monitor critical components such as wings, fuselage, and landing gear throughout the entire life of an aircraft. They monitor continuously to catch micro-damage that could lead to catastrophic failure if left unattended. These are strong data acquisition systems that allow for predictive maintenance, where upcoming issues can be identified and fixed before they are critical. Lastly, the complexity and strict safety needs of aerospace engineering also accentuate the need for high-performance SHM solutions to ensure that efficiency and safety are optimized in plane design and maintenance processes. This proactive strategy towards structural integrity management is critical in an industry where the risks are very high, thus underscoring SHM's position at the core of ensuring safety (Ahmad et al., 2015). Besides civil and aerospace applications, SHM technologies are also very important in the energy industry, particularly in the monitoring of wind turbines and the health of power generation plants. SHM systems enable energy operators to see mechanical equipment in real-time, such as blades and towers, and identify wear and tear early. The recording of information on environmental condition and performance parameters enables energy operators to maximize performance and extend the life of their assets by years. Through the integration of sensor networks and cloud computing technologies, data can be analyzed more efficiently and faster decision-making processes achieved when maintenance is needed. This use of SHM not only contributes to operational performance but also aids in energy industry sustainability practices by minimizing downtime and preventing costly failure, thereby endorsing the value of data-driven approaches in improving overall energy infrastructure (Liu et al., 2019).

Role of Sensors in Structural Health Monitoring (SHM) Systems

Sensor deployment in SHM systems plays a core function in contributing to the long-term integrity of infrastructure. These sensors are the primary data acquisition tools, capturing significant data on the structural status and performance parameters. By using various sensor types, like accelerometers, strain gauges, and temperature sensors, SHM systems are able to effectively monitor static and dynamic loads, material degradation, or environmental factors. The data gathered plays a crucial role in the subsequent analysis, primarily via data algorithms, to facilitate decision-making processes in real-time regarding maintenance and safety. Sensing technologies integrated are advanced to provide a very robust structure for damage detection at an early stage, which can reduce repair costs substantially and extend the service life of structures. Early detection of anomalies facilitated by these sensors contributes to the attainment of structural resilience in an increasingly challenging scenario. Data-driven software is central to processing and analyzing data picked up by sensors within SHM systems. Such software employs the gained data to evaluate the integrity of structures through sophisticated modeling techniques. Data-driven approaches are preferred as they are capable of not only handling large datasets but also providing timely insights to aid decision-making processes. However, such approaches have the drawback of requiring high-performance computing power due to their intricate nature. Increased computing capability and algorithmic improvement have been paramount, along with providing efficient means of handling large volumes of data collected from sensor installations. These algorithms can accurately detect, find, and recognize damage, which is of immense significance in the infrastructure safety management. The coupling between sensors and data-driven analytics brings out the integral ability of SHM systems to be able to deliver decisions in real-time (Cherta et al., 2020). Energy efficiency also plays a crucial role in the scenario of sensor networks for SHM applications. The availability of techniques such as Dynamic Voltage Scaling (DVS) shows an aggressive approach to energy consumption management while guaranteeing sensor functionality. SHM sensors are typically deployed in remote or hard-to-reach locations, where battery life is a relevant concern. Via energy minimization, DVS can allow sensor networks to work for extended durations, with sustained monitoring without frequent maintenance. One such example is in sensor network field experiments when DVS techniques were employed to reduce power usage without loss of time synchronization loss across the network (Abdelzaher et al., 2011). That compromise between performance and efficiency is just one such example of how sensor technology is evolving to help overcome the issues of long-term SHM deployment to ultimately contribute to more sustainable and resilient infrastructure management approaches (Figure 3.2).

Figure 3.2. Types of sensors used in structural health monitoring (SHM).

Source: https://www.sixense-group.co.uk/wp-content/uploads/sites/2/2020/09/eversense-shm-hero-image.jpg.

3.1. TYPES OF SENSORS USED IN STRUCTURAL HEALTH MONITORING (SHM)

In SHM, the selection of sensors is a central aspect to the correct identification of data regarding the integrity of a structure. Among the various sensors used, accelerometers are prominent due to their ability to measure vibration and dynamic response under various loads. Their responsiveness to minute fluctuations in structural response makes them important to detect potential damage over time. Accelerometers allow real-time monitoring and instant detection of deviations in a structure's performance, important for being able to make timely maintenance and repair decisions. In addition, they can be paired into data-driven algorithms to detect damage, hence optimizing the effectiveness of the SHM method. With ongoing advancement in SHM, applications of advanced sensor technologies such as accelerometers promise greater accuracy and reliability in monitoring structural performance (Cherta et al., 2020). Strain gauges are also a vital type of sensors used in SHM systems. Strain gauges measure the deformation or strain of a material to uncover information on the response of the structure under various conditions. Strain gauges prove to be extremely helpful for the identification of areas of critical stress concentration, aiding in early warning of impending failure sites.

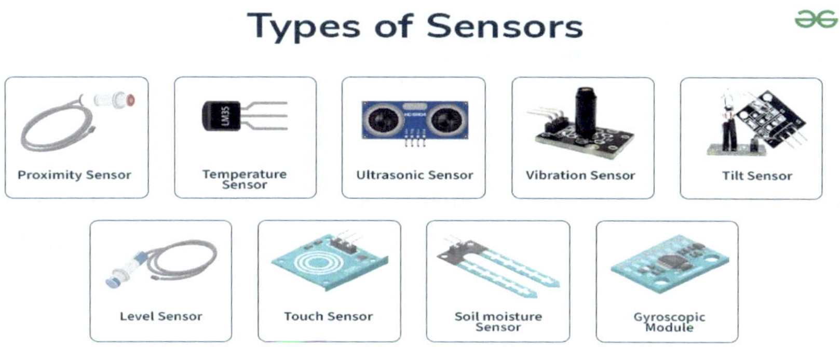

Figure 3.3. Types of sensors used in structural health monitoring (SHM).

Source: https://media.geeksforgeeks.org/wp-content/uploads/20240605014033/Types-Of-Sensors.webp.

In association with sensors such as accelerometers, they assist in providing a more accurate image of a structure's condition. The addition of strain-gauge data to analytical models enhances damage detection schemes, particularly with the application of data-driven techniques. These sensors not only help in identifying potential issues but also contribute towards developing predictive maintenance strategies, reducing the likelihood of catastrophic failure (Basto et al., 2017). Environmental condition sensors also find their application in SHM, fulfilling the demand for consistent data on external conditions influencing structural performance. Such sensors monitor the temperature, humidity, and moisture content parameters, which can significantly impact the integrity of a structure. For instance, temperature variations may lead to material thermal expansion or contraction that can lead to stress and therefore cause damage. With the incorporation of environmental data into SHM systems, engineers will better understand the nature of how a structure responds to the surrounding environment. This combined point of view serves to increase predictive models and improve determinations' accuracy regarding a structure's lifespan. As the diverse sensor types merge, they enhance the capacity to make informed data-driven decisions, ensuring prolonged safety and longevity of built environments (Cherta et al., 2020) (Figure 3.3).

3.1.1. Strain Gauges and Their Applications

Strain gauges are principal sensors of Structural Health Monitoring (SHM), providing critical information about deformation endured by materials loaded with stress. These sensors operate on the principle that when a conductive material gets deformed by compressing or tensile forces, its electrical resistance changes. This built-in property makes them measure mechanical strain in terms of quantifiable electrical signals. Strain gauges are applied to real-world applications for measuring the health of structures such as bridges, buildings, and pipelines, where minor changes in strain can signal failure. Their use enhances the safety and longevity of infrastructure, which is a tip of the iceberg in their application in aggressive management of public assets (Brandon et al., 2012). The greatest thing about strain gauges is that they are adaptable to different environments and conditions. They are adjustable in terms of shape, size, and materials and therefore are used in various industries like aerospace, automotive, and civil engineering. For instance, in aviation, the strain gauges can be utilized as an integral component in testing aircraft components' performance under stress, and in civil engineering, to test the structural strength of aging buildings. Their application in this case not only positions them as such vital tools in engineering but also calls for growing demands for precision in monitoring systems. This efficiency is further boosted with integration into data acquisition systems to provide real-time monitoring and analysis of the strain data (Brownjohn et al., 2005). As the field of SHM advances, strain gauge applications are also continually evolving with upcoming wireless technology and smart materials. Emerging new technologies such as fiber optic strain gauges are being created to provide more accurate measurements over longer distances and challenging environments. Further, data collected with the help of these gauges can be combined with advanced algorithms and machine learning networks to predict structural behavior in the long run. Incorporation of strain gauges and advanced technologies is a revolutionary step in monitoring and maintaining structures, with consequences to save money and improve safety protocols in maintenance and building processes. The growing integration of strain gauges into large-scale SHM systems identifies them as still relevant and with potential as the world becomes more automated and smarter in dealing with infrastructure (Brandon et al., 2012).

3.1.2. Accelerometers for Dynamic Monitoring

Accelerometers move to the forefront in the context of structural health monitoring (SHM) in the dynamic structuring assessment. These sensors work to sense the forces of acceleration on an object in order to enable both static and dynamic load responses to be evaluated. By vibration pattern data, accelerometers help in finding potential structural defects and instabilities. This is a preventive measure critical to it as it helps in the detection of damage when it begins to form and thereby prevents catastrophic failures. For conditions subjected to cyclic dynamic loading, accelerometers are an integral part, rendering real-time data supporting maintenance and safety decision-making. Addition of accelerometers to SHM systems augments real-time monitoring of structures, thus enhancing overall safety and lifetime. The integration of accelerometers into SHM systems emphasizes the use of algorithms using data for effective damage detection. These algorithms analyze raw measurements from accelerometers to establish structural health of buildings, bridges, and other critical infrastructure. The quality and speed of data collected from sensors determine how well these algorithms do their work.

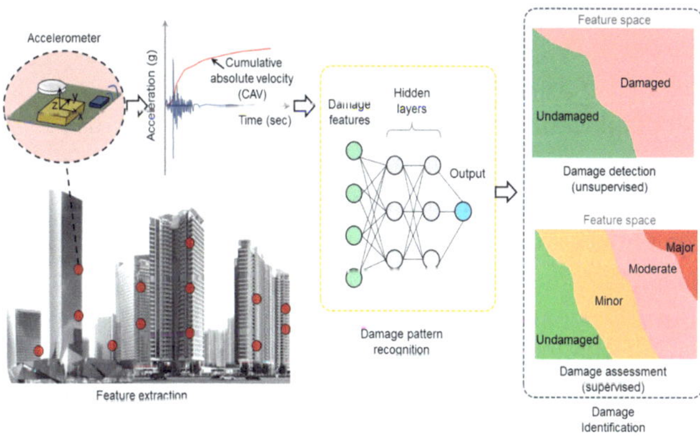

Figure 3.4. Accelerometers for dynamic monitoring.

Source: https://www.mdpi.com/applsci/applsci-11-05727/article_deploy/html/images/applsci-11-05727-g001.png

As described in newer publications, the use of data-driven techniques allows real-time monitoring and enhances damage localization and classification accuracy. With the implementation of advanced algorithms, engineers are able to demodulate dynamic response data taken by accelerometers and provide useful information regarding the structure's condition. Data-intensive approaches in SHM are a paradigm shift, requiring high-speed processing to maintain the computational loads of large-scale sensor data (Dong et al., 2010). Also, the integration of accelerometer technology into SHM systems allows new breakthroughs in intelligent structures. Such structures are gifted with self-sensing through knowledge derived from imbedded sensors, including accelerometers. By continuously assessing their own structure condition, these systems can generate automatic alarms or real-time adjustments to detected anomalies. Smart structure development is in full harmony with the advancement of data-driven algorithms, which offer effective real-time data processing and decision-making. As emphasized in recent studies, integrating accelerometers with sophisticated data analysis software enhances SHM performance, widening the boundaries of traditional engineering practice (Cherta et al., 2020). This integration not only improves safety and maintenance schemes but also supports the development of resilient structures in the future that can dynamically respond to environmental problems (Figure 3.4).

3.1.3. Displacement Sensors and Their Significance

Displacement sensors are significant contributors to structural health monitoring (SHM) as they provide precise measurements of relative motion in materials and structures. Such sensors provide the ability to measure deformations that may arise due to external loads, environmental conditions, or aging effects. Integration of displacement details into SHM structures by engineers helps determine the current state of a structure and thus facilitates prompt maintenance. Measurement of fine displacements is important as changes in fine displacements tend to reflect underlying issues that, if not treated, can lead to catastrophic failure. Displacement sensors, in this case, are not only data acquisition instruments but indispensable elements whose presence ensures safety and the longevity of infrastructure (Dong et al., 2010). Displacement sensors' use is not only restricted to monitoring because their use is part of predictive maintenance practice in various engineering fields. Civil engineering, for instance, has real-time displacement information that allows continuous monitoring of bridges and buildings subjected to dynamic loads. This is achieved with the capability of engineers to model potential failure modes and develop mitigation strategies on this basis. As sensors for displacement continue to advance, sophistication in the level of data analysis utilized may enable more informed decision-making. An example is the application of time-frequency analysis of ultrasonic waves proving to be successful in identifying internal flaws within material, such as in research on laser-generated ultrasound (Cojocaru et al., 2019). This indicates how integrating advanced sensing technologies is able to yield detailed information regarding the status of structures' health. Displacement sensors also aid in enhancing security measures by allowing structural irregularities to be easily identified with speed. Identification of abnormal displacement at an appropriate time can trigger countermeasures and, hence, avert catastrophes. In the domain of SHM, such sensors augment data acquisition practices by integrating multiple sources of signal to generate an integrated picture of structural condition. An integrated method seeks to promote material response to stressors in a historical context. Higher precision and reliability in displacement measurement offer engineers with better resources to customize their methods of maintaining infrastructure. Thus, the integration of displacement sensors within SHM systems is not just learning what currently exists; it is actually about ensuring structures that are significant in the future continue to be robust (Dong et al., 2010).

3.2. DATA ACQUISITION SYSTEMS (DAQ) IN STRUCTURAL HEALTH MONITORING (SHM)

The integration of DAQ within SHM systems has greatly improved the ability to assess the integrity of many structures on a regular basis. Traditional methods prefer to employ contact-type transducers, i.e., accelerometers, to measure dynamic responses, which must undergo significant physical handling for their installation and induce mass-loading effects on lighter structures. The combination of physical transducers with labor-intensive instrumentation issues poses challenges for large-area applications, where high spatial resolution over very large areas cannot always be assured. Otherwise, developments of recent non-contact measurement technology, e.g., digital video cameras, allow for rapid acquisition of high-density spatial information. Such developments allow improved structural behavior understanding, which facilitates improved damage identification through methods like Phase-based Motion Estimation (PME), e.g., related to a 2.3-meter wind turbine blade in the relevant literature (Mao et al., 2018). Moreover, the advent of open-source technologies has provided all with access to effective SHM strategies, particularly heritage buildings. The adoption of low-budget, open-source platforms enables researchers and experts to adopt constant static monitoring systems through calibrated components with little financial input. This practice is amenable to the prevailing trends toward sustainable structural monitoring practice with affordable yet reliable solutions for the idiosyncratic requirements of historic buildings. By application of these novel technologies, stakeholders can create holistic digital catalogs of cultural heritage and achieve preservation and active maintenance. Open-source platforms' flexibility provides space for various configurations to allow data output manipulation and testing, increasing the scope and scalability of SHM programs (Basto et al., 2017). As data acquisition in SHM evolves as a field, application of new technology and techniques remains the priority. Both traditional and novel approaches retain their own merits and limitations, which require an exquisite balance appropriate to individual structural conditions. While traditional DAQ systems may be best suited to provide exact data for well-endowed environments, the innovation of non-contact methods and open-platform approaches is poised to unlock the frontiers of SHM. With the exploitation of these technologies, scientists are able to better track a wide variety of structures, and consequently enhance the potential for timely intervention and the maintenance of structural long life. Eventually, continued investigation into these systems is an indispensable component of the advancement of SHM procedures and the maintenance of the safety and durability of infrastructure against future changes in environmental loads (Figure 3.5).

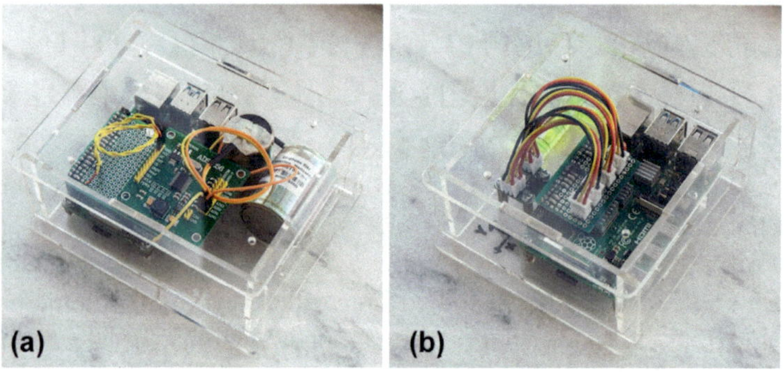

Figure 3.5. The integration of data acquisition systems (DAQ).

Source: https://pub.mdpi-res.com/sensors/sensors-24-04269/article_deploy/html/images/sensors-24-04269-g001.png?1719749918.

3.2.1. Components of Data Acquisition Systems (DAQ)

Data acquisition systems' components are the core of structural health monitoring (SHM) where they are tasked with collecting and examining data pivotal in deciding structural integrity. The central components of such systems are sensors, the primary tools used to quantify physical values such as strain, displacement, and acceleration. With the use of accelerometers, for instance, SHM systems can monitor dynamic behaviors that are capable of indicating structural damage. The success of these sensors, however, is based on the data acquisition system that supports them. Key components such as signal conditioners, which condition and filter incoming signals from sensors, are critical in maintaining the quality of the data collected. In addition, networking devices facilitate the transfer of data from sensors to data processing units and therefore enable real-time observation of structural responses to various loading conditions (Mao et al., 2018). The selection of appropriate sensors and associated technologies has a direct effect on the efficiency of data acquisition systems. While traditional contact sensors, for example, accelerometers, are often utilized due to their reliability, they are issues with regard to the time of installation and mass-loading sensitivity to low-mass structures. These limitations can pose hurdles to efficient monitoring, particularly for large structures where the installment of numerous sensors is much labor-intensive. Compared with conventional sensor approaches, being contact-based and incurring threats of hardware damage and disruption of structural operation, non-contact measurement techniques such as those performed using digital cameras are a cutting-edge option that can enhance measurement accuracy and lower risks. Utilizing Phase-based Motion Estimation (PME) to analyze recorded video data allows practitioners to obtain significant motion information from structures without the drawback of affixing physical sensors, thus enlarging SHM application scope (Mao et al., 2018). Furthermore, with the introduction of open-source, low-cost platforms, comes a revolutionary shift in the data acquisition system context of SHM, particularly in the context of heritage preservation. These platforms present a cost-effective method of incorporating cutting-edge technology into monitoring systems without the astronomical fees that come with traditional parts. With open-source, calibrated pieces, scientists can design flexible monitoring systems suitable for historic structures. Such an approach enables the creation of detailed digital records of the built environment, hence making conservation more feasible. It has, through recent studies, been demonstrated that the application of such readily accessible technologies allows for the use of adjustable monitoring systems that can easily fit the specific requirements of varying buildings, eventually leading to a more effective and precise assessment of structural condition.

3.2.2. Signal Conditioning and Processing

The significance of signal conditioning and processing in Structural Health Monitoring (SHM) cannot be overemphasized since it has a direct impact on the efficiency of data acquisition systems. Signal conditioning refers to a technique for manipulating sensor signals to improve the quality of the signals prior to their further analysis. It is a critical step in removing noise and disturbance that may obscure the useful information contained within the signal. Effective signal conditioning techniques such as amplification, filtering, and digitization produce an improved signal integrity such that presented data by sensors are a true representation of the structural health. The precision is not only crucial in real-time monitoring operations but also in decision-making during maintenance and repair, justifying the pivotal contribution of robust signal processing in SHM systems. Aside from initial signal conditioning, the processing step is where algorithms scan conditioned data to identify useful information about the integrity of structures. This is normally done through the use of data-driven algorithms, which have gained importance with their ability to identify useful information from vast amounts of data collected using sensors. Research into these algorithms plays a twofold function of damage detection and prognosis, significantly enhancing the efficiency of SHMs. Since data-driven strategy is based on real-time assessment, processors with high-performance are needed to execute complex calculations, as highlighted in Cherta et al. (2020). Therefore, ongoing growth of computing technology is necessary for unlocking the full potential of data processing and enabling development towards more accurate and efficient SHM systems. Furthermore, signal processing and conditioning apply not just in conventional buildings but also have a wide scope for their application in the monitoring of heritage buildings. Recently, it has been shown that low-cost open-source technologies are feasible in such a situation, particularly when the environment is tight on budget. With such new equipment, it is possible to make SHM more accessible for heritage buildings so that continuous monitoring in calibrated terms becomes achievable without requiring high financial inputs. This evolution emphasizes the dual benefit of inexpensive technologies: they enhance the preservation of cultural heritage while ensuring reliability and flexibility in data acquisition systems, as discussed in Basto et al. (2017). Thus, integration of signal processing techniques with emerging technologies will be fundamental in effectively maintaining the upkeep and protection of modern and historic infrastructures (Figure 3.6).

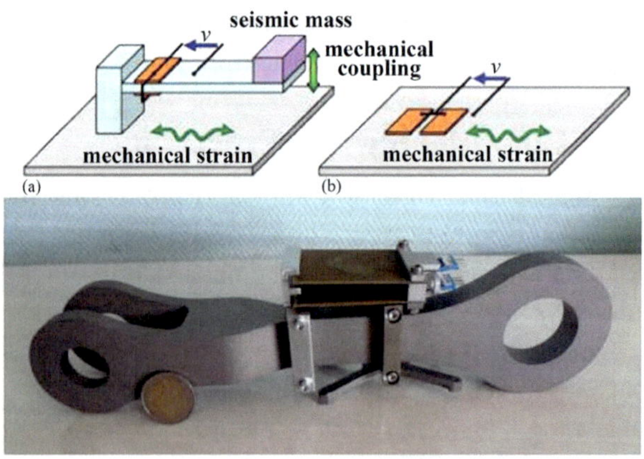

Figure 3.6. The role of signal conditioning and processing.

Source: https://cdnintech.com/media/chapter/69602/1512345123/media/F6.png.

3.2.3. Data Storage and Management Techniques

Data storage and management strategies are of paramount importance in the framework of Structural Health Monitoring (SHM) as they become crucial in handling the vast amounts of data generated by sensors. Data storage and management strategies not only ensure the quality and retrievability of data, but also enable rapid analysis and decision-making. Various storage media, ranging from cloud storage, have become of utmost importance in handling real-time data. Cloud computing makes consolidation of information from multiple sensors and devices convenient for access by many stakeholders. The combined procedure is particularly useful in SHM applications, where rapid recovery of data is critical to assessing structural integrity. By utilization of cloud technology, organizations can enhance collaboration and hasten responses to potential structural flaws, resulting in safer built environments.

Figure 3.7. Data storage and management techniques.

Source: https://neuroject.com/wp-content/uploads/2025/06/5-Leading-Sensor-Companies-in-Structural-Health-Monitoring-for-Construction-Main-Neuroject.jpg.

In addition to cloud storage, the employment of Building Information Modeling (BIM) adds to the management of sensor data in SHM. BIM gives a detailed model that integrates physical and functional attributes of structures and facilitates real-time monitoring. In this integration, new methods are used, such as the application of Radio Frequency Identification (RFID) and Long Range (LoRa) technologies for in situ tracking and analyzing performance data. The use of such technologies makes it possible to achieve accurate localization of prefabricated units, thus improving working efficiency and decreasing construction delays. As mentioned, the proposed strategy with new technologies can improve PC project performance and serve industry professionals by conveying sufficient required information (Liu et al., 2019). This integration of data acquisition and advanced modeling achieves greater visibility in perceiving structural health, something essential for making smart decisions. Also, secure data management procedures play a significant role in maintaining sensitive data gathered from sensor networks under their integrity and confidentiality. Having proper cybersecurity processes in place is a must, particularly with increasing data volumes and an increase in connected devices. According to the Office of Aeronautics and Space Technology report, good data management processes are necessary in various fields, including sensors and information systems (Beck et al., 2025). With the use of secure data encryption schemes, organizations can safeguard against unauthorized access and potential data breaches. Following this, through the combination of efficient storage strategies with stringent data security measures, not only is the dependency of SHM systems on reliable data enhanced but also confidence among stakeholders in structural safety (Figure 3.7).

3.3. SENSOR PLACEMENT AND CONFIGURATION

Sensor placement and layout are crucial in the effectiveness of structural health monitoring (SHM) systems. Proper sensor placement enhances the quality and relevance of data achieved, which is essential in the effective measurement of the structural integrity of various constructs. A sequenced approach to sensor placement mandates that a comprehensive analysis of the structures predicted performance and vulnerability to potential damage be conducted first. These include load-carrying parameters, material parameters, and environmental parameters that are to be considered. According to literature, data-based algorithms backed by quality data from sensors with appropriate locations can yield better results in damage detection processes, hence improving real-time decision-making as well as global structural health (Cherta et al., 2020). Due to this, the inclusion of engineering principles in the placement strategy ensures that the SHM system is efficient and effective. Besides, the sensor configuration plays a critical role in the capability and accuracy of data acquisition in SHM usage. Different sensors—such as accelerometers, strain gauges, and displacement transducers—have some properties that are appropriate for specific configurations for certain structures and anticipated stress conditions. For instance, accelerometers tend to be employed for the measurement of dynamic responses, whereas strain gauges play a very important role in more localized stress distribution determination. Best configurations may involve the employment of clusters of sensors in regions of importance with optimization of overlaps for minimizing redundancy of data. Such a complexity is possible that enables better understanding of structural behaviors while enabling real-time damage detection algorithms, as realized in the research on algorithmic performance for sensor data accuracy (Cherta et al., 2020). Such configurations have long-lasting impacts that go beyond such purely data acquisition and result in enhancing SHM system reliability. Finally, ongoing advancement in sensor technologies and data capture methods continues to shape best practices in sensor positioning and arrangement. In wireless sensor networks, for example, improvements ensure more flexibility with respect to sensor deployment at lower costs and lesser logistical hassles. Additionally, the application of machine learning methods ensures real-time optimization of sensor deployment strategies based on historical performance and future forecasting analytics. These new developments, while facilitating more wide-ranging monitoring, also open up prospects for sophisticated algorithms specifically designed for damage classification and localization. As addressed in the current discussion on SHM, both sensor arrangement and data-driven analytic approaches enhance the prospect of preventive maintenance and provision of safety in structural engineering. Last, the dynamic qualities of sensor technology require continuous reassessment of placement and configuration strategies to meet the demands of escalating infrastructure requirements (Figure 3.8).

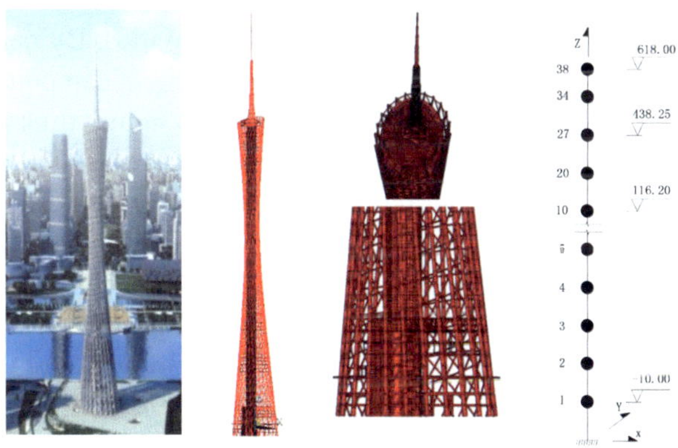

Figure 3.8. The strategic placement and configuration of sensors.

Source: https://encyclopedia.pub/media/item/202312/658d2d61b38babuildings-13-03129-g012-550.jpg.

3.3.1. Strategies for Optimal Sensor Placement

Effectiveness of sensor placement strategies in the case of Structural Health Monitoring (SHM) is a multi-sourced complexity problem that requires careful consideration of numerous factors. Optimum sensor location is critical since it directly affects the accuracy and reliability of data gathering for the purpose of determining structural integrity. The methods employed to determine the optimal positions of sensors are finite element modeling and optimization techniques, which guarantee the simulation and analysis of expected structural responses to various loading conditions. These computational approaches can identify the crucial points in a structure where damage is most likely to occur, thereby maximizing the efficiency of the sensor network. Furthermore, with data-driven algorithms gaining wide usage in SHM systems, understanding the behavior of individual sensors at chosen points maximizes the quality of the data, which eventually supports a more precise damage detection process (Cherta et al., 2020). Another important method is combining sensors with different sensitivity to different phenomena. Utilizing a variety of accelerometers, strain gauges, and acoustic emission sensors can provide a comprehensive estimation of the structural health because each kind of sensor excels in sensing certain signs of damage. For instance, accelerometers are highly effective in dynamic response measurements, while strain gauges provide useful information on static loads and stress distributions. Also, the placement strategy must address environmental conditions and operation factors that will be prejudicial to data. For optimal functioning, sensors must be positioned in locations that represent overall structural behavior, as opposed to local anomalies. By careful selection of sensor types and corresponding positions, researchers can construct a strong framework for effective data collection in SHM systems, thus enhancing the entire performance of monitoring systems (Delgado et al., 2010). Additionally, real-time data acquisition techniques as well as sensor positioning in optimal positions significantly enhance monitoring ability in SHM systems. Where sensors are strategically located, the continuous data accumulation allows for the early identification of potential damage, which is important in decision-making regarding maintenance and safety measures. Smart structures, embedded with sensors, assist in monitoring by allowing self-reporting of health metrics. The synergy between advanced analytics and data-driven algorithms coupled with strategically located sensors produces a synergistic effect to facilitate the detection of critical issues in advance before they turn into larger problems. Therefore, sensor placement optimization is not all about gaining optimum data quality but also ensuring the strength and long-term durability of structures through proactive maintenance practices (Cherta et al., 2020).

3.3.2. Factors Influencing Sensor Configuration

The sensor configuration in Structural Health Monitoring (SHM) is often dictated by a number of factors influencing their performance in data acquisition and analysis. The nature and type of structural component to be monitored are among the most important factors that dictate the type of sensor that will be used and its placement. Different structural components, such as beams, columns, or bridges, possess unique vibration and stress characteristics; thus, it is crucial to utilize sensors with the ability to properly capture these attributes. Temperature and humidity changes in environmental conditions can also affect the performance and lifespan of sensors. Sensor placement and configuration must take these factors into account to guarantee consistency in data despite the existence of different external factors (Brandon et al., 2012). Consequently, a strong sense of the specific structural elements and their interaction with environmental parameters is a deciding factor in designing ideal sensor configurations for SHM applications. The intended purpose for the monitoring system is the second critical parameter influencing sensor integration. Depending on whether the application is to monitor the onset of damage, assess the structural response, or record measurements for further long-term analysis, the sensors used and the arrangements employed might vary considerably. For instance, damage detection systems may require a denser array of accelerometers or strain gauges to ensure high sensitivity in capturing anomalies. Conversely, systems focused on overall structural response might prioritize fewer sensors that can capture broader trends over time. Furthermore, the data collection frequency becomes critical; certain applications may demand high-frequency sampling to detect transient events, while others might suffice with lower sampling rates. This delimitation in purposes calls for a tailored sensor design that is sensitive to the specific monitoring needs, thereby enhancing the efficiency and reliability of SHM systems (Cherta et al., 2020). Besides structural and operational parameters, financial considerations also have a major impact on sensor designs. Budget can limit the number and quality of sensors that are possible to deploy, and there may be cost-versus-benefit tradeoff in designing the monitoring system. Researchers and engineers typically must deal with the tradeoff between the amount and quality of data received and the project budget in total. As a result, new approaches like the use of multifunctional sensors or advanced algorithms that can work with fewer devices but still generate quality data are coming more and more into the focus. Identifying and evaluating such economic factors can lead to better sensor configurations that yield maximum performance at minimum cost. Overall, a combined approach to configuration must take into account the structural characteristics, performance requirements, and costs in order to yield a robust and efficient monitoring system in SHM (Brandon et al., 2012).

3.3.3. Case Studies on Effective Sensor Deployment

Sensor deployment in structural health monitoring (SHM) systems plays a central role in assessing the integrity of structures, particularly historical ones. An astute case study explains utilizing low-cost, open-source sensors to effectively monitor cultural heritage sites. Utilizing more ubiquitous digital modeling and fabrication technology, researchers have created workable alternatives for monitoring systems. The character of this new method emphasizes the possibility of using common, inexpensive monitoring systems to capture the intricacy of built cultural heritage. The generated information not only aids in short-term structural studies but also aids in safeguarding such valuable resources in the long run, enabling an environment where digital papers can be preserved in open-source format for access and research at a later time (Basto et al., 2017). In another significant case study of a wireless sensor network, focus was on optimal energy utilization towards real-time surveillance purposes. This research refers to the complexities of the realization of energy efficiency, particularly in DVS-based systems. While benefits of DVS are visible, the issue is maintaining synchronization in time between sensor networks and maintaining energy usage at a minimal level; without proper synchronization, information collected becomes unreliable. It was discovered that running a robust algorithm exclusively for energy saving had significant enhancements over typical DVS strategies. This innovation is vital in long-term SHM projects, most notably in applications where continuous and reliable data collection is demanded without affecting functional reliability (Abdelzaher et al., 2011). These case studies identify the mixed threats and innovative solutions part of the installation of sensors in SHM. Through the observation of open-source integration and the development of energy-efficient strategies, researchers are paving the way for more sustainable and effective monitoring techniques. The contribution of such studies extends beyond straightforward applications, enabling a transition towards more adaptable, reliable, and economical systems of monitoring. As the technology continues to improve, such research into the case studies will be essential to the advancement of methodologies, ultimately aiding the enhanced resilience and longevity of primary infrastructure as well as the preservation of the cultural heritage embodied in buildings of historical interest (Abdelzaher et al., 2011; Basto et al., 2017).

3.4. DATA ANALYSIS TECHNIQUES IN STRUCTURAL HEALTH MONITORING (SHM)

Data analysis techniques form the backbone in the area of SHM, primarily due to their capacity to process significant amounts of data collected by various sensors. Techniques are classifiable under two broad methodologies: data-driven algorithms and model-based methods. Machine learning and statistical algorithms make use of the power of machine learning and statistical methods to derive patterns from the sensor data to enable real-time damage prognosis, detection, and localization. Timely response is crucial in the maintenance of the safety and integrity of the structures, particularly the structures that are subjected to uncertain loads such as earthquakes or extreme weather conditions. Through the application of cutting-edge data-driven techniques, SHM can make a transition from traditional monitoring to predictive maintenance strategies and, consequently, minimize risks associated with structural failure as well as enhance infrastructure asset lifespan (Cherta et al., 2020). Data-driven algorithms also demand the availability of high-performance computing capacity as massive processing is necessary for real-time data analysis. These algorithms allow thorough investigation of structures, in addition to mere detection, including classification and prognostic capability.

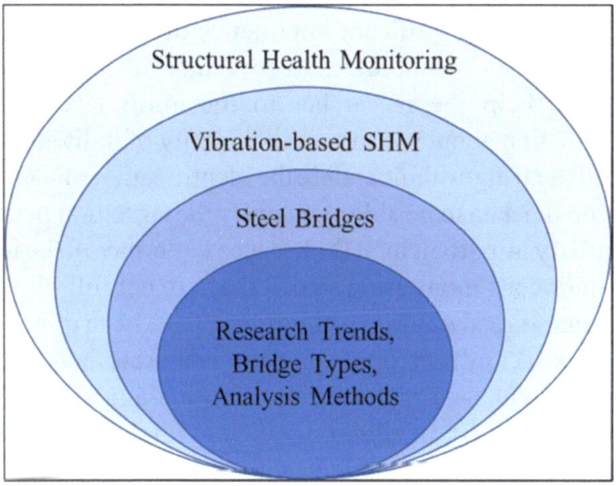

Figure 3.9. Data analysis techniques in structural health monitoring (SHM).

Source: https://ars.els-cdn.com/content/image/1-s2.0-S2090447923003908-gr1.jpg.

The synergy of their combined application ensures holistic determination of structural conditions and, therefore, informed decision-making by engineers and stakeholders regarding the timing and extent of maintenance interventions. As data acquisition technology advances, low-cost, open-source platforms are increasingly utilized. These technologies democratize entry into SHM systems, allowing greater numbers of applications, from critical infrastructure monitoring to the preservation of historic buildings. Although traditional approaches remain viable, the promise of these new approaches reinforces the necessity for ongoing research and development in the SHM field (Basto et al., 2017). In conclusion, continuous improvement of data analysis techniques in SHM has a proportional effect on the usability and efficiency of sensor technologies for infrastructure upkeep. By bridging the gap between data acquisition and applicable interpretation, the techniques enhance monitoring systems as a whole. The union of data-driven algorithms with emerging sensor technologies provides a robust tool for engineers to ensure structural integrity in various forms of architecture. With evolving research, a general understanding of these approaches will be key to guiding subsequent research in SHM, eventually yielding safer and more resilient built spaces. Consequently, the coming together of new data analysis approaches will remain a major point of focus for both scholarship and industry professionals (Cherta et al., 2020) (Figure 3.9).

3.4.1. Statistical Methods for Data Interpretation

Statistical methods are of significant importance in interpreting data collected through the use of sensor technologies in structural health monitoring (SHM). Statistical methods help the researcher to transform raw sensor data into meaningful information about the structural integrity of individual components. For instance, statistical algorithms enable the identification of damage by pattern recognition within the data to be able to quantify the structural performance over time. It is particularly important in SHM because faults have to be identified early enough using continuous monitoring so that they are not full-blown failures. As proposed in the literature, data-driven algorithms have been quite popular owing to their ability to extract information out of large datasets that are acquired from sensors and thereby enhance real-time decision-making for maintenance and repairs (Cherta et al., 2020). In addition, efficient implementation of statistical methods in SHM is further combined with robust models that are capable of extracting intrinsic variability present in environmental and operational conditions. Traditional damage detection techniques tend to struggle with the type of variability posed by such heterogeneity, potentially leading to biased output and false positives or negatives. Recent methodology, as suggested by Hajrya et al. (2014), calls for the use of advanced statistical methodology, i.e., independent component analysis and matrix perturbation, which can provide a more precise basis for evaluating structural health. By employing these methods, engineers can develop analytical thresholds that minimize statistical assumption dependence, significantly improving detection reliability and confidence in assessments. Finally, implementation of smart structural systems via multifunctional sensors makes statistical methods not only necessary but also more sophisticated. These smart structures also contain sensors within their structure to monitor their health regularly, giving real-time data that can be statistically analyzed for quick interpretations. Using sophisticated statistical approaches enhances the capacity to distinguish normal operating variations and critical damage indications, driving innovations in SHM technology. And as the technology continues to advance, the combination of sensor technology and robust statistical techniques will certainly define the future direction of infrastructure safety and maintenance (Hajrya et al., 2014). Therefore, research investment in understanding these interactions is central to the evolution of effective SHM practice.

3.4.2. Machine Learning Applications in Structural Health Monitoring (SHM)

The evolution of machine learning technologies has revolutionized SHM with the enhancement of data acquisition and interpretation systems. Through the traditional approach, SHM utilized model-based techniques that employed significant information concerning the material behavior of the structure and its response to varied loads. With the deployment of data-driven algorithms, the movement is towards the deployment of methods that learn directly from empirical observations captured by sensors. As debated in Cherta et al. (2020), they allow for real-time damage detection, localization, and prognosis through time-series sensor data. This development not only improves the accuracy of damage identification but also facilitates ongoing monitoring without needing extensive model approximations, making machine learning an invaluable asset in enhancing structural safety and reliability. Machine learning-based algorithms, particularly in vibration-based SHM, have significant advantages in the monitoring of large and complex structures. Traditional SHM methods, through a sparse array of accelerometers and other contact sensors, are prone to suffer from effects such as mass-loading effects and labor-intensive installation processes. Machine learning can potentially harness non-contact measurement techniques, such as video-based motion analysis, to provide high-resolution spatial data acquisition without compromising the integrity of the structure. The methods presented in Mao et al. (2018) illustrate how Phase-based Motion Estimation (PME) facilitates the isolation of subtle structural movement from video recordings, thus enhancing the capacity for damage detection. These technologies not only minimize installation issues but also provide a scalable solution for monitoring large infrastructures, providing a paradigm shift in the effectiveness of SHM systems. The combination of machine learning and advanced sensor technology produces more integrated data acquisition in SHM. Combining these technologies allows for the development of intelligent structures with sensors integrated into them that provide continuous data input and thus generate immense volumes of data that can be utilized in predictive maintenance. Algorithm-based analytics of data can systematically analyze such information to identify patterns or outliers, allowing for more anticipatory decision-making. Furthermore, as Cherta et al. (2020) highlight, the efficacy of such algorithms can be substantially improved by incorporating heterogeneous data from various sources, so that a structure's health is better understood. This convergence of machine learning and sensory data capture not only generates innovation in the use of SHM but also leads to more sustainable infrastructures that prioritize safety and longevity (Figure 3.10).

Figure 3.10. Machine learning applications in structural health monitoring (SHM).

Source: https://www.bitsathy.ac.in/wp-content/uploads/AI-in-Construction.jpg.

3.4.3. Real-Time Monitoring and Alert Systems

The use of real-time monitoring and alert systems in Structural Health Monitoring (SHM) systems has revolutionized civil infrastructure management. Real-time monitoring of structural health helps to ensure safety and durability of key assets. Real-time monitoring is possible through the application of advanced sensors and data acquisition technology, which can detect deviations from normal operating conditions, such as anomalies in stress or strain, and trigger warning to operators and maintenance personnel. In this way, proactive maintenance allows stakeholders to address potential failures before they turn into full-fledged issues, thereby extending structures' lifespan and overall safety. In addition, deployment of such systems is increasingly being recognized as essential to upkeep both old infrastructure and new construction, leading to wiser choices made using timely and accurate information (Brownjohn et al., 2016). Alert systems and real-time monitoring are particularly valuable to use in pavement management applications, wherein they enable preliminary detection of distresses that can erode structural performance. Because infrastructure managers are faced with traffic loading and environmental factors leading to progressive deterioration, the application of SHM systems would be capable of significantly enhancing maintenance schemes. By employing sensors that continuously monitor information concerning pavement performance, agencies will be able to develop more effective rehabilitation programs, and ultimately optimize resource allocation. The transition from conventional wired sensors to new technologies like micro-electromechanical systems (MEMS) provides greater flexibility and life in pavement monitoring, solving safety issues and saving costs in infrastructure management (N/A, 2016). In addition, data from real-time monitoring systems are also important in the analytics and decision-making processes of SHM. Modern warning systems not only provide immediate feedback on building condition but also aid in long-term data collection and trend analysis. With this wealth of real-time and historical information, operators and engineers can develop predictive maintenance schedules, permitting timely interventions based on facts rather than assumption. Moreover, the application of data mining techniques enhances overall structural health and performance characterization, leading to enhanced management strategies and policy-making. The ongoing development of real-time monitoring technologies positions them as key components of contemporary SHM, with the goal of making infrastructure systems failure-proof (Brownjohn et al., 2016) (Figure 3.11).

Figure 3.11. Real-time monitoring and alert systems.

Source: https://b2bblogassets.airtel.in/wp-content/uploads/2022/03/iot-application-in-weather-monitoring-system-scaled.jpg.

3.5. CHALLENGES IN SENSOR TECHNOLOGY FOR STRUCTURAL HEALTH MONITORING (SHM)

The use of sensor technology in SHM systems is accompanied by a range of challenges that need careful scrutiny and innovative solutions. Sensor calibration and reliability across different environments are some of the prominent concerns. Sensors should have the ability to provide consistent performance under different conditions, from intense temperatures to varying humidity levels, which significantly impact their accuracy and longevity. When monitoring ancient structures, as outlined in recent research, the use of low-cost and open-source sensors can help bridge some of the budget constraints, but at the cost normally being accuracy and lifespan. Scientists have studied their potential for use, observing that although they are highly promising for more widespread use, effective integration into monitoring systems is subject to close examination and possibly further steps to ensure consistency of operation under the particular conditions of heritage sites (Basto et al., 2017). Data collection and processing is another major challenge in using sensor technology for SHM. Sensors generate large amounts of data that must be processed in order to make informed decisions regarding the structure of the systems being monitored. Utilization of algorithms based on data, which are preferable as they can provide real-time perspectives, introduces computational burdens and data management. These algorithms typically require highly capable processors to deal with the large amount of data in an effective manner, thereby increasing cost and system complexity. Additionally, sensor reliability and data quality are not always consistent, and variation exists in damage detection and interpretation, which is evidenced by research into damage identification processes. The effectiveness of SHM systems depends in great measure on the balance of performance among data collection methods and analytical models to convert such data (Cherta et al., 2020). Further, challenges surrounding sensor technology for SHM are compounded by interoperability and standardization issues. With all the various types of sensors and data structures available, creating a system that provides seamless communication between sensors remains a challenge. There are applications for SHM that need customized solutions, resulting in fragmentation that hinders interoperability across different platforms. Moreover, there is a lack of internationally recognized standards that can cause discrepancies in data interpretation and evaluating sensor performance. Addressing these challenges is essential to future SHM technology advancement, particularly as the field moves toward the integration of increasingly advanced and complex monitoring systems capable of providing high-fidelity data and actionable information. It can only be done through combined efforts and hardware and software innovations that productive solutions to these challenges can emerge to enhance the overall efficiency of structural health monitoring (Basto et al., 2017).

3.5.1. Environmental Factors Affecting Sensor Performance

Environmental conditions are at the core of the operation of sensors utilized in Structural Health Monitoring (SHM) systems. Temperature fluctuations, humidity, and exposure to potential corrosive substances have significant effects on the validity and performance of sensors. For instance, elevated temperatures can lead to thermal drift in sensors, hence yielding erroneous readings that nullify the integrity of data being collected. Similarly, high humidity can cause condensation, which in turn can cause damage to the electrical parts of sensors, further reducing the reliability of their performance. With SHM relying heavily on data-driven algorithms for real-time decision-making about structural safety and performance, any degradation of sensor performance by virtue of environmental factors can lead to erroneous results and potentially unsafe consequences, putting a premium on environmental robustness at the design level of sensors (Cherta et al., 2020). Vibration and mechanical load are also critical environmental factors that could adversely affect the performance of sensors, particularly those deployed in dynamic monitoring applications such as bridges and high-rise buildings. Sensors should be resistant to traffic and wind load-induced vibrations in normal operation; however, they could become sensitive to excessive vibration, leading to misinterpreting the structure's behavior over time. Under circumstances where high levels of seismic activity are recorded, uncalibrated sensors which ignore such dynamic changes will give erroneous data regarding the structural integrity. This introduces complexity in the damage identification process of SHM systems, and strong algorithms must be used which would eliminate the noise and give accurate readings of the monitored structures (Abdelzaher et al., 2011). Moreover, external electromagnetic interference is a significant factor that may influence the integrity of data within wireless sensor networks implemented for SHM. As such sensors are usually deployed inside urban and industrial environments, they are subject to numerous sources of electromagnetic radiation which may cause interference in the transmission and reception of signals. Such interference will result in loss or corruption of data, and data-driven algorithms will find it difficult to process and translate inputs. To enhance the performance of sensors in such circumstances, scientists are working on the incorporation of shielding techniques and error correction codes. By concentrating on electromagnetic interference and other environmental conditions, sensors used in SHM applications can significantly improve their reliability and quality to effectively enhance their operating lifespan and have a positive influence on the effectiveness of structural health monitoring initiatives as a whole (Cherta et al., 2020).

3.5.2. Calibration and Maintenance Issues

Sensor calibration in Structural Health Monitoring (SHM) is a significant issue that has a major influence on the accuracy of data obtained for damage detection. The sensors must be calibrated with a great deal of care so that they register accurate and identical readings; slight differences can lead to huge errors in structural health determination. Calibration is actually the procedure of setting up the outputs from the sensor to conform to defined standards or reference levels, thereby ensuring compensation for any drift or deviation occurring over time. Calibration cannot be stressed enough; it directly impacts the decision-making process of SHM. When the sensors are poorly calibrated, the data created can mislead decision makers and engineers into taking the wrong actions or overlooking repair that is necessary. Moreover, the complexity of calibration always has to be constantly addressed, once again encouraging the need for periodic maintenance techniques to ensure sensor accuracy in the long term. SHM sensor maintenance needs to be treated with care so as not to degrade that conceals the integrity of structural analysis. Regular checking, maintenance, and cleaning of sensors are essential tasks that contribute to ensuring their efficiency. Over time, exposure to environmental factors like moisture, temperature fluctuations, and mechanical stress can lead to the breakdown of sensor components, generating less precise data. Formulating a detailed schedule for maintenance that suits each sensor's individual working environment can prevent these hazards. In addition, the integration of data-driven algorithms in literature requires high-speed processors to execute the high-level computation required for real-time analysis (Cherta et al., 2020). The algorithmic solutions rely on not just the precision of the collected data but also on the unflappable dependability of sensor performance, thus maintenance is a necessity. In the realm of SHM, in regards to sensor selection and field installation, minimizing the issues of calibration and maintenance is a critical parameter to guarantee productive outcomes. A report such as INE/AUTC 12.30 highlights how sensitive the process of selecting sensors needs to be coupled with the process adopted in successful field installation that directly impacts calibration as well as the maintenance regimen to be adopted (Brandon et al., 2012). By adhering to stringent standards and orderly practices, practitioners can optimize their SHM systems, where the sensors installed are not only sufficient for the job but also able to feed precise data throughout the working life. Proper calibration and good maintenance are thus the fundamentals to achieving successful damage detection and analysis ultimately leading towards enhancing the reliability of structural assessments and guaranteeing the safety of structures under surveillance via SHM technologies (Figure 3.12).

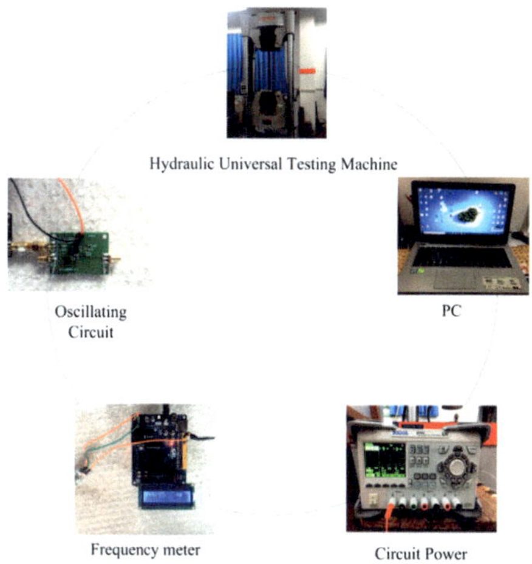

Figure 3.12. Calibration and maintenance issues.

Source: https://pub.mdpi-res.com/sensors/sensors-19-02782/article_deploy/html/images/sensors-19-02782-g006-550.jpg?1607316142.

3.5.3. Data Integrity and Security Concerns

In the Structural Health Monitoring (SHM) domain, security and data integrity of sensor-acquired information are critical concerns that potentially impact safety as well as operational performance. With the dependence on accurate and reliable information for making decisions related to maintenance and safety, any corruption or breach in data can lead to disastrous occurrences, like structural collapse. With improved technology, the utilization of more sophisticated sensors that monitor persistently boosts the volume of data to be validated and secured. The integration of sensors with advanced data acquisition systems thus must be equipped with robust security features to prevent unauthorized access or tampering. Furthermore, with pending threats like cyber-attacks, there has to be an active defense to ensure that the data integrity is not breached during its entire life cycle. SHM system security transcends protecting information at the moment, but also includes ensuring the privacy of those whose information may be unintentionally collected by these sensors. For instance, bio-sensors in new health care applications can potentially collect sensitive personal health data, raising ethical concerns over data-sharing and privacy rights. The trend in interconnected devices is a grave threat to security, as evidenced in recent advances in microelectronic and biosensor technologies. To ensure that the collected data is not just accurate but also safe, new strategies must be employed, including encryption, secure data transfer protocols, and anonymization of data. Because these matters are so important, the sector is encouraged to prioritize offering these protections more in SHM technologies, particularly where public safety and personal privacy are concerned. Addressing the data integrity and security issues in SHM ultimately necessitates a multi-pronged approach involving collaboration between engineers, cybersecurity experts, and policymakers. Such collaborative measures might generate standards and best practices governing handling data across a variety of applications. For example, findings from the research on high-energy battery technologies emphasize the importance of safety analysis and risk management systems. Research findings probing thermal accidents associated with lithium-ion batteries could similarly evolve the protocols necessary for the handling of risks in SHM applications. By actively identifying potential vulnerabilities, professionals can design counter-strategies that reduce these risks and increase the dependability of SHM systems. Continual sensor technology development brings challenges and opportunities in equal measure, underlining the necessity of ongoing dialogue and innovation to ensure data integrity and security in SHM (Figure 3.13).

Figure 3.13. Data integrity and security concerns.

Source: https://wallstreetmojo-files.s3.ap-south-1.amazonaws.com/2023/05/Data-Integrity-Meaning.png.

3.6. FUTURE TRENDS IN STRUCTURAL HEALTH MONITORING (SHM) AND SENSOR TECHNOLOGY

The SHM environment is on the cusp of radical change with advancements in sensor technology and data acquisition. Directions for the future are towards an increasing trend of more integrated autonomous systems for facilitating real-time data collection and analysis to significantly improve the responsiveness of maintenance processes. Perhaps most importantly, the availability of integrated non-destructive evaluation (NDE) methods has become critical, especially for complex structures like Composite Overwrapped Pressure Vessels (COPVs). The success of such systems relies on the creation of quantitative accept-reject criteria, which must be in line with specific damage levels proven to cause failure. In the absence of such structures, the promise of SHM may remain untapped, impacting high-stakes missions, especially in space exploration, where structural integrity plays a significant determinant of mission success and the safety of personnel (Nichols et al., 2025).

Figure 3.14. Future trends in structural health monitoring (SHM) and sensor technology.

Source: https://www.variohm.com//images/content/meta/903_meta_1200x630.jpg.

Simultaneously, developments in remote sensing technology, particularly synthetic aperture radar (SAR), represent a paradigm shift in surveillance. Modern developments in SAR allow for high-resolution temporal resolution, observing nanoscale deflections of megastructures over long periods and thus enhancing the performance of SHM systems. The systems can function even in harsh conditions where traditional in situ measurements are unable to be done or are even prohibitive. All-weather and multi-temporal characteristics of SAR systems make them especially suitable for SHM applications, where structures such as dams or cultural heritage buildings can be inspected remotely. SAR methods are not only employed for detection and localization of damage but further apply to assessment and prediction, which is a complete approach towards structural integrity management (Biondi et al., 2016). Besides, the use of machine learning and artificial intelligence for SHM is leading future trends. These technologies enable increased interpretation of vast data sets due to sensor networks, allowing for more accurate assessments of structural condition. With the use of algorithms combining historic data, present conditions, and predictive models, engineers are able to anticipate future issues even before they arise. This predictive capability is also critical for optimizing maintenance schedules and extending the working life of primary infrastructure. These data analysis advances, combined with robust sensor architectures, are opening up the way for a more proactive and unified SHM philosophy that can combat hazards and deliver confidence across engineering disciplines as buildings become increasingly intricate and require advanced monitoring strategies (Figure 3.14).

3.6.1. Advancements in Sensor Materials and Designs

Sensor material and design advancements have significantly enhanced the performance potential of data acquisition systems in structural health monitoring (SHM). The new era of advanced materials such as nanocomposites and smart polymers offers greater sensitivity and responsiveness, which are essential for real-time analysis of structural integrity. The integration of these new materials allows sensors to react more to environmental change and mechanical stress and ultimately to gather more precise and timely information. Engineers and scientists now possess tools that not only sense minute changes in structural condition but also predict potential failures, hence maximizing safety and minimizing maintenance expenditures in infrastructure management. These advancements reveal the key role material science and engineering have in developing more efficient and reliable SHM solutions that address existing infrastructure demands and sustainability goals. Sensor performance in SHM has also been enhanced by incorporating MEMS technology. MEMS sensors offer enormous advantages compared to traditional sensing methods, primarily through their compact size, lower power requirements, and enhanced performance characteristics. Advancements in MEMS technology have resulted in increased reliability and longevity of sensor systems, particularly for highway infrastructure monitoring. In addition to providing real-time continuous monitoring, sensors also introduce cost savings in lifecycle expenses for operating and maintaining structural systems. Using MEMS technology, engineers can integrate a more robust and responsive data acquisition system that can efficiently cater to environmental stresses and conditions and sustain the safety and advancement of infrastructure resilience against natural hazards (Buss et al., 2013). Wireless sensor networks combined with sophisticated sensor materials are a revolutionary means of integrating SHM. This strategy makes it possible to include sensing capabilities in building materials during the manufacturing process, thus speeding up data collection activities. The emergence of off-the-shelf MEMS devices has prompted a new generation of affordable, low-power sensing systems that can enable long-term real-time monitoring of critical infrastructure such as bridges and roads. Moreover, the use of advanced communication technology, like radio-frequency identification (RFID), also enhances monitoring capability by supporting remote tracking of structure states without invasive installations (Dong et al., 2010). All these developments are a deep orientation toward sustainable infrastructure decisions, closing the gap between high-tech developments and building practices to create safer, more resilient surroundings.

3.6.2. Integration of Internet of Things (IoT) in Structural Health Monitoring (SHM) Systems

Incorporation of IoT into SHM systems is a significant leap towards real-time data analysis and collection. By utilization of an integrated network of sensing entities, IoT enables structural integrity monitoring in real time, enabling engineers to detect potential failure prior to its occurrence as cataclysmic events. Smooth connectivity of the IoT technologies allows for data to be forwarded in real-time from an array of sensors, such as strain gauges, accelerometers, and thermal sensors, to centralized data processing systems. Real-time analysis is critical in guaranteeing the safety and reliability of structures, specifically critical infrastructures like bridges and skyscrapers, where failure could lead to disastrous consequences. In this regard, advanced technologies such as Radio Frequency Identification (RFID) and Long Range (LoRa) communication have emerged as pivotal aspects in developing IoT-based SHM systems. RFID enables precise on-site tracking of individual prefabricated components, whereas LoRa technology offers effective long-distance data transmission without high energy consumption. According to recent studies, integrating these technologies facilitates precise structural performance monitoring of parts as they are installed and throughout their lifecycle. Information collected by these methods can be conveniently stored and processed by cloud-based Building Information Modeling (BIM), which provides stakeholders readily available, real-time information regarding the structural condition and structural performance indicators of the monitored assets (Liu et al., 2019). Moreover, employment of sensor data fusion in IoT-driven SHM systems holds deep meanings. Utilizing inputs from multiple sensors and sources, the system enjoys a comprehensive understanding of structural behavior in various states. Not only does this paradigm shift favor decision-making between engineers and construction staff, but it also helps promote more sustainable construction and maintenance practices. Matching sensor information with pre-defined frameworks is revealed to have significant potential in enhancing deployment and management of these systems in smart city applications. As the field advances for sensor data fusion, ongoing research into applying it in IoT architecture remains critical in easing data accuracy and reliability shortcomings (Perera et al., 2015) (Figure 3.15).

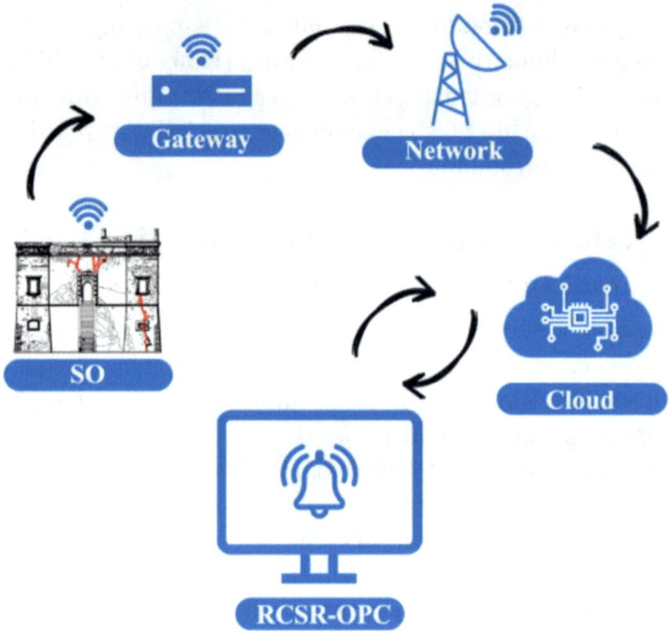

Figure 3.15. The integration of the Internet of Things (IoT) into structural health monitoring (SHM) systems.

Source: https://ars.els-cdn.com/content/image/1-s2.0-S0950061821008527-gr1.jpg.

3.6.3. Predictive Analytics and Its Implications

With the advent of predictive analytics, there has been a revolutionary change within the Structural Health Monitoring (SHM) sector. By employing advanced sensors and data-acquisition systems, professionals are able to capture high-resolution structural integrity information in real-time. All of these abilities are supported by the development of the Internet of Things (IoT), which enables seamless communication between interlinked devices. Machine learning programs embedded within them enable analysis of vast amounts of data to identify patterns that may be predictive of potential structural failure prior to its occurrence. Consequently, predictive analytics not only enhances pro-active infrastructure management but also alleviates the likelihood of catastrophic structural failures and makes possible on-schedule maintenance and repair, thus ensuring safety and reliability in construction processes (Baaziz et al., 2015). The implications of predictive analytics to SHM also extend to economic dimensions. By shifting away from traditional maintenance work, which relies on reactive repair and scheduled inspection, organizations can leverage data-driven knowledge to transition towards preventive and condition-based maintenance strategies. Not only does this reduce operation costs associated with infrastructure maintenance but it also maximizes the lifespan of assets. With ever-more sophisticated predictive models, industries realize better utilization of resources and reduction in downtime and, as a result, significant cost savings. Moreover, the use of predictive analytics allows decision-makers to allocate maintenance budgets more effectively, investing where structures are most at risk. This practice maximizes financial planning, in addition to facilitating public confidence in the safety and performance of infrastructure (Baaziz et al., 2015). In addition to economic value, use of predictive analytics for SHM advances environmental sustainability. Structural health monitoring allows organizations to prevent premature deterioration of infrastructures that are accountable for waste generation and resource loss. For example, predicting potential failure in marine structures can guarantee more sustainable ocean management by ensuring that harmful marine detritus and pollution are detected and cleaned up early. In addition, with industries becoming more ecologically conscious, the combination of predictive analytics and environmental concerns can result in novel techniques that anticipate and tackle problems such as ocean deoxygenation or invasive species control. As a consequence, the combination of predictive analytics, data gathering, and environmental sustainability highlights the various implications of these technologies in SHM.

SUMMARY

The integration of sensors and data acquisition systems in SHM has transformed how we assess and confirm the structural integrity of structures. As highlighted across this essay, correct application of these technologies allows for real-time data acquisition, analysis, and decision-making of the structural conditions. The flexibility of data-driven algorithms, particularly those adopted together with sensor technology, enhances the ability for detecting, locating, and characterizing potential damage in structures. These capabilities are their own reward, not only in terms of performance and security but also in terms of assisting in the extension of critical infrastructure longevity, which ultimately means cost savings and optimal resource utilization. With continuous research and development, the effectiveness of these systems of monitoring will, in the future, improve further and give more sophisticated evaluations that can react to the myriad of issues raised by different kinds of structures and situations, and thus reinforce their place in current engineering practices. Additionally, the new developments in low-cost, open-source platforms have huge potential for the widespread application of SHM, and particularly the monitoring of historic structures. This democratization of technology is a milestone in the industry because it shatters precedents for using high-end monitoring systems at non-deterrent prices, which otherwise remain costly on the pocket. The research discussed here verifies the efficacy of such cost-effective solutions and demonstrates that the use of such platforms in SHM makes preservation of cultural heritage more potent. Including such systems not only adds to the structural integrity of these buildings but also generates an online repository of digital information that can aid future generations in their conservation and understanding of their cultural worth. By doing so, the possibility of enabling sustainable methods of heritage preservation becomes more feasible, being in the service of greater social goals. Lastly, if we consider the current scenario of Sensors and Data Acquisition in SHM, we can see that the integration of technology and engineering principles holds unprecedented promise. The intense interaction between sensor technology and data-centric algorithms enables proactive maintenance and safety with paradigms where intricacies could be resolved prior to the development into catastrophic failure. A sustained commitment to research and innovation will eventually drive the advancement of SHM methods, leading to smarter, more responsive systems for managing infrastructure. By promoting interdisciplinarity between researchers, engineers, and conservators, the field will be in a position to continue to advance, providing evolutionary dividends that enhance structural safety, conserve cultural heritage, and increase resilience against environmental adversity. This change will have the methods employed now not only address present issues but also adequately equip us to address those of the future.

Advances in sensors and data acquisition devices have revolutionized SHM, particularly for heritage structures. Present research emphasizes the potential of low-cost and open-source platforms in monitoring the latter, which have been historically constrained by meager budgets and technical challenges. With the help of emergent digital modeling and fabrication technology, researchers can generate cost-effective and innovative monitoring systems that could be efficiently incorporated into the heritage environment. Not only does it ensure proper resource deployment, but it also makes cultural heritage conservation easier as ongoing monitoring is ensured. The conclusions of this research identify a revolutionizing shift towards democratization of SHM practice, thus enhancing proactive maintenance and evidence-based preservation methods in the heritage sector (Basto et al., 2017). Data acquisition methods in SHM are generally categorized into two vast techniques: data-driven algorithms and model-based analyses. The first utilizes data gathered by sensors to assess structural integrity, providing real-time feedback about the health of the structures being monitored, which is most vital in supporting timely decision-making processes. The second focuses on developing models that capture the structural content, aiming at high accuracy in reading the data. Although both approaches have their strengths, data-driven approaches are increasingly becoming favored due to their malleability and capability to effectively process large amounts of data. The high computational demands necessary to run such data-driven algorithms indicate the need for powerful processing facilities to address the complexity that comes with dealing in real-time monitoring applications. Thus, the disparity in methodology captures the ongoing evolution of SHM technologies and their potential for applicability (Cherta et al., 2020). The advent and utilization of sensors in SHM provide significant opportunities for enhancing our structural behavior and performance knowledge in the long term. With the application of some sort of low-cost hardware and sophisticated algorithms, scientists can identify damage, locate problem areas, and even predict future performance of structures. Such multi-dimensional analysis is crucial for extending life for existing structures, minimizing risks, and ensuring safety levels at all times. Besides this, the incorporation of intelligent sensor technologies aids the realization of structures into adaptive and self-sensing dynamic systems to react against environmental conditions. The current research activity in this field not only augments the current knowledge field but also offers the basis for developing integrated SHM solutions capable of addressing current as well as future structural engineering problems effectively (Basto et al., 2017; Cherta et al., 2020).

REFERENCES

1. Abdelzaher, T. F., Agha, G. A., Mechitov, K. A., et al. (2011). Dynamic voltage scaling techniques for energy efficient synchronized sensor network design. https://core.ac.uk/download/4834374.pdf (accessed on 2nd October 2025).
2. Agis Cherta, D., Gomez Vargas, R. C., Pedraza, C., et al. (2020). Damage identification in structural health monitoring: A brief review from its implementation to the use of data-driven applications. https://core.ac.uk/download/288625955.pdf (accessed on 2nd October 2025).
3. Ahmad, S. W., Jennions, I. K., Perinpanayagam, S., et al. (2015). A carrier signal approach for intermittent fault detection and health monitoring for electronics interconnections system. https://core.ac.uk/download/42144211.pdf (accessed on 2nd October 2025).
4. Anstrom, J., Bosch, M. S., Cleary, T., et al. (2016). Safety of lithium nickel cobalt aluminum oxide battery packs in transit bus applications. https://core.ac.uk/download/129532663.pdf (accessed on 2nd October 2025).
5. Baaziz, A., Kleiner, A., Feigelson, E. D., Waller, M. A., Boyd, S., Vandermerwe, S., & Zhou, Z. (2015). *Big Data and the Internet of Things*. http://arxiv.org/abs/1503.07092 (accessed on 2nd October 2025).
6. Baj-Rossi, C., Burleson, W., Carrara, S., et al. (2012). Fully integrated biochip platforms for advanced healthcare. https://core.ac.uk/download/286729684.pdf (accessed on 2nd October 2025).
7. Basto, C., Chacón Flores, R. A., & Pelà, L. (2017). Open-source digital technologies for low-cost monitoring of historical constructions. https://core.ac.uk/download/81578722.pdf (accessed on 2nd October 2025).
8. Beck, S. M., Breckenridge, R. A., Clark, L. G., et al. (2025). *Technology for the Future: In-Space Technology Experiments Program, Part 2*. https://core.ac.uk/download/pdf/42817270.pdf (accessed on 2nd October 2025).
9. Biondi, F., Bürgmann, R., Cigna, F., et al. (2016). Structural health monitoring of engineered structures using a space-borne synthetic aperture radar multi-temporal approach: From cultural heritage sites to war zones. https://core.ac.uk/download/84588322.pdf (accessed on 2nd October 2025).
10. Brandon, P., Hulsey, J. L., & Xiao, F. (2012). Structural health monitoring and condition assessment of Chulitna River Bridge. https://core.ac.uk/download/286607640.pdf (accessed on 2nd October 2025).
11. Brownjohn, J. (2016). Structural health monitoring of civil infrastructure.

https://core.ac.uk/download/43096639.pdf (accessed on 2nd October 2025).

12. Brownjohn, J. M. W., Moyo, P., & Omenzetter, P. (2005). Identification of unusual events in multi-channel bridge monitoring data. https://core.ac.uk/download/30435671.pdf (accessed on 2nd October 2025).
13. Buss, A., Ceylan, H., Ceylan, H., et al. (2013). Highway infrastructure health monitoring using micro-electromechanical sensors and systems (MEMS). https://core.ac.uk/download/38934593.pdf (accessed on 2nd October 2025).
14. Cojocaru, C., Delgado Prieto, M., Selim, H. M., et al. (2019). Laser ultrasound inspection based on wavelet transform and data clustering for defect estimation in metallic samples. https://core.ac.uk/download/187265561.pdf (accessed on 2nd October 2025).
15. Dong, Y., Liu, H., & Song, R. (2010). *Bridges Structural Health Monitoring and Deterioration Detection: Synthesis of Knowledge and Technology*. https://core.ac.uk/download/286607584.pdf (accessed on 2nd October 2025).
16. Hajrya, R., & Mechbal, N. (2014). Perturbation analysis for robust damage detection with application to multifunctional aircraft structures. https://core.ac.uk/download/29138092.pdf (accessed on 2nd October 2025).
17. Liu, Z., Mbachu, J., & Zhao, L. (2019). Development of intelligent prefabs using IoT technology to improve the performance of prefabricated construction projects. https://core.ac.uk/download/231841646.pdf (accessed on 2nd October 2025).
18. Mao, Z., Niezrecki, C., Poozesh, P., et al. (2018). Vibration-based damage detection in wind turbine blades using phase-based motion estimation and motion magnification. http://arxiv.org/abs/1804.00558 (accessed on 2nd October 2025).
19. Mujica Delgado, L. E., Rodellar Benedé, J., & Vehí Casellas, J. (2010). A review of impact damage detection in structures using strain data. https://core.ac.uk/download/46605571.pdf (accessed on 2nd October 2025).
20. N/A. (2015). *Launching the Grand Challenges for Ocean Conservation*. https://core.ac.uk/download/84911929.pdf (accessed on 2nd October 2025).
21. N/A. (2016). *Development of a Wireless MEMS Multifunction Sensor System and Field Demonstration of Embedded Sensors for Monitoring Concrete Pavements, TR-637, 2016*. https://core.ac.uk/download/76640544.pdf (accessed on 2nd October 2025).

22. Nichols, C., Saulsberry, R., & Waller, J. (2025). *Smart Composite Overwrapped Pressure Vessel–Integrated Structural Health Monitoring System to Meet Space Exploration and International Space Station Mission Assurance Needs*. https://core.ac.uk/download/pdf/10566765.pdf (accessed on 2nd October 2025).

23. Perera, C., Wang, M., Zhang, M., Strazdins, P., Jayaraman, P. P., Shyamsundar, R. K., & Ranjan, R. (2015). City data fusion: Sensor data fusion in the Internet of Things. http://arxiv.org/abs/1506.09118 (accessed on 2nd October 2025).

CHAPTER 4

WIRELESS SENSOR NETWORKS FOR SHM

Contents

INTRODUCTION

The incorporation of Wireless Sensor Networks (WSNs) in Structural Health Monitoring (SHM) revolutionized the way we monitor and maintain infrastructure. SHM plays a significant role in establishing the integrity and safety of engineering structures such as bridges, buildings, and dams. The inspection of these structures was previously manual-based, which not only consumed a lot of time but was also prone to human error. Latest advances in wireless sensing and communication have facilitated a more structured process for real-time monitoring of the health of structures. This has facilitated continuous capturing and analysis of data, which is important for timely decision-making and intervention. With WSNs, engineers can significantly enhance the detection of possible damage and be in a position to take preemptive measures, hence extending the lifespan of critical infrastructure. The damage detection methodology in SHM generally involves two main strategies: data-driven algorithms and model-based approaches. The former invokes physical models of structural systems to simulate their behavior, whereas the latter uses real sensor measurements to determine the present state of a structure. More often than not, preference nowadays is shifting towards data-driven tools because of their efficiency in handling the large volumes of information gathered with the help of sensors and because of their ability to adapt to changing requirements. Such algorithms not only facilitate real-time damage detection but are also conducive to predictive analytics, allowing for enhanced planning and resource allocation in maintenance activities. The significant computational cost of data-driven algorithms, according to Cherta et al. (2020), necessitates sophisticated processing power; however, insights derived from their application often outpace current technical barriers. Furthermore, the continually improving wireless technology and sensor design enable the development of SHM further. Today's sensors can measure a variety of parameters, such as vibrations, temperature variations, and strain. The multifunctionality of sensing capabilities extends data resolution and the accuracy of data obtained from WSNs, hence increasing the precision of damage diagnosis and prognosis. Emphasis on developing intelligent structures with these technologies results in more durable infrastructure systems. Combined with the results introduced by data-driven algorithms mentioned above, this is a big leap towards more resilient health-monitoring solutions. With the development of research, WSNs are to be incorporated into SHM and reduce expenditures, increase efficiency, and establish new standards in structural safety and reliability to achieve the goals of sustainable engineering practice (Fan et al., 2011).

Definition of Wireless Sensor Networks (WSNs)

WSNs are a technological breakthrough of the current era that allows the collection and transmission of information in any environment. Made up of geographically remote autonomous sensors, WSNs are designed for the monitoring of physical or environmental conditions such as temperature, humidity, and vibration. The networks allow real-time data collection, which is vital for applications ranging from environmental monitoring to SHM. One of the fundamental features of WSNs is their ability to be deployed in environments where traditional monitoring is going to be impossible, significantly improving the scalability and effectiveness of the monitoring systems. By employing low-power wireless communications, WSNs offer an affordable and versatile solution for various industries that must continuously monitor physical states to ensure operational integrity and security. For SHM, WSNs are increasingly significant because they potentially can offer a complete picture of structures' states over time. This technology involves the application of numerous sensors to detect changes in parameters that are an indicator of structural deterioration. For example, by the implementation of WSNs, real-time vibration monitoring of critical infrastructure such as bridges and railroads is achievable. Multi-point capability for collection and transmission allows analysis on a comprehensive scale with immediate intervention, which increases maintenance plans significantly.

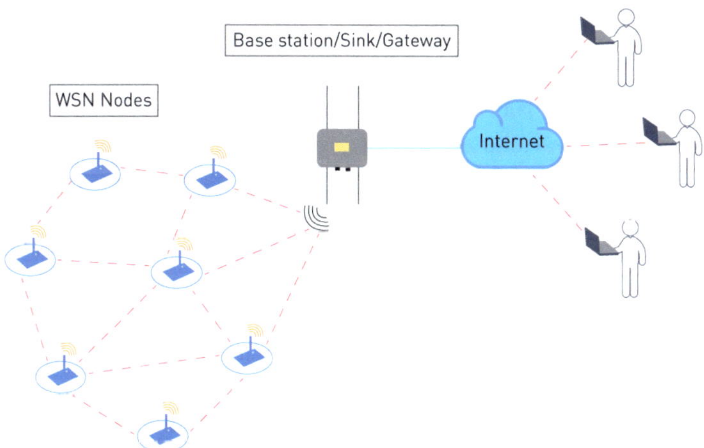

Figure 4.1. Wireless sensor networks (WSNs).

Source: https://media.monolithicpower.com/wysiwyg/Educational/Diagram_Recreation_-_Sensors_Chapter_4_Fig3-_960_x_650.png.

Technological advances in sensors and communications have enabled WSNs to become ever more capable, such that they are now feasible for employment in sophisticated environments where traditional monitoring would often be too unwieldy or too expensive. The advancements emphasize the potential of WSNs to allow active maintenance and extended longevity of critical infrastructure. Their effectiveness in operation largely depends on their design and the requirements of the application. For example, in cases where high precision of data is a strict necessity, such as in vibration monitoring of helicopter rotor blades, WSNs must be constructed to overcome many challenges, such as power supply, communication, and processing of data (Basu et al., 2013). Interaction among these factors is most crucial in determining the capability of WSN applications for SHM. In addition, with industries embracing the Industrial Internet of Things (IIoT), the use of WSNs in their operations is becoming essential. The evolution of advanced communications technologies such as LTE and 5G further supports WSN technologies to enable efficient data transmission and real-time monitoring crucial in maintaining operational safety and efficiency in such applications as railways (Castedo et al., 2017). Lastly, later developments in the technology of WSNs will result in revolutionary improvements in structural monitoring and maintenance methods in various industries (Figure 4.1).

Overview of Structural Health Monitoring (SHM)

The primary discipline of SHM is a preventive approach for maintaining the integrity and safety of structures through constant testing and data acquisition. Through monitoring various parameters, such as stress, strain, and vibration, SHM enables engineers to identify potential problems before they develop into full-blown issues. The use of high-tech solutions, such as wireless sensor networks, has transformed SHM with real-time capture of data without any constraints from wire-based infrastructure. This not only improves the accuracy of the monitoring process but also provides flexibility so that sensors may be placed in inaccessible or hazardous locations. Thus, SHM becomes a crucial tool for guaranteeing structural reliability and eventually saving public safety and reducing maintenance cost. In addition, adoption of wireless technology in SHM makes energy efficiency and sustainability in structural monitoring applications possible. WSNs typically adopt energy harvesting techniques, which allow them to operate autonomously without relying on conventional sources of power. For instance, recent developments have led to the development of autonomous wireless nodes that receive power from their environment, such as seismic accelerations during earthquakes. Through piezoelectric materials, such nodes can effectively convert mechanical vibrations into electrical power, thus enhancing the longevity and operational effectiveness of SHM systems. As outlined in the literature, energy harvesting has the potential to do much to ease the logistics burden of battery-powered sensor maintenance and hence improve the overall efficiency of SHM systems and pave the way to more sustainable practice (Marano et al., 2017). Apart from energy efficiency, the complexity of SHM technologies may also be seen in their ability to use sophisticated computational models to analyze data. Incorporation of wireless sensor networks facilitates rapid collection of huge amounts of data, which can be processed with the aid of advanced algorithms, hence resulting in more effective maintenance and repair decision-making processes. For example, a method based on the theory of random vibrations for energy harvester performance analysis has been proposed. The model allows for the simulation of ground acceleration due to seismic activity, enhancing the predictive capability of SHM systems. These advances put forward the dynamic dependence between data analysis and structural monitoring and position SHM at the forefront of modern infrastructure management and resilience practice (Bontempi et al., 2017). Through these developments, SHM not only emerges but also becomes a major contributor to upholding the sustainability and safety of built structures (Figure 4.2).

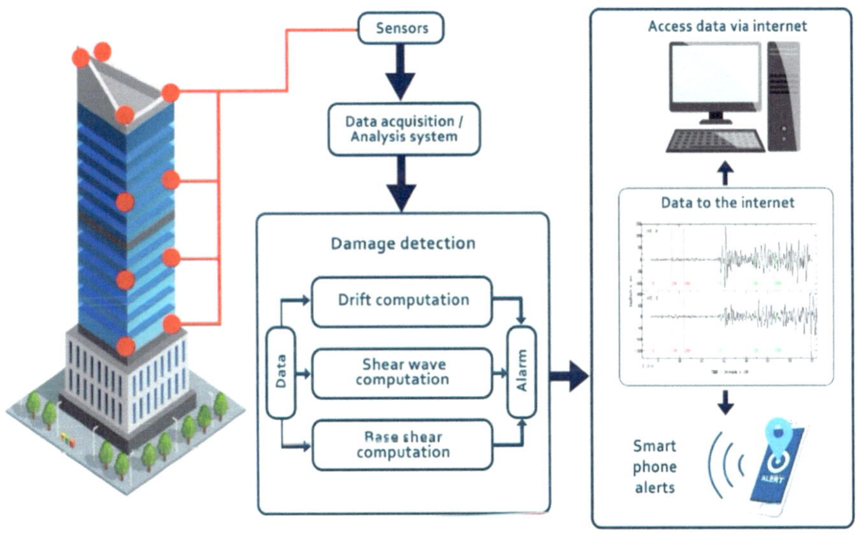

Figure 4.2. The critical domain of structural health monitoring (SHM).

Source: https://www.mdpi.com/buildings/buildings-11-00263/article_deploy/html/images/buildings-11-00263-g001.png.

Importance of Integrating Wireless Sensor Networks (WSNs) in Structural Health Monitoring (SHM)

The integration of WSNs in SHM presents a groundbreaking method that upholds the reliability and efficiency of structural health monitoring. With an array of sensors mounted in a structure, WSNs enable the on-going monitoring of real-time data for parameters including vibrations, temperature shifts, and strain. With such an abundance of information, timely structural condition assessments are possible, which is extremely important to detect possible damage in its initial stages. As literature has explored, data-driven and information-based algorithms using the data gathered by such sensors can provide decision-making answers in real time, taking aim at the heavy computational cost of data processing and enabling safer infrastructure management by active intervention (Cherta et al., 2020). The use of WSNs thus revolutionizes the methodological approach of SHM from reactive to proactive, thereby guaranteeing significantly increased safety in engineering practice. In addition, current demand for alternative sources of energy, e.g., wind energy, fuels the importance of integrity of supporting infrastructure. Precisely, the rise in onshore wind turbine installations underscores the significance of advanced monitoring systems.

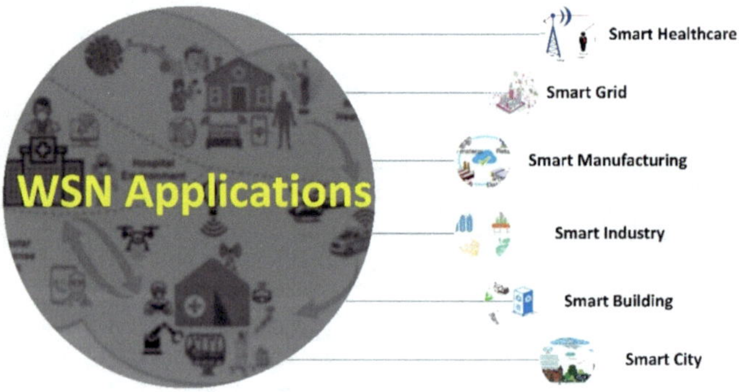

Figure 4.3. The integration of wireless sensor networks (WSNs) in structural health monitoring (SHM).

Source: https://media.springernature.com/lw1200/springer-static/image/art%3A10.1007%2Fs12083-025-02070-0/MediaObjects/12083_2025_2070_Fig13_HTML.png.

WSNs with a large number of sensors can easily monitor performance and integrity of the structures to meet demands of the dynamic loads found in wind energy utilization. As stressed by studies, a cost-effective custom-designed structural health monitoring system is vital in ensuring the reliability and safety of these renewable energy resources (Currie et al., 2012). This paradigm demonstrates that integrating WSNs and SHM not only enhances safety but also promotes environmental and sustainability goals, making it an essential consideration in modern engineering practice. Furthermore, the operational advantages contributed by WSNs extend beyond those of increased near-term safety to encompass long-term economic benefits. The deployment of WSNs in SHM can minimize maintenance costs through enabling predictive maintenance methods, where possible modes of failure can be diagnosed and eliminated before they become propagated failures. This preventive solution not only saves resources but also minimizes downtime associated with structural repair. As reflected in the literature, algorithmic integration of data-based algorithms with sensor-derived real-time data plays the crucial role in the development of smart structures to self-monitor through damage detection, localization, and prognosis (Cherta et al., 2020). Summarily, the use of WSNs in SHM is not just a new technology but a necessary evolution towards stronger and more efficient infrastructure management systems, capable of addressing the challenges facing modern engineering requirements (Figure 4.3).

4.1. FUNDAMENTALS OF WIRELESS SENSOR NETWORKS (WSNS)

WSNs are an innovative advancement in the practice of structural health monitoring (SHM). With the use of distributed sensing technology, WSNs can constantly gather data from various structural elements and provide near real-time judgments on structural integrity. Constant information flow allows for early fault detection and hence enhanced safety and reliability of infrastructure. The union of advanced communication technologies and new sensor techniques, such as Micro-Electro-Mechanical Systems (MEMS), makes it possible to utilize small and efficient sensors that can easily be installed in areas that are inaccessible. As a result, WSNs not only make monitoring more efficient but also provide for an anticipatory approach to maintenance operations in a way that necessary interventions are planned to prevent catastrophic failures, as evidenced in the context of SHM for heritage buildings (Barontini et al., 2017) and innovative engineering demonstrator systems (Aguilar et al., 2011). Operational WSN concepts entail a variety of functionalities crucial to the technology, including data processing, communication, and data collection. A WSN node is typically an embedded processor, memory, and a communication module that facilitates efficient data collection and transfer. Connecting nodes is achievable through several protocols to facilitate data transference back to a centralized processing unit for analysis. This distributed approach not only enhances system reliability but also ensures robustness despite node failures. Furthermore, the application of nature-inspired optimization techniques offers various advantages, particularly concerning energy efficiency, which is necessary for extending the operating life of battery-powered sensors in distributed networks. These features are core to facilitating effective deployment of WSNs in SHM applications, as they align with the necessity of deploying cost-effective and sustainable structural monitoring approaches (Barontini et al., 2017). While replete with potential for revolutionary changes, WSN application in SHM has its challenges. Signal interference, sensitivity thresholds, and data synchronization remain principal obstacles to effective monitoring. For example, some commercially available MEMS sensors can suffer from low sensitivity or resolution, which will degrade the damage detection and assessment accuracy (Aguilar et al., 2011). Additionally, handling complexity in joining data streams of multiple sensors can make it challenging to interpret the structural health, where high-level algorithms must be used to effectively combine and analyze the data. Overcoming these shortfalls is crucial in order to improve the role of WSNs in this context. Ongoing research and development activity aimed at optimizing sensor technology and algorithmic systems must be undertaken to allow wireless sensor networks the potential to deliver on their promise of maintaining the integrity of civil infrastructure and contributing to the longevity of built heritage (Figure 4.4).

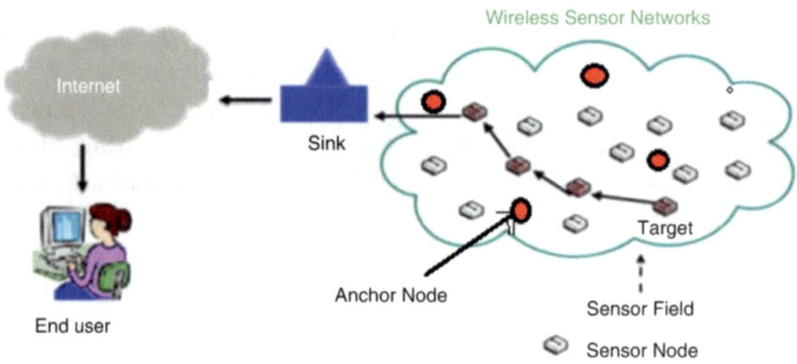

Figure 4.4. Wireless sensor networks (WSNs).

Source: https://media.springernature.com/lw685/springer-static/image/chp%3A10.1007%2F978-3-030-74061-0_1/MediaObjects/501267_1_En_1_Fig1_HTML.png.

4.1.1. Components of Wireless Sensor Networks (WSNs)

WSNs play a tremendous role in the advancement of Structural Health Monitoring (SHM) systems through the composition of various fundamental elements that facilitate efficient data gathering and transmission. The nodes constitute the core of a WSN that has hardware that is necessary to measure structural parameters such as strain, temperature, and vibration. The nodes gather meaningful data that reflect the structure's condition and performance of the asset in consideration. Once the data is obtained, the data are communicated to a central processing node, referred to by terms like sink or base station, through wireless communication protocols. Reliability and accuracy of data obtained from sensor nodes such as these are essential for effective SHM. Hence, the choice and application of sensor types and configurations become important to achieve improved coverage and fidelity of data in testing environments. Another crucial point of WSNs is linked to the communication infrastructure that can transmit data from sensor nodes to processing devices with ease. Various communication protocols, such as Zigbee, LoRa, and Wi-Fi, are used for various applications and different performance requirements in SHM. For instance, LoRa technology has particularly been renowned to provide long-range communication using low power and hence is suitable for monitoring widespread infrastructures like bridges or buildings where sensor nodes may be extremely dispersed. In addition, the integration of RFID technology with WSNs can enhance operation efficiency of monitoring systems in that it enables precise location tracking of prefabricated units, thus facilitating effective on-site management during construction. Such capability is vital in delivering the analyzed and improved structural performance promptly and dependably, particularly under real-time data needs in settings (Liu et al., 2019). Data analysis and processing mechanisms constitute another core element of WSNs, particularly in SHM. The data obtained from sensor nodes must be subjected to sophisticated analysis to provide useful information regarding the structural conditions. It is here that data-driven algorithms come into play, using the collected data to establish damage, identify issues, and predict impending performance issues. These algorithms derive significant benefits from instant analysis of real-time data, making timely decisions that can minimize risks associated with structural failure. In the majority of cases, such an ability relies on robust computational power due to the vast volume of data generated. Yet, the development of smart structures increasingly facilitates the integration of sophisticated data-driven methods that can effectively monitor and assess infrastructure integrity and accommodate multiple forms of data, ultimately resulting in enhanced structure safety and longevity in surveilled networks (Cherta et al., 2020).

4.1.2. Communication Protocols Used in Wireless Sensor Networks (WSNs)

For WSNs employed in Structural Health Monitoring (SHM), communication protocols are of utmost priority to facilitate successful data transmission from sensor nodes to the processing unit. They allow for various tasks such as data gathering, data aggregation, and sending it on networks as well as addressing issues regarding energy consumption, latency, and reliability. For instance, light-weight communication protocols are particularly important in WSNs because of the restricted power supply of sensor devices. Energy-limited protocols, such as the Media Access Control (MAC) protocol, help in minimizing collisions and maximizing the utilization of the channel to enable sustainable operation of the network over long monitoring periods. These processes are required for gathering information in real time for SHM applications, where timeliness and precision of information are very significant in order to evaluate structural condition and make sound maintenance choices. The integration of communications protocols within WSNs also emphasizes the necessity of precision and timeliness of data for continuous monitoring systems. Based on the specific application in question, different protocols are customized to support different levels of data transmission rates and reliability. For example, Transport Control Protocol (TCP) is generally utilized where data delivery is to be guaranteed, whereas TCP may not be suitable for low-power sensor networks due to overhead. In contrast, User Datagram Protocol (UDP) offers a less heavy option at the cost of reliability, with speed being given greater importance than loss, which can be applied in applications where losing data is tolerable. The choice of protocol is ultimately an issue of the specific needs of the SHM system, emphasizing performance versus energy savings as the top goal for its most efficient network operation. Advances introduce new ways through which communication protocols for WSNs can be enhanced in SHM applications. Dynamic Voltage Scaling (DVS) is one of the technologies that have emerged as primary means to reduce sensor node energy consumption without sacrificing timekeeping performance. The application of DVS in WSNs illustrates the manner in which power management can greatly prolong the lifespan of monitoring systems during use, an aspect that is important for sustained data gathering over prolonged durations. Interestingly, pioneering open-source technologies have also emerged with a future in supporting low-cost, yet efficient, monitoring solutions for heritage buildings. These technologies pair advanced communication protocols that have the potential to enhance functionality while financially viable, thereby facilitating greater use of continuous SHM systems across varying construction environments. This union of technological advances heralds a new future of structural monitoring as sustainable and affordable, promising significant benefits to the preservation of built cultural heritage (Abdelzaher et al., 2011; Basto et al., 2017).

4.1.3. Types of Sensors Commonly Used in Wireless Sensor Networks (WSNs)

The effectiveness of WSNs when applied to Structural Health Monitoring (SHM) relies highly on the types of sensors used in such networks. Various sensors are employed to carry out particular tasks based on the parameters being measured, such as displacement, strain, temperature, and humidity.

Displacement sensors, for instance, are crucial in detecting movement within structural members so that any movement is identified and responded to in real time. Strain gauges, on the other hand, detecting the deformation of materials, are similarly crucial in monitoring stress response in structures subjected to loaded state. Such sensors support the collection of real-time data, central to the delivery of safety and integrity throughout the life of a structure, in accordance with the data-driven approaches emphasized in the literature for the use of sensors within SHM (Cherta et al., 2020). Other functions within SHM are facilitated by environmental sensors employed for monitoring the conditions of the environment such as temperature and humidity. Such sensors provide vital context to structural data by measuring external environmental conditions that can potentially affect the performance of a structure. For instance, thermal expansion or contraction is caused by changes in temperature, and this can stress materials.

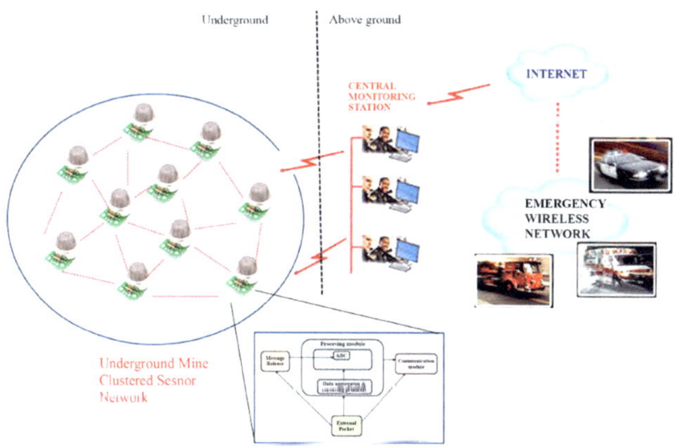

Figure 4.5. Types of sensors commonly used in wireless sensor networks (WSNs).

Source: https://www.elprocus.com/wp-content/uploads/2014/03/38.jpg.

By constantly collecting such environmental data, WSNs provide the overall investigation of how external factors cause potential issues in structures over time. Further, technological innovation in sensors has enabled them to be integrated with processing capabilities to carry out preliminary analysis prior to data transmission. This feature lightens the data load of the communication network and enhances the efficiency of information handling and decision-making in SHM (Liu et al., 2019). In addition, integration of next-generation sensor technologies such as Radio Frequency Identification (RFID) and Long Range (LoRa) into WSNs has revolutionized traditional monitoring techniques in construction and structural health. RFID technology enables real-time location tracking of prefabricated building materials on a construction site while strain sensors are utilized for performance monitoring. These technologies, when integrated with cloud computing, enhance the accessibility of data such that stakeholders in the project are able to access vital information via Building Information Modeling (BIM) on a decentralized level. Hence, this multi-dimensional approach not only increases the reliability of data but significantly enhances onsite decision-making and project performance. These advancements highlight the way that modern sensor technologies are central to the conversion of WSNs into effective platforms for effective SHM, which promotes more intelligent and dynamic structural monitoring approaches (Liu et al., 2019) (Figure 4.5).

4.2. APPLICATIONS OF WIRELESS SENSOR NETWORKS (WSNS) IN STRUCTURAL HEALTH MONITORING (SHM)

The application of WSNs in SHM has revolutionized the way engineers and researchers assess and guarantee the integrity of structures such as bridges, dams, and buildings. Through a sensor network that is distributed, WSNs enable the gathering and examination of real-time information, hence supporting more enhanced decision-making and resource utilization. Modular Monitoring System (MMS) is a good example of this revolution by offering a tailorable modular structure that accommodates mesh network topology, adding to coverage and reliability for various SHM applications. This flexibility is indicative of the considerable leap in monitoring technologies to accommodate tailored solutions that can address the particular needs of various structures at the expense of cost-effectiveness and operational efficiency without compromising overall structural safety (ANGELETTI et al., 2015). In damage identification, WSNs' ability to support data-driven algorithms cannot be overemphasized. Data-driven algorithms make use of sensor data to examine the condition of a structure at the time, enabling interventions when necessary. The two major ways of doing damage identification are data-driven methods that are sensor-dependent and model-based methods that are model-dependent. Nonetheless, algorithms driven by data emerge as the prevailing techniques with potential to provide real-time information on structural condition and are thus the go-to for dynamic SHM systems. These techniques involve enormous computation burdens that call for the inclusion of high processing capacity in WSNs and highlight further the importance of sophisticated detection techniques in maintaining infrastructure integrity and safety (Cherta et al., 2020). The advantages of employing WSNs for SHM are more than mere data collection and analysis, as they also facilitate the development of smart structures that can self-monitor without the need for human interference. Technological innovations in sensor technology and data analysis procedures have pushed SHM to more intricate systems with predictive analytics and self-diagnostics. With the help of information obtained from WSNs, buildings can not only detect damage but even anticipate future failures before they happen, which improves reliability and reduces maintenance expenses. The shift to such smart systems is a sustainable strategy in infrastructure management, where round-the-clock monitoring results in preemptive action, making crucial structures operational and safe for public consumption for long durations (ANGELETTI et al., 2015; Cherta et al., 2020) (Figure 4.6).

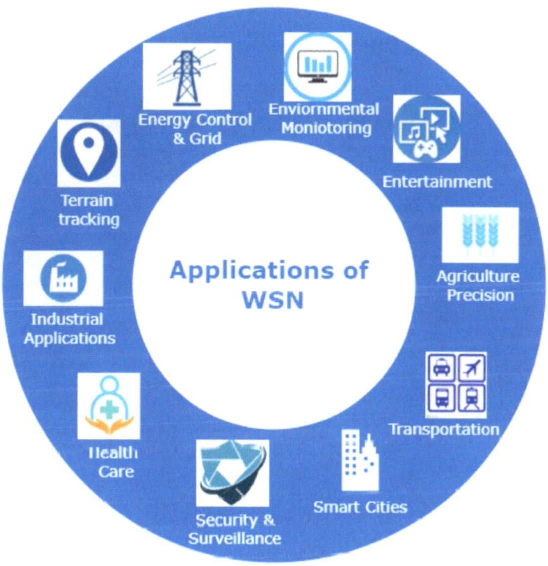

Figure 4.6. Applications of wireless sensor networks (WSNs) in structural health monitoring (SHM).

Source: https://www.mdpi.com/sensors/sensors-22-02087/article_deploy/html/images/sensors-22-02087-g002.png.

4.2.1. Monitoring Bridges and Infrastructure

The integration of wireless sensor networks (WSNs) significantly enhances bridge and other critical infrastructure monitoring capacity. The networks facilitate real-time data acquisition and transmission, which is essential for

SHM usage. WSNs can monitor and assess structural response to environmental excitation, such as temperature changes and vibrations, vital in maintaining structural integrity, via the implementation of various sensors. As technology advances, data-driven algorithms have emerged as key solutions in the processing of the tremendous amount of data gathered by such sensors. Such algorithms allow for rapid identification of damage or defects in structures, hence ultimately informing maintenance and overall safety improvements. Through such active monitoring of bridges, infrastructure systems become stronger, the risk of abrupt failure decreases, and the lifespan of key assets is increased (Cherta et al., 2020). In the interests of SHM, the development of smart structures is a significant consideration in altering infrastructure monitoring. Sensing functionalities can be embedded into construction materials during production so that engineers can create buildings that self-monitor their health without any involvement by humans.

Figure 4.7. Monitoring bridges and infrastructure.

Source: https://besgroup.com/app/uploads/2023/10/Infrastructure-Monitoring-Bridges-Forth-rail-bridge-min-scaled.jpg.

This new technology has generated interest considering the advancement in micro-electromechanical systems (MEMS), which ensure improved reliability and performance over traditional monitoring methods. Application of MEMS technology in infrastructure enables monitoring of different parameters, such as stress and strain, in real-time conditions continuously. Hence, smart structures enable accurate feedback about their performance and safety, thus enabling the detection of probable flaws at an early phase. This shift to smart infrastructure finds a place in the arena of sustainable practices, ultimately resulting in reduced maintenance costs and enhanced safety for the infrastructure's users (Buss et al., 2013). Wireless sensor networks are an exemplary shift in the maintenance and monitoring of infrastructure. More specifically, the integration of MEMS and WSNs offers cost-effective long-term monitoring solutions answering the complex issues of traditional monitoring systems. In addition to the economic benefits, these technologies enhance safety aspects against natural hazards and normal wear and tear of bridges and roads. Worth mentioning specifically here is that experiments with wireless concrete monitoring, based on radio-frequency identification (RFID) technology combined with MEMS sensors, demonstrate real-world implementation of this technology in practice. The implications of such innovations trigger the necessity for additional development and research in this field, along with the integration of these technologies into various infrastructure projects to reap their potential in guaranteeing structural health (Buss et al., 2013) (Figure 4.7).

4.2.2. Assessing the Health of Buildings and Dams

Structural health monitoring of buildings and dams is a crucial aspect of the subject, and assessing the health of these structures keeps them safe and long-lasting. With increasing urbanization and age of buildings, there has never been a more critical need for high-efficiency assessment methods. Wireless sensor networks are emerging as central technologies in this field with their capacity to monitor buildings in real time and provide immediate responses to structural changes. Such sensor networks, as exemplified by initiatives like ShakeNet, make possible rapid deployment of monitoring instruments after earthquake activity, thus enabling valuable feedback on structural response to aftershocks. Through the integration of high-resolution vibration sensors and advanced data acquisition techniques, such networks have the capability not only to enhance safety but to make maintenance of older structures economically viable, very much in accord with contemporary practice in civil engineering (Govinda et al., 2015). The growth in SHM triggered by wireless sensor networks is particularly groundbreaking for gigascale structures like dams. Such structures require continuous monitoring to detect potential weaknesses that lead to catastrophic failure. Wireless accelerometers incorporated in the structural component allow continuous vibration and load response monitoring even in harsh environments. In situations where wired systems would be too costly or inconvenient in other situations, wireless technologies offer a more feasible option. In addition, the data acquired can support our understanding of dam performance over different load states and environmental factors. Based on actual time data, engineers are able to make sound judgments concerning maintenance planning, ultimately leading to optimized operating lifespan for principal structures and safeguarding public safety (Brownjohn et al., 2016). Besides monitoring, the use of wireless sensor networks exceeds because they are a part of active building and dam health management. Information gained through the application of SHM can be critical in informing design choices and guidance for retrofitting activities. Increasing awareness among stakeholders of the importance of structural integrity will drive the demand for advanced monitoring systems. Ongoing developments in data collection technologies and communication systems will continue to augment the competence of wireless networks to analyze and interpret intricate datasets efficiently. This advancement represents a significant shift towards augmenting data-driven practice in civil engineering, wherein the use of wireless sensor networks will be regular practice in determining the health of key infrastructures, making them safer and more reliable for generations to come (Govinda et al., 2015) (Figure 4.8).

Figure 4.8. Assessing the health of buildings and dams.

Source: https://www.building.govt.nz/assets/Uploads/managing-buildings/building-safety/dam-safety-timeframes__FillWzcyNCw0MDhd.jpg.

4.2.3. Real-Time Data Collection and Analysis

The evolution of wireless sensor networks (WSNs) has transformed real-time data acquisition and analysis, especially in the domain of Structural Health Monitoring (SHM). With the use of a network of connected sensors, these networks are able to monitor continuously vital data about a structures integrity and performance. For instance, Radio Frequency Identification (RFID) and Long Range (LoRa) technologies facilitate real-time collection and transmission of data from various structural components, expanding the scope to monitor conditions at critical stages, such as installation. Real-time data not only facilitates instant response to potential defects in the structure but also adds vigor to proactive maintenance practices. The pairing of Building Information Modeling (BIM) and cloud computing enables easy access to such data, providing the stakeholders with the current status of the well-being of the structure, hence better decision-making and optimal resource utilization. Also, the functionality of real-time data capture systems can greatly increase construction and prefabrication project efficiency. For example, in a practical application, the use of strain sensors and LoRa technology demonstrated that live tracking of prefabricated members could result in accurate location evaluations and performance assessment of the structure in the construction site. As explained in recent publications, such a capability allows on-site project teams to obtain relevant information, thereby enhancing accountability and the efficiency of operations. With the assistance of cloud-based BIM systems, the experts in the construction sector are able to see and examine real-time information, which not only assists moment-by-moment troubleshooting but also facilitates more general overall judgments throughout the building's life.

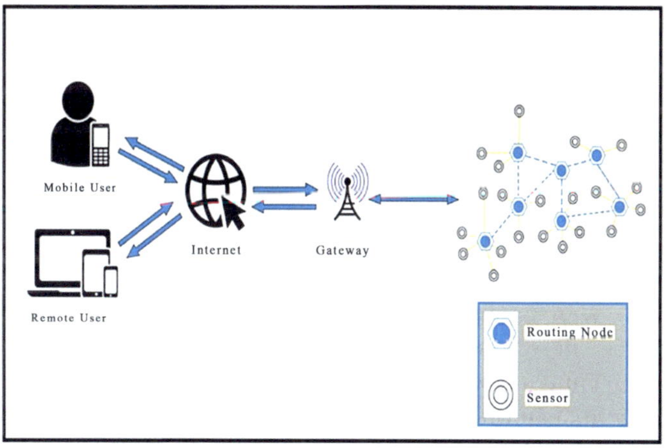

Figure 4.9. Real-time data collection and analysis.

Source: https://media.springernature.com/lw685/springer-static/image/chp%3A10.1007%2F978-3-031-26564-8_18/MediaObjects/516225_1_En_18_Fig1_HTML.png.

Thus, the combination of technology and analytics of data is one of the main drivers of enhancing the degree of SHM. Finally, the question of how to maintain energy efficiency in wireless sensor networks is most critical to the sustainability of SHM systems. Techniques such as Dynamic Voltage Scaling (DVS) have been explored to minimize energy spending while maintaining performance reliability. By dynamically controlling the CPU frequency, such techniques can extend the lifespan of sensor networks while enabling high-fidelity monitoring activities. Utilizing such techniques would help to optimize SHM system sustainability. However, one critical matter is time synchronization, which is fundamentally important in determining reliable data in SHM. With recent research findings, while DVS has the potential to significantly reduce energy usage, possible loss of synchronization should receive serious consideration. Research efforts have attempted to mitigate such issues, making sure that energy-efficient systems are able to work efficiently without compromising data integrity, thereby ensuring the reliability of overall real-time data acquisition in structural monitoring applications (Abdelzaher et al., 2011) (Figure 4.9).

4.3. ADVANTAGES OF USING WIRELESS SENSOR NETWORKS (WSNS) FOR STRUCTURAL HEALTH MONITORING (SHM)

WSNs integration within SHM systems offers a variety of advantages that significantly enhance the reliability and efficiency of infrastructure monitoring. One of the primary strengths of WSNs is real-time data collection and analysis. This aspect enables timely decision-making by providing immediate feedback regarding the structural safety of bridges and buildings. Such rapid communication is highly critical at the time of adverse conditions or emergencies in order to provide rapid responses to any faults found. As indicated by latest studies, WSNs with modular designs can be easily configured to meet the specific requirements of various SHM applications to achieve better adaptability and performance in real-world environments (ANGELETTI et al., 2015). This flexibility is an indicator of the technologically sophisticated nature of WSNs, thus rendering them a highly sought-after choice within the realm of structural monitoring. Moreover, WSNs provide enhanced coverage and dependability through their mesh network topology, by which interlinked sensor nodes are able to communicate smoothly over long distances. This topology not only helps overcome difficulties posed by harsh environmental conditions but also provides redundancy in data transmission, reducing the likelihood of data loss. Such dependable data communication links ensure that critical structural health data are continuously monitored and reported accurately. Further, energy-efficient design practices utilized in WSNs prolong the lifespan of sensor systems deployed on buildings, thus lowering the cost of maintenance and downtime (ANGELETTI et al., 2015). Consequently, cost-effectiveness brings another attractive aspect to WSNs as they enable organizations to make better use of resources without compromising structural safety. The progression from traditional SHM techniques to data-driven techniques facilitated by WSNs makes it possible to perform more precise damage identification and assessment in structures. Data-driven techniques take advantage of the abundance of information obtained by sensors to detect, localize, and classify damage types. The ability to process and analyze in situ real-time data from structures allows for more informed repair and maintenance decision-making. In addition, the advancement of intelligent structures with WSNs allows for an integrated view of health performance indicators, thus enabling proactive measures in place of reactive solutions. Altogether, WSNs' advantages in SHM not only augment safety features but also call for the shift to smart infrastructures that can self-sense and respond adaptively, and this shifts the field of civil engineering (Figure 4.10).

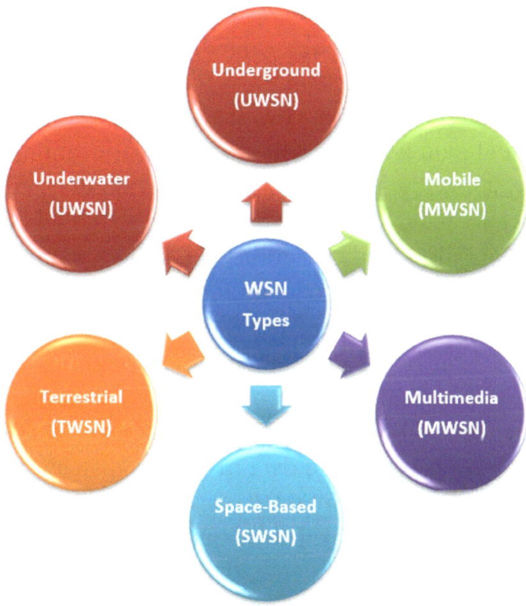

Figure 4.10. The integration of wireless sensor networks (WSNs) in structural health monitoring (SHM) systems.

Source: https://pub.mdpi-res.com/sensors/sensors-22-05094/article_deploy/html/images/sensors-22-05094-g004.png?1657169942.

4.3.1. Cost-Effectiveness Compared to Traditional Methods

The application of wireless sensor networks (WSNs) in structural health monitoring (SHM) is a cogent case for cost-effectiveness compared to traditional approaches. Traditional monitoring procedures are often time-consuming and involve lengthy manual inspections and wired, static sensors that are not only time-consuming but also prone to human errors and unexpected costs. In contrast, WSNs take advantage of automated data collection, which does away with the need for frequent on-site physical visits. This is not only cost-effective but also saves the cost of labor, allowing for improved resource allocation. Integration of recent technologies like LoRa and RFID supports this affordability by enabling real-time data gathering and transferring, supporting the monitoring of multiple structures simultaneously without significant increases in costs (Liu et al., 2019). In addition, WSNs support greater long-term economic benefits through minimized maintenance and operational expenses associated with traditional methods. The traditional monitoring is typically done by routine manual checking and expensive hardware to ascertain the structural condition of buildings. Applying data-based algorithms, as pointed out in certain studies, WSNs enable real-time tracking and early damage detection. The proactive approach can lead to interventions at a lower stage, ultimately bringing enormous savings on repair and maintenance. Cloud-based platform implementation also enhances these systems by providing readily available and comprehensive project data to well-versed decision-making as well as better management of the projects. Flexibility and scalability of WSNs can ensure sustainable construction operations with consequent short-term and long-term economic advantages (Cherta et al., 2020). Besides direct cost reduction, wireless sensor network deployment can result in better project scheduling and performance. Using traditional techniques, latencies are typically experienced due to the need for manual assessment and processing of data collected in the field. WSNs, on the other hand, provide real-time data to project managers and engineers in real-time, accelerating verification processes and allowing for swift reaction to any irregularities detected. This real-time timeliness can cut down on project times and overall construction standards, thereby saving on costs for extended timelines. Constant real-time observation, WSNs can also serve as an excellent quality assurance system during the lifecycle of a structure, providing a cheaper approach in total than conventional methods of observation (Liu et al., 2019).

4.3.2. Enhanced Data Accuracy and Reliability

The application of wireless sensor networks (WSNs) in structural health monitoring (SHM) is at the heart of providing enhanced data accuracy and reliability. WSNs facilitate continuous capturing of vital data regarding the state of structures via a network of sensors. Live monitoring enables engineers to detect problems such as material fatigue or structural weakening at the earliest stage. Data-driven algorithms play a vital part in handling this information overload to facilitate enhanced analyses that result in well-enlightened decision-making in repair and maintenance work. The application of data-driven approaches compared to traditional model-based approaches is primarily due to their ability to utilize vast amounts of data using numerous sensors, thus improving the overall accuracy of damage identification processes in SHM systems, as noted in Cherta et al. (2020). These advances indicate the growing relevance of WSNs in maintaining the integrity and safety of infrastructure. The precision of data obtained from wireless sensor networks is critical to their operation in SHM schemes. Modern technologies, such as the integration of Radio Frequency Identification (RFID) and Long Range (LoRa) technologies, not only enhance data collection but also enhance the reliability of data transmission across construction sites. In practice, these technologies help track structural components through the assessment of their physical performance criteria so that the information gathered is reliable and actionable. Cloud-based solutions further improve data integrity so that sensor information is easily incorporated into Building Information Modelling (BIM) models. These advancements constitute an integrated platform that allows project teams to view real-time information, and thus enhance process management and timely intervention during the construction process, as the evidence in Liu et al. (2019) suggests. In order to ensure maximum benefit is derived from better data accuracy and reliability, it is necessary to overcome challenges that wireless sensor networks may have in SHM applications. Sensor drift, data noise, and external interference pose risks to the validity of the gathered information. These risks are countered by robust calibration techniques and advanced filtering strategies. Data-driven algorithm optimization continuously ensures that provided information is current and correct, ultimately driving better predictive analytics and risk analyses. This cyclical process not only improves the precision of diagnostic outcomes but also builds confidence among stakeholders in the decision-making process. Through the use of technology to improve data accuracy, WSNs can potentially transform SHM practice to ensure safer structures and effective resource allocation for maintenance needs, as highlighted in the data-driven algorithms for damage detection in SHM applications review contained in Cherta et al. (2020).

4.3.3. Ability to Monitor Remote or Inaccessible Structures

In the case of Structural Health Monitoring (SHM), priority goes to the ability to monitor remote or inaccessible structures. Traditional methods of monitoring do not provide in situations where access is not good due to geographical reasons, environmental conditions, or security concerns. Wireless Sensor Networks (WSNs) provide an alternative solution to the problem through the ease of sensor deployment in locations that are hard for human experts to reach. WSNs can monitor structural integrity, environmental conditions, and possible threats on a continuous basis and thus ensure adequate monitoring of critical infrastructures such as bridges, dams, and offshore platforms. This makes it possible to realize hazards to safety and reliability in real-time and accordingly perform maintenance or intervention. The operational efficiency of WSNs in the field is due to the fact that they can function autonomously, using low-power communication technologies.

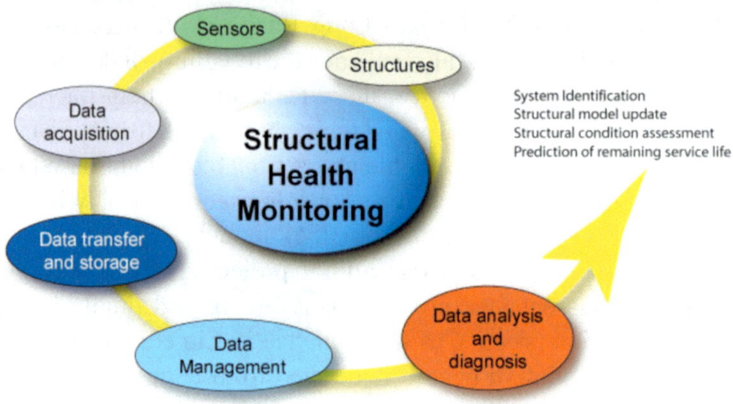

Figure 4.11. In the context of structural health monitoring (SHM).

Source: https://ars.els-cdn.com/content/image/3-s2.0-B9780128035818100396-f10039-01-9780081005330.jpg.

The sensors are strategically positioned in pre-decided locations to provide a comprehensive overview of the state of structures without human intervention at regular intervals. Therefore, WSNs are capable of collecting important data over lengthy periods, monitoring real-time fluctuations in structural behavior that might escape notice otherwise. For instance, in the protection and tracking of shipping containers, WSNs provide the ability to monitor vast logistics operations and thereby enhance security solutions by preventing loss and theft (Flavio et al., 2012). Such capability to provide constant monitoring is crucial in crowded areas where human observation cannot be sufficient. Moreover, advances in sensor technology and communication standards have significantly improved the reliability of data collected from WSNs deployed in adversarial or far-flung locations. The networks are now capable of easily transmitting data across long distances, and this further widens their application. With data fusion and remote diagnostics among other approaches, engineers receive accurate and timely information that is critical to decision-making processes related to structural upkeep and safety assessments. The application of WSNs in SHM operations ultimately encourages an innovative approach to infrastructure management, thus reducing risks and enhancing the resilience of critical systems (Fan et al., 2011). As the need for reliable monitoring solutions continues to rise, WSN application in monitoring distant structures is likely to become more dominant (Figure 4.11).

4.4. CHALLENGES IN IMPLEMENTING WIRELESS SENSOR NETWORKS (WSNS) FOR STRUCTURAL HEALTH MONITORING (SHM)

Deployment of WSNs in SHM involves several issues worthy of careful consideration. Of primary concern is the dependency on data-driven algorithms, which are crucial to handle information obtained via sensors placed in the field. While these algorithms facilitate real-time decision-making, their deployment is accompanied by high-performance processors due to the computational requirements posed when handling huge datasets. This requirement can limit the feasibility of applying WSNs to limited environments, where hardware expenses and power consumption are major considerations. Another aspect is that the high technology involved in data-oriented approaches further emphasizes the complexity of the overall system, with a higher maintenance and running cost that could be beyond what conventional monitoring operations can keep up with (Cherta et al., 2020). Another pressing concern is the reliability of the sensor network itself. WSNs are susceptible to various environmental parameters, such as temperature, humidity, and electromagnetic interference, which have a tendency to adversely affect the precision and longevity of the sensors. Moreover, the vulnerability of wireless communication channels to signal loss and interference poses serious reliability issues for data transmission. Any disruption in data flow can result in partial or erroneous interpretations of a structure's health and thereby violate the purposes of SHM. It is necessary to design robust and adaptive communication protocols that can effectively support these environmental conditions without sacrificing the integrity and accuracy of data exchanged between sensors (Liu et al., 2019). The requirement for seamless integration across various technologies also makes it difficult to implement WSNs in SHM systems. The use of newer technologies like Building Information Modeling (BIM) and Radio Frequency Identification (RFID) can add more functionality to sensor networks, but with that comes another layer of complexity. The problem is, how to make sure that multiple systems talk to one another properly, creating full data collection and analysis. There is a misalignment in various technologies that destroys the performance and efficiency of the SHM system overall. With ongoing research, it is critical that new developments remain aimed at the interoperability of WSNs with existing infrastructure to guarantee that they benefit to their full capacity in structural monitoring and maintenance (Liu et al., 2019) (Figure 4.12).

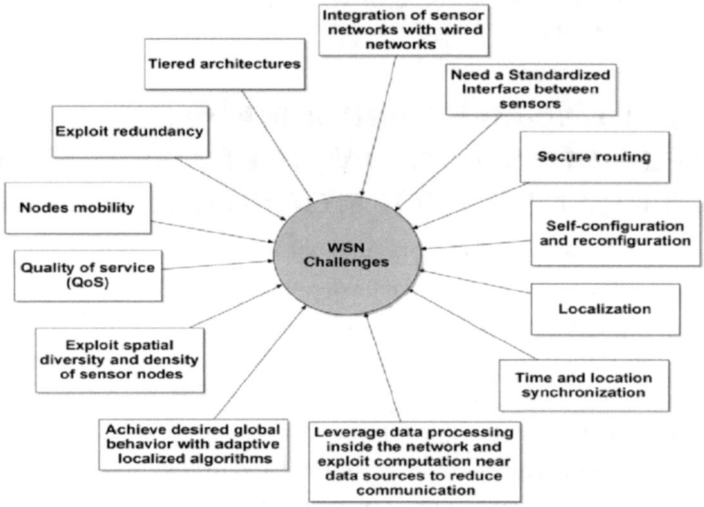

Figure 4.12. Challenges in implementing wireless sensor networks (WSNs) for structural health monitoring (SHM).

*Source:*https://www.researchgate.net/profile/Mohamed-Elhoseny/publication/315794085/figure/fig7/AS:614134123073556@1523432393656/Current-Challenges-and-Research-Directions-of-WSN.png

4.4.1. Energy Consumption and Battery Life Issues

In the application of wireless sensor networks (WSNs) for Structural Health Monitoring (SHM), battery life and power usage acquire prime concern status. As such networks consist of resource-constrained sensors that are predominantly battery-operated, it is crucial to adhere to energy-conserving practices that prolong the operational period. It becomes clear from understanding the dynamic relationship between energy demands and usage of SHM applications that there is an urgent need for innovation. The introduction of energy-consumptive security technologies such as encryption and authentication, though important in data integrity, has the byproduct of exacerbating battery drain. Thus, the challenge is to balance the introduction of security mechanisms with the necessity to guarantee efficient energy use, pushing researchers to look for new approaches that optimize longevity without compromising security (Udoh et al.,

2019). To counter both energy efficiency and security issues, it is necessary to look into methods such as duty cycling, which manages the power status of the sensors in order to conserve battery life. Duty cycling protocols such as Sensor MAC, Timeout MAC, and Tunable MAC have proven effective in maximizing energy consumption by minimizing active transmission times. Comparison of such protocols across various network situations gives information about how they impact energy savings and the system throughput in general. For instance, simulations that measure quantifiable values like transmit power and throughput may give one information regarding the efficiency of each mechanism for duty cycling. Lastly, the aim is to develop new protocols that not only effectively conserve energy but also enhance resilience to forthcoming denial of sleep attacks, which have severe consequences on sensor longevity (Udoh et al., 2019). Moreover, integration of new energy management techniques is a vital contribution to solving battery life issues in WSNs for SHM. As the systems become more complex with increased deployment, innovative solutions to energy are the need of the hour. The research area has now featured smart algorithms that dynamically adjust sensor activity based on environment and operation requirements. These advancements have great potential to optimize the energy consumption and extend the lifetime of the sensor nodes. Moreover, as technology continues to develop, there is greater emphasis on creating systems that are energy drain attack resilient and effectively hold back battery resources. This equilibrium between security and energy efficiency encapsulates the evolution of wireless sensor networks based on realizing that an integrated methodology will be essential in SHM applications in the future (Figure 4.13).

Figure 4.13. Energy consumption and battery life issues.

*Source:*https://blog.nordicsemi.com/hs-fs/hubfs/Blog_pics/Tech_Explained/2017-06/shutterstock_637062103-029243-edited.jpg?width=559&height=315&name=shutterstock_637062103-029243-edited.jpg.

4.4.2. Data Security and Privacy Concerns

Data security and privacy concerns have become fundamental in the area of Wireless Sensor Networks (WSNs) for Structural Health Monitoring (SHM) as these types of networks are prone to use a series of networked sensors that gather sensitive information regarding the structure's health of buildings and infrastructure. As the data collected could include stress, strain, and vibration pattern data, the absence of functional security systems represents a severe risk. It would be possible for criminals to hijack such data or intercept it during transmission, and the consequences for public safety would be catastrophic. Ensuring that this sensitive information is safeguarded against unlawful tampering and access is imperative. As underscored in the literature, security of sensitive data, including structural data, against illegal exposure is crucial for ensuring confidence in WSN technologies and their applications in SHM (Baig et al., 2014). CPS and IoT intricacies also make security and privacy in WSNs for SHM even more challenging. The seamless incorporation and monitoring through these technologies tend to result in the dynamic reconfiguration of network components, rendering traditional security protocols inadequate. Such a changing landscape negates the authority of design teams over system components, with components evolving as a function of user context or environmental requirements. Information security carried across these networks thus becomes paramount, requiring advanced protocols for guaranteeing integrity and privacy. As is seen in recent analysis, the combination of sophisticated data analytics and real-time changes affects user privacy and the commodification of susceptible information (Azorin-Lopez et al., 2018). These innovations require a nuanced understanding of the threats with the incorporation of WSNs in susceptible monitoring scenarios. To address the pressing concerns of data security and privacy in WSNs for SHM, it is of crucial significance to undertake security frameworks with customization that consider the unique nature of the sensor networks. There have been various methodologies proposed, including traditional key-based and paired key-based paradigms, to offer secure communication among nodes. These methods are applied to ensure the security of data transmission and prevent unauthorized access, thus establishing a secure environment for SHM implementation. Additionally, utilizing encryption techniques and the regular rotation of security measures can efficiently overcome data breach vulnerability. With the growing complexity of WSNs and their deployment as a constituent part of larger IoT networks, an active security and privacy strategy will be necessary in maintaining the integrity of structural monitoring data, thereby ensuring the safety and reliability of the infrastructures they are intended to safeguard (Figure 4.14).

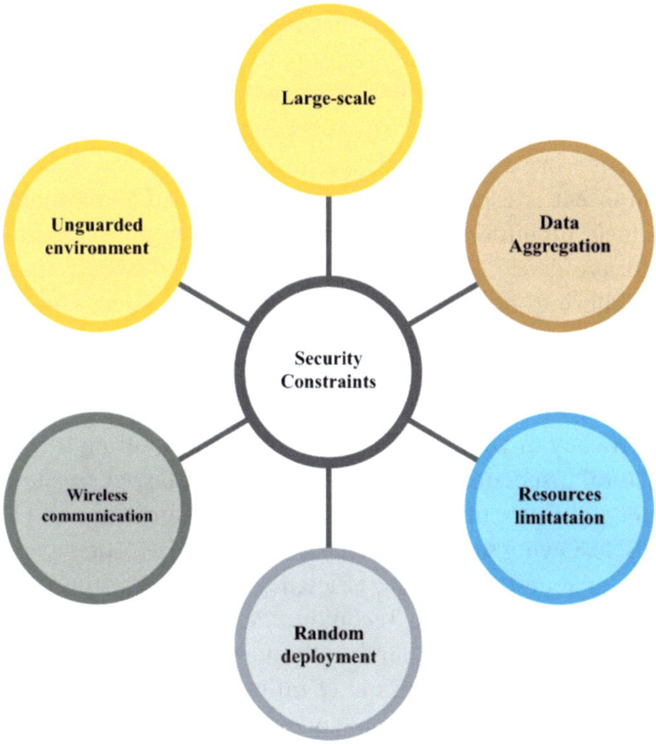

Figure 4.14. Data security and privacy concerns.

Source: https://media.springernature.com/lw685/springer static/image/art%3A10.1007%2Fs11277-020-07213-5/MediaObjects/11277_2020_7213_Fig2 HTML.png.

4.4.3. Environmental Factors Affecting Sensor Performance

In wireless sensor networks (WSNs) for structural health monitoring (SHM), environmental factors can directly affect the performance of sensors. The performance of working sensors relies on factors such as temperature, humidity, and electromagnetic interference. For instance, extreme temperatures may alter physical properties of sensor materials, thus influencing their response rate and accuracy. Similarly, excess moisture can result in condensation that can compromise sensor integrity and measurement accuracy. Such environmental conditions must be properly controlled and accounted for during the design of WSNs, as they technically establish the quality of data collected, enabling accurate assessment of a structure's condition. Consequently, adequate precautions that account for these environmental conditions are important to enhance the robustness of sensor networks utilized in SHM applications. Besides temperature and humidity, environmental noise, especially in urban environments, is another major threat to sensor performance. Vibrations and acoustic noise from localized construction, traffic, or natural occurrences can cause interference in sensor signals, resulting in incorrect readings or signal loss. Such an effect is especially worrisome in dynamic environments where around-the-clock monitoring is important to allow for immediate damage detection and avoidance. To address these issues, advanced algorithms that include noise filtering and data preprocessing should be used. Through effective decoupling of genuine structural responses from extraneous signals, such algorithms can enhance the reliability of data-driven approaches to SHM. Thus, understanding and mitigation of environmental noise impacts can significantly increase the efficacy of wireless sensor networks in capturing true and relevant data for structural inspection and maintenance. Also, power consumption is the key environmental factor that directly affects the performance and longevity of sensors in WSNs. Owing to the fact that the majority of sensor networks are used in unreachable or remote locations, efficiency of energy is the focus to enable prolonged monitoring activities. Techniques such as dynamic voltage scaling (DVS) can reduce power consumption through enhanced CPU performance. However, CPU frequency changes may influence time synchronizations and overall data integrity, especially during longer monitoring periods. Performance and energy efficiency must be balanced carefully in a manner that does not cause data loss or sensor failure. Using advanced energy management strategies, aided by effective data acquisition algorithms and robust processing algorithms, can significantly enhance the operational life of the sensors, hence resulting in more reliable SHM systems in various environments (Abdelzaher et al., 2011; Cherta et al., 2020).

4.5. CASE STUDIES OF WIRELESS SENSOR NETWORKS (WSNS) IN STRUCTURAL HEALTH MONITORING (SHM)

Application of WSNs in SHM has been well established by series of case studies that prove their potential to enhance construction quality and safety. One of the best examples is the employment of prefabricated construction (PC) projects, where Radio Frequency Identification (RFID) and Long Range (LoRa) communications technologies have been adopted. Sensors there monitor real-time data on structure performance while installing the PC components. The application of a cloud-based Building Information Model (BIM) allows project teams to access such critical information from anywhere, with the objective of achieving efficiency and making informed decisions. The system test on real construction sites indicates the potential of WSNs in providing precise data and monitoring capabilities, thereby reflecting the high value that WSNs add to the success of PC projects with respect to SHM (Liu et al., 2019). Another key aspect of WSNs in SHM is the utilization of data-driven algorithms to detect damage, significantly enhancing condition assessment of a structure. Case studies of data-driven approaches show that sensor data have a major contribution towards real-time monitoring and decision-making. These algorithms examine data from various sensors, allowing for comprehensive analysis and accurate detection, localization, and prognosis of damage. The literature proves that data-driven methods are favored due to their potential for working with actual-time data quickly compared to model-based methods that rely heavily on prior structural models. As a result, the immense computational needs of these algorithms necessitate the utilization of robust processing units to ensure timely and efficient outcomes in structural assessments (Cherta et al., 2020). In addition, the complementarity of sensor technology developments and the possibilities of WSNs are also being put to the test in certain case studies, further solidifying their application in SHM. For instance, integrating strain sensors in a WSN system allows for continuous stress and patterns of deformation measurement in buildings. These sensors pick up critical data that allow for early signs of impending failure or structural defects. Coupled with machine learning procedures, the evaluation becomes predictive to enable proactive maintenance strategies that could extend the lifespan of critical infrastructure. This development sets the stage for a future where WSNs not only perceive, but also enhance the robustness of structures through data-driven maintenance actions, eventually contributing to safer built environments and fewer costs for unplanned repairs (Cherta et al., 2020; Liu et al., 2019).

4.5.1. Successful Implementations in Urban Infrastructure

Wireless sensor network application in urban infrastructure promises enormous scope to develop structural health monitoring (SHM). By the distributed nature of such networks, cities can gather complete data on the status of their infrastructure in real-time. Successful applications have been characterized by good network designs, which ensure efficient communication between sensors placed on different structures. Another significant factor in how these networks operate optimally is monitoring the quality of network communication both prior to and following deployment. As discussed under assessment models for long-term network performance, it is essential to achieve high data throughput rates as well as sensor topologies optimization in a manner that minimizes communication failure to help increase the efficiency of SHM operations in urban areas (Baccour et al., 2016). Besides, the urban environment, which is typically exposed to variable external conditions and stresses, requires a dynamic management approach in their SHM systems. This need has fueled the development of innovative frameworks such as utility computing models (UCM). The models allow for enhanced urban planning through dynamically adapting monitoring systems with real-time feed from sensors disseminated in metropolitan areas. This strategy has been demonstrated through the SHM-UCM model, using a geo-informatics system to combine geometrical and environmental data for structure health evaluation and prognosis of possible issues. Case studies, such as the Qatar University application, have demonstrated the efficacy of this approach in optimizing monitoring of current structures and supporting strategic planning for future urban infrastructure developments (Al-Hitmi et al., 2019).

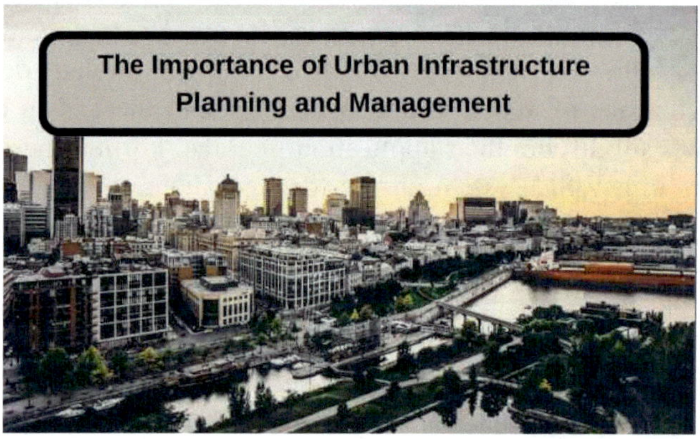

Figure 4.15. Successful implementations in urban infrastructure.

Source: https://www.constructionplacements.com/wp-content/uploads/2023/03/Building-Sustainable-and-Resilient-Cities-The-Importance-of-Urban-Infrastructure-Planning-and-Management.jpg.

Additionally, the successful deployment of such systems in urban infrastructure highlights the need for interdisciplinary collaboration to promote SHM technology development. The optimum deployment of WSNs must be optimized by interdisciplinarity among urban planners, engineers, and data scientists to facilitate cities making effective integration of sensor data into practical insights. As cities continue to expand and deal with problems of climate change and congestion, the potential to conduct real-time assessments of infrastructure condition becomes increasingly relevant. With good planning and innovative strategies like advanced sensor networks and models, cities can become more resilient against infrastructural collapse, thus safeguarding the health of citizens while fostering sustainable development for future generations. The deployment of effective SHM systems will not only streamline maintenance but also lead to a proactive approach in urban infrastructure management (Figure 4.15).

4.5.2. Lessons Learned from Failed Projects

Learning from failed projects in the realm of Wireless Sensor Networks (WSNs) for Structural Health Monitoring (SHM) offers significant insights into the complex dynamics of technology adoption. One valuable lesson is the instrumentation system design, especially based on the lessons learned in the course of NASAs Habitation Demonstration Unit (HDU) endeavor. The challenges of subsystem integration highlight the value of forethought and vision in planning projects. For instance, by revising previous designs that were successful in different scenarios while considering the unique demands of SHM, project managers are better able to anticipate pitfalls, hence reducing the likelihood of system failures. Further, these experiences stress the use of rapid development processes so that the teams can progressively improve their work based on field trials, thus improving the dependability and robustness of the WSN systems in future deployments (Hafermalz et al., 2011). Furthermore, all failed projects typically stress the importance of proper collaboration and communication of multi-disciplinary teams. Referring to the HDU, lessons learned indicate that effective coordination between engineers, scientists, and technologists is key to successful project execution. When communication between teams fails, there are risks of misunderstandings of project goals and requirements that lead to less efficient solutions, as seen with various SHM projects using WSNs. Open channels of communication assist in the sharing of knowledge and experience so that teams can readily address emerging problems as they occur. Such an observation is particularly critical in technology innovation, where learning from past projects can inform later work. Disjointed team failure is normally followed by disparate system components that do not operate harmoniously with one another, again highlighting the need for coordinated collaboration under project frameworks (Dankanich et al., 2025). In addition, performance analysis of failed projects underscores the significance of strict risk assessment processes before planning. By rigorous evaluation and understanding of potential failure modes, WSN projects for SHM can pre-empt weaknesses. For instance, the lessons from NASAs research expeditions, where research advances have shaped the direction of their strategic initiatives, come to mind. As it is not always possible to preclude threats, project teams are forced to plan for backups using lessons from past failure experiences. The HDUs focus on a standards-compliant wireless sensor node is an explicit instance of how experience gained in earlier projects can lead to innovative solutions in high-risk environments. Building on these findings can significantly enhance the deployment of resilient WSNs, leading ultimately to enhanced resilience in SHM approaches (Dankanich et al., 2025).

4.5.3. Comparative Analysis of Different Wireless Sensor Network (WSN) Technologies

Comparing different WSN technologies reveals numerous alternatives tailored to the needs of applications such as SHM. Each technology has certain strengths and weaknesses, which affect its applicability to specific environments and applications. For instance, Wireless Body Area Networks (WBANs) have been designed for medical applications and allow for ongoing monitoring of physiological parameters non-invasively. The networks' architecture is usually made up of various inbuilt elements that allow for real-time data transfer to health monitoring systems, thereby improving patient care. Besides, as shown in the National Integrated Network for Patient Monitoring in Benin design, careful consideration of system qualities and functional requirements is crucial in designing practical monitoring solutions effectively addressing challenges in healthcare such as minimizing hospitalization time and enhancing coordination of nursing staff (Ahouandjinou et al., 2020). Another technology that has greatly contributed in the area of WSNs is the Internet of Things (IoT), which collates a range of sensors from various areas to create dynamic networks with real-time capabilities of fusing data. The trend facilitates effective deployment in areas like city monitoring and environmental monitoring, where large volumes of data oftentimes referred to as big data are generated and transmitted. Such data management calls for high-level models to be able to analyze and extract meaningful insights, particularly as the data may be structured or unstructured. In response to challenges data management in IoT paradigms might pose, the use of advanced database models, such as graph databases together with traditional relational databases, becomes imperative for maintaining performance and scalability in WSN applications. These considerations highlight the importance of selecting appropriate technologies that are capable of handling large data rates without trading off responsiveness and accuracy (Küçükkeçeci et al., 2017). Lastly, the comparison of these WSN technologies highlights the role of communication protocols and energy efficiency in determining the working effectiveness of sensor networks. Protocols such as Zigbee, LoRa, and MQTT all have varying power consumption levels, range, and data transmit capacity, significantly impacting their applications in real-world scenarios. Zigbee, for instance, is opted for where low power and short-range applications with high networking requirements are required. LoRa, however, is applied where applications require connectivity to be maintained across expansive regions. Therefore, an appropriate selection of WSN technology is not only a matter of physical qualities but also of the manner in which these technologies interact with particular demands of SHM systems. These factors aid the development of improved networks capable of adequately providing for diverse applications while their operational lifespan and efficiency for monitoring infrastructure condition are maintained.

4.6. FUTURE TRENDS IN WIRELESS SENSOR NETWORKS (WSNS) FOR STRUCTURAL HEALTH MONITORING (SHM)

WSN implementation for SHM is poised to significantly enhance infrastructure efficiency and reliability. With environmental concerns gaining greater significance, power-efficient sensor technology is becoming a new priority, particularly in energy harvesting methods. Developments such as piezoelectric devices are an example, as they can capture environmental forces to create power, minimizing the need for power sources outside the system. Not only do these sensors gather data on infrastructure health, but also they contribute to sustainability efforts by conserving energy use in buildings and infrastructures. All of these developments point to the dual benefit of WSNs; they enhance monitoring efficiency while promoting green practice, which ultimately lead to more intelligent infrastructure solutions well-aligned with contemporary sustainability goals (Bontempi et al., 2017). In addition, the integration of leading-edge communication technologies into WSNs is revolutionizing SHM strategies. Directions for the future are to focus on using Internet of Things (IoT) platforms, making data exchange and real-time analysis more convenient. With the installation of Internet-based sensors, stakeholders are able to obtain crucial information in real time, facilitating timely decisions as well as intervention actions. Not only does such interconnectedness enhance the efficiency of monitoring, but also predictive maintenance capacity, which facilitates anticipatory repair. Evolution of low-power, energy-efficient communication protocols will further make such networks robust, ensuring reliability and durability of long-term data transmission. These technological advancements pave the way for boundless reformation of SHM by WSNs and at times even more safety and life of infrastructure in various industries (Ahouandjinou et al., 2020). Lastly, the employment of machine learning and artificial intelligence (AI) in the field of WSNs for SHM is a revolution in data analysis and interpretation. The capacity of AI algorithms to accommodate large amounts of data supports more sophisticated predictive analytics, which can identify slight structural changes that may indicate concealed issues. Since WSNs keep generating huge volumes of data through sensors of different types, the application of such technologies can assist in pattern identification and anomaly detection within structural response better. Apart from enhancing the accuracy of health monitoring, this also reduces the manpower required for routine data analysis. As the technologies are further developed and integrated into WSN systems, they hold the potential to transform SHM practices by being responsive, efficient, and informative and, at the end of the day, safeguarding public safety and investment in infrastructure (Ahouandjinou et al., 2020).

4.6.1. Advances in Sensor Technology and Materials

Technological advancements in materials and sensors have also been highly influential in the use of structural health monitoring (SHM), which is highly reliant on accurate and efficient data gathering for real-time processing. Advanced sensor systems incorporated in application make it simpler to record critical details regarding structural integrity, hence facilitating proactive maintenance methods. Higher-end sensors, often equipped with data-driven algorithmic features, facilitate advanced damage localization and detection by applying stringent data analysis methodologies. As established from the latest literature, data-driven algorithms help in transitioning towards better than decision-making procedures of SHM applications with the utilization of real-time data produced on various structural components. The enhancement reveals that data-driven approaches are the preferred method as major contributor to the timely detection of structural anomalies, which further provides a vital backbone for safety and reliability improvement in infrastructure management (Cherta et al., 2020). Advancements in wireless sensor networks (WSNs) over recent years have also introduced new technology and new materials that further the role of SHM systems. One such evolving role is incorporating energy harvesting techniques, so that sensors become everlasting without the frequent need for replacement of batteries. By harnessing vibration energy sources, such systems exhibit an extremely high degree of operational autonomy, as described in recent research on low-power wireless communications. Continuing monitoring with such energy-efficient structures augmented by good data transmission protocols, the reliability of data collected is thus secured. Even the utilization of smart materials used in sensor construction promises to make responsiveness better through more sophisticated networks of sensors that would be able to adapt to the dynamic environment and structural response (Giuliano et al., 2012). These sensor technology advancements not only enhance data gathering and durability in SHM but also pave the way for even better advances in the future. Enhanced sensor capabilities pave the way for more advanced algorithms that can provide insight into damage prognosis, advancing the applications of SHM systems from simple monitoring to predictive maintenance methods. With growing sophistication of SHM structures, attempts to develop intelligent networks capable of self-diagnosis and offering necessary maintenance data are increasingly relevant. Continued development of sensor materials along with altered analysis techniques places the field of wireless sensor networks at the forefront of structural engineering. This wide-ranging understanding of sensor developments ultimately gives rise to safer, more advanced infrastructure management systems that are crucial to the rapidly evolving technology landscape of the current era (Cherta et al., 2020) (Figure 4.16).

Figure 4.16. Advances in sensor technology and materials.

Source: https://www.mdpi.com/sensors/sensors-20-06783/article_deploy/html/images/sensors-20-06783-g001.png.

4.6.2. Integration with Internet of Things (IoT) and Smart City Initiatives

The huge urbanization in which the world is presently grappling across the globe necessitates innovative solutions to the issues of modern-day cities. Smart city initiatives have also emerged as a response to mitigate the unprecedented pressures placed on urban infrastructures by population growth. These initiatives are primarily based on the integration of wireless sensor networks (WSNs) and the IoT to get the most out of the effectiveness and efficiency of the utilization of resources. By utilizing a myriad of sensors that are connected to the IoT, cities become more efficient in monitoring environmental statuses, traffic movement, and building conditions. Here, WSNs are the basis for information gathering, providing timely data to inform decision-making and enable sustainable city development. As explained, the basis of such initiatives is the seamless integration of technology in a manner that enables spaces to endure and recover from challenges brought about by urbanization (Akyildiz et al., 2013). The applications of WSNs in intelligent city environments extend beyond data gathering; they encompass the whole analytical framework of instruments and organizational collective practices required for effective resource management. Based on the analytics derived from data gathered through IoT devices, urban policymakers and city planners can devise strategies to minimize traffic jams, energy consumption, and public health hazards. According to smart city practice study, organizations must create environments that foster innovation and leverage analysis insights that originate from WSN data. The collaboration of these initiatives, in addition to enhancing the decision-making process, promotes an enlightened citizenry that can effectively contribute to urban development projects (Stahl et al., 2017). Hence, WSNs are the key to building smart urban ecosystems that react smartly to the needs of their populations. Moreover, the integration of WSNs with IoT is considered pivotal towards enabling more sustainable and resilient urban systems. As cities increasingly use smart technology, the use of sensors for infrastructure status monitoring, such as bridges or roadways, is imperative to ensuring public safety and longevity. Through live monitoring, WSNs can detect structural defects or environmental hazards and provide for prompt action that prevents catastrophic failure. As an extension of smart city initiatives, the immediate data flow from such networks promotes proactive action rather than reactive response, thus enhancing the resilience of the entire urban infrastructure. Lastly, the integration of wireless sensor networks into IoT frameworks encourages an enhanced recasting of city planning, toward wiser decision-making and evolving sustainable strategies required for future urban resilience (Akyildiz et al., 2013).

4.6.3. Potential for Machine Learning (ML) and Artificial Intelligence (AI)in Data Analysis

The advent of AI and ML has considerably transformed data analysis, particularly for wireless sensor networks (WSNs) in structural health monitoring (SHM). AI and ML can analyze enormous amounts of data generated by sensors that are mounted on infrastructure such as buildings and bridges to identify anomalies and predict potential failures. With the use of advanced algorithms, ML can learn from historical data and recognize patterns that might otherwise go unnoticed.

As such, these prediction models can potentially enhance maintenance schedules and resource allocation immensely, thus potentially reducing the expense of structural repair and ensuring public infrastructure safety. The integration of AI into WSN for SHM, apart from automating data analysis, increases decision-making speed and precision, therefore allowing interventions in a timely fashion when there are faults. Moreover, ML and AI facilitate the correct fusion of heterogeneous data types obtained through several sensors in a WSN. Such capability is required for extensive monitoring and analysis as different sensor modalities, for instance, temperature, strain, and vibration, can provide data that must be interpreted in multifaceted manners. With mechanisms such as data fusion and feature extraction, AI systems can combine information so that there can be a comprehensive view of the structural performance and integrity over time. This holistic approach is paramount in preventive maintenance strategies, where stakeholders are able to respond to conclusions from multi-dimensional data sets, thus enhancing the resilience of critical infrastructure. Moreover, with continuous AI algorithm improvements, less opportunity for error exists, enabling high quality in data-driven decision-making in SHM structures. In addition to improving predictive analysis and data integration, the application of AI in WSNs also extends to addressing problems related to data ownership and privacy. Because wireless sensors produce an ongoing flow of sensitive health data—especially in healthcare and pervasive healthcare settings—the potential for data abuse rises. AI can mitigate this by generating rigorous data management policies that safeguard patient information but make it possible to analyze and share information effectively. This dual focus on security and analytics points to the importance of ethical AI frameworks function in alignment with regulatory standards. As discussed, ongoing innovation in technologies through sensors and machine learning extends realms of opportunities to enhance the effectiveness of SHM with effective data analysis practices, transcending practical challenges while improving health outcomes and infrastructure safety (Baumgartner et al., 2019; Hosseinpour et al., 2019).

SUMMARY

It is important to know how they can be used for transformation in improving the security and life of infrastructural facilities. As organizations increasingly rely on clean sources of energy, such as wind turbines, structural resilience and strength cannot be more important. WSNs offer a unified platform through which information is collected and processed in real time and analyzed so that stakeholders can formulate timely information about infrastructural frameworks. The incorporation of advanced sensing technologies supports the integration of bespoke systems to accommodate specific problems posed by evolving load regimes, as in recent studies on onshore wind turbine foundation design (Currie et al., 2012). The new paradigm can help reduce the financial and human cost of catastrophic failure and promote a preventive maintenance culture. Besides this, the use of WSNs in SHM is not merely a technical breakthrough; it is a perfect transition towards infrastructure management. Engineers can achieve proactive monitoring of building health with the help of these networks, and they can identify anomalies that could indicate potential failure ahead of time. Early detection is significant as it enables informed decision-making in the maintenance process, resulting in optimal resource allocation. The enhanced capability for constant monitoring is groundbreaking, particularly in the case of systems that possess complex load patterns where traditional methods may fall short. For instance, the development of new condition-based monitoring systems particularly for onshore wind turbines captures the necessity of WSNs' contribution to respond to the dynamic demands of modern energy infrastructures (Currie et al., 2012). Such systems demonstrate the necessity of adaptive technology in aid of contemporary safety and operational objectives. On balance, the use of wireless sensor networks in structural health monitoring is not only a coming together of engineering and technology but also an ongoing devotion to safeguarding public safety and maximizing infrastructure efficiency. The ongoing refinement of sensor technology and data processing is creating opportunities for increasingly sophisticated monitoring systems that are adaptable over a wide range of environments. With the demand for sustainable energy sources on the rise, insights from these networks will be immensely valuable in reducing the danger of structural failure. By embracing these innovations, stakeholders can make decisions that establish rigorous monitoring processes in place to enhance infrastructures' integrity and the future of managing energy. Deploying these in an organized way can enable robust early warning systems to be designed and guarantee a more resilient infrastructure that can stand nature and time's tests, as underscored in current research efforts in the discipline.

Studies on WSNs in the context of SHM have also opened the curtain on some of the most critical advancements in the technology. Most relevant among these is the adoption of new technologies such as RFID and LoRa, which have revolutionized the monitoring of prefabricated construction sites. The system facilitates automatic acquisition, analysis, and reporting of real-time data associated with prefabricated units. Through the utilization of sensor networks and cloud-based BIM platforms, project stakeholders are able to access critical project data remotely, hence improving decision-making and operational performance. The results of the application of such technologies have been found to be more efficient and trustworthy when it comes to data collection, which has enabled one to properly localize and follow construction items throughout their installation phases, hence showing the promise of today's WSN applications in SHM (Liu et al., 2019). With technological advancement, the methodology in detecting damage in structures has progressed a great deal with the implementation of data-driven algorithms. The algorithms have become increasingly significant due to their ability to perform real-time analysis of sensor inputs, thus becoming a preferred choice over traditional modeling methods. With the ability to facilitate real-time monitoring, data-driven methods allow for enhanced capability in real-time decision-making, addressing potential structural vulnerabilities before they develop into significant issues. In addition to this, these algorithms provide a broad base of functionalities such as damage detection, localization, and prognosis, thus enhancing the total robustness of SHM systems. This focus on data-driven techniques emphasizes the shift towards appropriate utilization of computer resources to facilitate high-performance processing needed for large-scale analyses and to ensure increased safety and lifespan of structures (Cherta et al., 2020). Moreover, the integration of wireless sensor networks into SHM systems is not only technological progress but also an industrial practice paradigm shift regarding maintenance and construction. The ability to have integrated streams of data in real time facilitates a proactive approach in assessing structural integrity as compared to traditional reactive ones. The implications of such a move are drastic, heralding a future where predictive maintenance and immediate availability of data become the norm in construction management. Such a capability matters to aid sustainable construction methods and mitigate wastage, as seen through effective monitoring of prefabricated components in installation phases. Lastly, the ongoing evolution of WSNs in SHM underscores the importance of adapting such technologies for purposes of promoting innovation and enhancing the performance of built structures as a whole (Cherta et al., 2020; Liu et al., 2019).

REFERENCES

1. Abdelzaher, T. F., Agha, G. A., Mechitov, K. A., et al. (2011). Dynamic voltage scaling techniques for energy efficient synchronized sensor network design. https://core.ac.uk/download/4834374.pdf (accessed on 2nd October 2025).
2. Agis Cherta, D., Gomez Vargas, R. C., Pedraza, C., et al. (2020). Damage identification in structural health monitoring: A brief review from its implementation to the use of data-driven applications. https://core.ac.uk/download/288625955.pdf (accessed on 2nd October 2025).
3. Aguilar, R., Alves, M., Gandra, P., et al. (2011). Prototype WSN platform for performing dynamic monitoring of civil engineering structures. https://core.ac.uk/download/55615281.pdf (accessed on 2nd October 2025).
4. Ahouandjinou, M. H., Houessouvo, R. C., Jossou, T. R., et al. (2020). New intelligent network approach for monitoring physiological parameters: The case of Benin. https://core.ac.uk/download/305119894.pdf (accessed on 2nd October 2025).
5. Akyildiz, I. F., Alvarado, J., Ashton, K., Banerjee, A., Cabral Pinto, R., Chou, W., et al. (2013). Sensing as a service model for smart cities supported by Internet of Things. http://arxiv.org/abs/1307.8198 (accessed on 2nd October 2025).
6. Al-Hitmi, M. A. E., Crescini, D., Manouer, A. B., Tahir, A., Tariq, H., & Touati, F. (2019). Geographical area network–structural health monitoring utility computing model. https://core.ac.uk/download/322372049.pdf (accessed on 2nd October 2025).
7. Angeletti, F., Colesanti, U. M., Paoli, M., et al. (2015). Wireless sensor networks in structural health monitoring: A modular approach. https://core.ac.uk/download/54527370.pdf (accessed on 2nd October 2025).
8. Azorin-Lopez, J., Bradley, D., Hehenberger, P., et al. (2018). Mechatronics & the cloud. https://core.ac.uk/download/228178466.pdf (accessed on 2nd October 2025).
9. Baccour, B. S., Chipcon, H., Hill, J., Hollar, J., Song, J., et al. (2016). Probabilistic assessment of high-throughput wireless sensor networks. https://core.ac.uk/download/79711706.pdf (accessed on 2nd October 2025).
10. Baig, Z. A. (2014). Securing the Internet of Things infrastructure–Standards and techniques. https://core.ac.uk/download/41537076.pdf (accessed on 2nd October 2025).
11. Barontini, A., Boscato, G., Ceravolo, R., Conde, B., de Castro, P., Farrar,

C., et al. (2017). An overview on nature-inspired optimization algorithms for structural health monitoring of historical buildings. https://core.ac.uk/download/154276373.pdf (accessed on 2nd October 2025).

12. Basto, C., Chacón Flores, R. A., & Pelà, L. (2017). Open-source digital technologies for low-cost monitoring of historical constructions. https://core.ac.uk/download/81578722.pdf (accessed on 2nd October 2025).
13. Basu, B., Das, K., Havinga, P. J. M., et al. (2013). Wireless sensor network for helicopter rotor blade vibration monitoring: Requirements definition and technological aspects. https://core.ac.uk/download/pdf/18296029.pdf (accessed on 2nd October 2025).
14. Baumgartner, C., Deserno, T., Hsu, W., et al. (2019). Advancing artificial intelligence in sensors, signals, and imaging informatics. https://core.ac.uk/download/323076975.pdf (accessed on 2nd October 2025).
15. Bono, F., & Renaldi, G. (2012). Analysis of current and potential sensor network technologies and their incorporation as embedded structural system. https://core.ac.uk/download/38627565.pdf (accessed on 2nd October 2025).
16. Bontempi, F., Gkoumas, K., & Petrini, F. (2017). Piezoelectric vibration energy harvesting from airflow in HVAC (heating, ventilation and air conditioning) systems. https://core.ac.uk/download/141696016.pdf (accessed on 2nd October 2025).
17. Brownjohn, J. (2016). Structural health monitoring of civil infrastructure. https://core.ac.uk/download/43096639.pdf (accessed on 2nd October 2025).
18. Buss, A., Ceylan, H., Ceylan, H., et al. (2013). Highway infrastructure health monitoring using micro-electromechanical sensors and systems (MEMS). https://core.ac.uk/download/38934593.pdf (accessed on 2nd October 2025).
19. Castedo, L., Fernández-Caramés, T. M., & Fraga-Lamas, P. (2017). Towards the Internet of smart trains: A review on industrial IoT-connected railways. https://core.ac.uk/download/95054341.pdf (accessed on 2nd October 2025).
20. Currie, M., Quail, F., & Saafi, M. (2012). Development of a robust structural health monitoring system for wind turbine foundations. https://core.ac.uk/download/9558383.pdf (accessed on 2nd October 2025).
21. Dankanich, J. W., & Morris, H. C. (2025). *Marshall Space Flight Center Research and Technology Report 2018*. https://core.ac.uk/download/pdf/161997280.pdf (accessed on 2nd October 2025).

22. Fan, W., Hulsey, J. L., McLean, D., et al. (2011). Smart FRP composite sandwich bridge decks in cold regions. https://core.ac.uk/download/286607559.pdf (accessed on 2nd October 2025).

23. Giuliano, A., Marsic, V., Pozzi, M., et al. (2012). Performance testing of a low power consumption wireless sensor communication system integrated with an energy harvesting power source. https://core.ac.uk/download/19542667.pdf (accessed on 2nd October 2025).

24. Govinda, R., Hao, S., Kohler, M. D., et al. (2015). ShakeNet: A portable wireless sensor network for instrumenting large civil structures. https://core.ac.uk/download/33126295.pdf (accessed on 2nd October 2025).

25. Hafermalz, S., Kennedy, K. J., Rojdev, K., et al. (2011). The Habitat Demonstration Unit project: A modular instrumentation system for a deep space habitat. https://core.ac.uk/download/pdf/10562678.pdf (accessed on 2nd October 2025).

26. Hosseinpour, F., Ilic, A. S., Jakobik, A., Jarynowski, A., Krzyszton, M., Marks, M., Moldovan, D., et al. (2019). Medical data processing and analysis for remote health and activities monitoring. https://core.ac.uk/download/224993779.pdf (accessed on 2nd October 2025).

27. Isnin, N., & Junaidi, N. H. (2020). *Katakan Tidal Pada Rasuah*. https://core.ac.uk/download/12007855.pdf (accessed on 2nd October 2025).

28. Küçükkeçeci, C., & Yazıcı, A. (2017). Big data model simulation on a graph database for surveillance in wireless multimedia sensor networks. http://arxiv.org/abs/1708.03878 (accessed on 2nd October 2025).

29. Liu, Z., Mbachu, J., & Zhao, L. (2019). Development of intelligent prefabs using IoT technology to improve the performance of prefabricated construction projects. https://core.ac.uk/download/231841646.pdf (accessed on 2nd October 2025).

30. Marano, G. C., Maruccio, C., Quaranta, G., et al. (2017). Energy harvesting from earthquake for vibration-powered wireless sensors. https://core.ac.uk/download/98349523.pdf (accessed on 2nd October 2025).

31. Stahl, E., & Dupuis, N. (2017). *Trends in Smart City Development*. https://core.ac.uk/download/83111937.pdf (accessed on 2nd October 2025).

32. Udoh, E., & Udoh, E. (2019). An energy aware and secure MAC protocol for tackling denial of sleep attacks in wireless sensor networks. https://core.ac.uk/download/305119264.pdf (accessed on 2nd October 2025).

CHAPTER 5

IOT PLATFORMS AND CLOUD INTEGRATION

Contents

INTRODUCTION

The development of the Internet of Things (IoT) has initiated a revolutionary shift in how devices get connected and communicate. As more devices get loaded with sensing and actuating functionalities, there has been a clamor for robust IoT platforms that can make them integrate into smooth systems. Such platforms serve as fundamental middleware connecting the most crucial connectivity challenges between heterogeneous devices, including sensors, actuators, and other smart technologies. In order to achieve mass adoption and satisfaction of various stakeholders like device manufacturers and application developers, IoT platforms need to be thoroughly evaluated on their usability and readiness. One recent study discusses the capability of various platforms in the context of heterogeneous technology support, ability to process data, and completeness of the overall IoT system (Mazhelis et al., 2016). This discussion emphasizes the need for constant optimization and improvement to keep pace with evolving needs of the IoT environment. The mutual synergy between cloud technology and IoT platforms provides unparalleled potential to enhance different industry operations. Although devices in IoT generate tremendous volumes of data, the cloud serves as a master point of storage, processing, and analytics. However, heterogeneity of the IoT devices provides a significant barrier to service developers who need to navigate an environment of varied connectivity protocols and communication patterns. Developers have challenges when trying to create applications that peacefully coexist across different systems. In response to this intricacy, current cloud-based platforms aim to provide simplified integration, allowing seamless interaction among devices regardless of the underlying technology (Demeester et al., 2015). Such platforms not only facilitate ease of development of multi-domain services but also enhance the overall functionality and range of IoT systems, eventually achieving the potential for greater efficiency. Security and privacy issues, as ever, remain high on the agenda when the issue of IoT platforms and cloud integration arises. With IoT systems becoming ever more ubiquitous, interests in ownership, security, and compliance of data increase exponentially. Effective platforms ought to include robust security features to safeguard sensitive information that moves across devices and the cloud while not compromising end-users to maintain any level of control over their data. The current situation indicates vast gaps in the application of security and privacy protocols in most prevailing IoT platforms. This necessitates stakeholders to convene in establishing standardized procedures that not only fortify security models but also empower user trust (Mazhelis et al., 2016). As IoT platforms evolve, building a secure exchange of data environment will be fundamental in realizing the complete potential of cloud integration in the context of IoT.

Definition of IoT (Internet of Things)

IoT is a revolutionary architecture which connects a large number of physical systems and devices with each other using the internet, enabling them to exchange, collect, and process data. With the connectivity, there is effective communication among a large amount of intelligent devices such as sensors, appliances, and even transport systems, changing their traditional modes of operation. By integrating novel technologies, IoT offers the foundation for creative applications that enhance efficiency and user experience. This can be noticed very much in Smart City initiatives, where IoT functionality is utilized to create optimum city life with enhanced service delivery as well as quality of life for residents. The common application of the technologies highlights the importance of IoT to today's development, particularly in the context of urbanization and increasing demands for sustainability in the infrastructure of society (Chaves et al., 2017). The heart of IoT's definition lies in the fact that it is not only about connectivity but the deployment of embedded systems that interact with their environment. These systems integrate sensors and actuators that gather real-time data, make judgments, and execute functions autonomously.

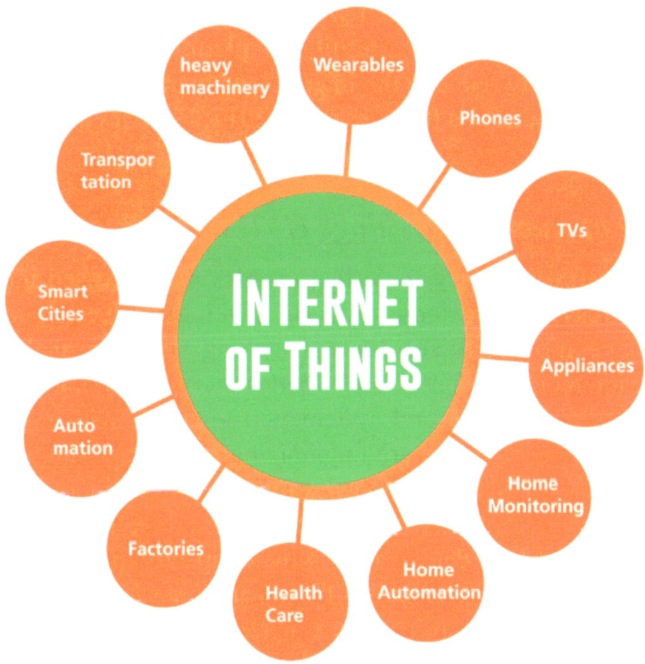

Figure 5.1. The Internet of Things (IoT).

Source: https://images.theengineeringprojects.com/image/main/2021/03/What-is-IoT-14.png.

As urban areas around the world grapple with climate change, pollution, and overpopulation, IoT's significance is even greater. By embracing smart technology, cities can effectively monitor and manage resources, conserve energy, and engage citizens more meaningfully. The functioning of these devices and the data generated are key to initiatives toward harvesting the advantages of urban life, consistent with the vision of Smart City deployments, where IoT is a facilitator of enhanced government and more efficient sustainability (Fischer et al., 2018). The IoTs definition notion extends beyond devices being connected; it is a vision for future intelligent systems contributing to informed decision-making. Within this consolidating context, IoT platforms facilitate the organization and management of devices so that data can be gathered and processed securely. Cloud integration is most important to allow this infrastructure to scale and have the data in one place for analysis. It is the unification of IoT and cloud technology that allows organizations to access infinite amounts of data being generated by connected devices. Apart from this, it supports machine learning algorithms and analytics integration with insights into fueling efficiency and innovation across industries. As a result, IoT platforms, when integrated with cloud systems, not only redefine the interactions but also optimize operational frameworks across industries (Chaves et al., 2017; Fischer et al., 2018) (Figure 5.1).

Importance of Internet of Things (IoT) in Modern Technology

Today, IoT is a revolutionary force that enables the increased integration of technology across various industries. By making devices connected, IoT is changing operational efficiencies and making real-time analyses of data possible. The interconnectedness enables better decision-making through instant overviews of user activities and system performance. With the unimpeded flow of information, industries are maximizing operations, minimizing costs, and enhancing customer satisfaction. Furthermore, with increasing instances of smart devices being all-pervasive in business life, the reliance on IoT platforms is more evident, demonstrating their capacity to promote innovation and competitiveness within business. This change not only provides value to the business world but also affects the experiences of consumers by enabling customization based on personal needs and preferences. The application of IoT in existing infrastructures is best exemplified in the creation of smart grids, where traditional power infrastructures are re-engineered to facilitate bi-directional energy transfer. The employment of IoT devices in smart grids is fundamental to monitoring and controlling energy consumption in real time, addressing the defects of traditional power grids, which were not interactive or prone to instant feedback. By facilitating better energy distribution reliability and efficiency, IoT ensures a more sustainable energy system to meet changing demands and environmental considerations. As stated, IoT technology supports solutions that address the pressing problem of energy management and thus constitutes an essential component of modern technological infrastructure. Thus, effective usage of IoT ensures optimized energy use among consumers and producers, paving the way for a green future in terms of energy utilization (Copeland et al., 2019). Moreover, expansion in IoT technology is not restricted to energy management alone but is implemented in other sectors, as emphasized by its role in the construction of smart cities. These urban spaces utilize IoT platforms to improve infrastructure performance, improve public services, and maximize transport systems. Actuators and sensors in urban infrastructure monitor traffic patterns, air quality, and waste management, with data-driven urban planning and governance as the outcome. The ability to obtain and analyze massive amounts of data in real time enables a proactive approach in dealing with urban challenges. Moreover, cloud integration of such IoT-capable systems also makes them more scalable and accessible, hence making it easy for municipalities to deploy smart solutions. The convergence of IoT and cloud integration facilitates the development of strong urban ecosystems that ultimately depict the very essence of the importance of IoT in today's tech innovation as an enabler of smarter life (Ahmed et al., 2019).

Overview of Cloud Integration in Internet of Things (IoT)

The integration of cloud computing into the IoT platform fundamentally transforms data management, analysis, and real-time application. Through cloud technology, IoT devices are able to export vast amounts of data to the server where scalable storage and processing capacity required to handle the complexities of multi-perspective applications are possible. Cloud integration allows IoT platforms to facilitate the unimpeded sharing of information between devices in different environments, like creating the required infrastructure to fuel smart city projects. These platforms not only optimize operation efficiencies but also promote city services and urban life quality, as exemplified in current studies on smart city software platforms based on enabling technologies such as Cloud Computing and IoT (Chaves et al., 2017). In addition, the problems caused by modern urbanization—climate change, transport issues, and pollution—demand new solutions that cloud-based IoT systems are equipped to. Smart cities are emerging as rich case studies for the demonstration of the potential of converging technologies, predominantly through networked devices that have the capability to compile and process data. An example of this is the use of sensors on urban infrastructure to enable the monitoring and response to urban problems in real time. The most significant role that IoT technologies have played as enablers towards this vision can be viewed as they are the infrastructure for building responsive and adaptive cities, emphasizing the contribution of cloud platforms in ensuring these outputs (Fischer et al., 2018). In conclusion, integration of clouds in IoT not only aids in data management but also aids in effective communication channels between the various stakeholders involved in smart city systems. This networked approach encourages public-private collaborative efforts between stakeholders and public authorities to ensure that urban matters are addressed through evidence-based decision-making. Moreover, with continuous advancements in cloud platforms, they continue to support the growth of IoT applications, which drive innovation from industries ranging from transportation to public safety. The future of smart cities is rosy, with ongoing innovation in cloud integration that helps cities revamp their infrastructures towards improved sustainability as well as increased citizen engagement (Chaves et al., 2017; Fischer et al., 2018) (Figure 5.2).

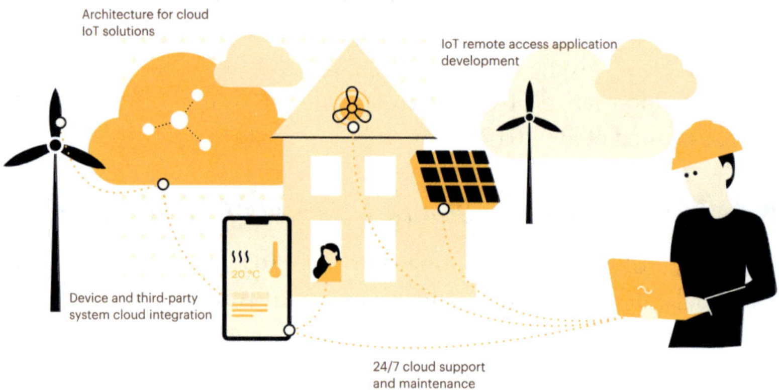

Figure 5.2. The Internet of Things (IoT)-cloud-integration-applications-illustrations.

Source: https://proekspert.com/wp-content/uploads/2022/09/iot-cloud-integration-applications-illustrations.svg.

5.1. OVERVIEW OF INTERNET OF THINGS (IOT) PLATFORMS

IoT platform development is the major enabler in facilitating cloud integration, which increases the efficiency of smart city initiatives. The platforms consolidate technologies to manage the huge volumes of data generated by thousands of networked devices. They leverage Cyber-Physical System tools, Big Data analytics, and Cloud Computing frameworks, which collectively offer robust data processing and real-time analysis. As cities grow in terms of population, the need for new ways to overcome issues such as climate change and mobility becomes increasingly urgent. IoT platforms, thus, are the prime component of delivering solutions to the issues in cities by facilitating mass data collection and engagement between the different stakeholders, including public authorities and citizens, to bring the vision of smart cities into reality. In addition, IoT platform deployment squarely contributes to the establishment of smart models of governance for cities. With the use of advanced ICT, municipal leaders can boost infrastructure management and citizen engagement with a concentration on effective interaction among a number of municipal departments. Successful deployment in this case relies on the achievement of significant technological challenges exemplified by the integration of sensors and actuators within city infrastructure.

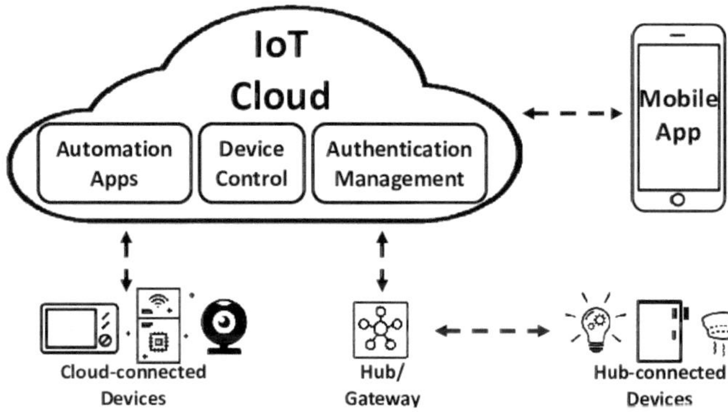

Figure 5.3. The evolution of Internet of Things (IoT) platforms plays a pivotal role in facilitating cloud integration.

Source: https://www.researchgate.net/publication/338292181/figure/fig1/AS:842376625012736@1577849646880/Overview-of-IoT-Platform-Architecture.png.

According to existing research, there is a requirement for a formalized methodology encompassing full awareness of IoT system functional and non-functional requirements. This model not only assists in refining the architecture of IoT platforms but also assists city planners in making informed investment decisions and technology deployment, thereby enhancing urban resilience and quality of life for residents, as in Chaves et al. (2017). In addition, advancements like the European Union's MONICA illustrate the ways IoT platforms can transform urban landscapes with mass deployments. The project is intended to transform urban landscapes by utilizing interconnecting systems to monitor, collect, and analyze data at public events, showing a practical application of IoT technology. Urban municipal governments, such as Hamburg, are also a part of this revolution through conducting field trials of sensor applications that are new and for creating a vertically integrated system. These are not only indispensable steps in fixing contemporary urban problems but also contribute significantly towards creating synergies among various entities to leverage the overall performance of smart city systems. By demonstrating the possibility of versatility and adaptability of IoT platforms, these types of projects leave room for future innovations in urban management, thereby contributing to the overall goals of smart city development as outlined in Fischer et al. (2018) (Figure 5.3).

5.1.1. Definition and Purpose of Internet of Things (IoT) Platforms

Concept of IoT platforms is a part and parcel of the broader scenario of integration of cloud into modern technology landscapes. An IoT platform is essentially a complete bundle of software applications designed to facilitate connectivity and management of different physical objects embedded with sensors and communicative capabilities. IoT platforms are an intermediary agency between connected devices and provide important functions such as data acquisition, device management, and analytics. The ability of IoT platforms makes it possible for businesses to exploit real-time information from heterogeneous sources to improve operations, decision-making, and ultimately create better systems. Besides, the integration of IoT platforms with cloud solutions enables scalability, which helps organizations to adjust dynamically to fluctuating demand, thus driving innovation across various industries, ranging from manufacturing to smart cities, cementing their root position in current digital ecosystems. Continuing on into the position of IoT platforms, it is apparent that they are designed not only to offer connectivity for devices but also to take advantage of user interactions as well as system abilities. By orchestrating heterogeneous devices, IoT platforms enable users to interact with technology naturally and in context. This user-centric approach is vital as it projects system behavior onto user intention as well as requirements and makes interactions meaningful and pertinent. Contextualization capability allows IoT systems to adapt dynamically to the activities of users and provide an unobtrusive experience. Such adaptability is essential in the case of smart cities because the convergence of IoT platforms leads to enhanced public service and quality of life for citizens, indicating the potential of such technologies in transforming urban environments (Chaves et al., 2017). In addition, the potential of the IoT platform is also maximized by placing it in context with integrating with the cloud. The cloud provides the backend infrastructure required in data processing and storage of the big amounts of data collected from IoT devices. The combination ensures not only an improved efficiency in handling data but also facilitates richer analytics capabilities, for example, provision of predictive insights that support anticipatory decision-making. By making use of cloud resources, IoT platforms can tap into high-performance computational capabilities such as machine learning and big data analytics and ultimately improve their functionality and value proposition. The more devices that are added to the IoT ecosystem, the more evident it becomes that strong platforms that can both manage interoperability and scalability are critical. The role of IoT platforms is central here, serving as a go-between between consumer demand and technology, and addressing the unique challenges of a networked digital world (Davidsson et al., 2017) (Figure 5.4).

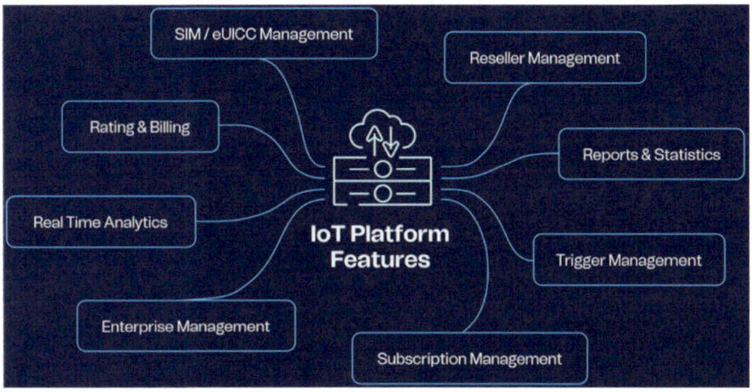

Figure 5.4. The concept of Internet of Things (IoT) platforms.

Source: https://iot.telenor.com/wp-content/uploads/2025/02/key-iot-platform-functions.png.

5.1.2. Key Features of Internet of Things (IoT) Platforms

The IoT platforms market is characterized by a number of key features necessary for their effective operation and integration into the broader digital infrastructure. One of the primary traits is the support for heterogeneous sensing and actuating technologies, enabling the interaction and communication between different devices. Because of the pervasiveness of connected products, platforms ought to cater to multiple protocols and standards in order to accommodate various components provided by various providers operating collaboratively. Moreover, the platforms should exhibit superior data processing and sharing ability in the efforts to enhance the synergies among devices. This facilitates real-time analysis and data-driven decision-making, which is essential in designing responsive smart environments as envisioned in the current literature on IoT platforms (Mazhelis et al., 2016). Therefore, the interoperability and adaptability of such platforms are the foundation of their functionality, making them a prerequisite for contemporary IoT applications and integration. In addition to device compatibility, another critical requirement of IoT platforms is their emphasis on data ownership, security, and privacy. As sensitive information is collected from diverse devices operating in personal homes, public spaces, and industrial settings, the significance of this requirement increases. IoT platforms must also ensure transparent data handling procedures where users retain control over their data. The data ownership implications also translate to the development of trust among stakeholders and end users; therefore, platforms are charged with having strong security measures in place to protect data integrity and confidentiality. Recent assessments indicate weaknesses in most current solutions in terms of handling these issues, pointing to a critical need for improvement in this space (Mazhelis et al., 2016). In the end, effective data governance improves the entire reliability of IoT platforms to create a secure environment for developers as well as users. Another aspect of complete IoT platforms is application support through dedicated marketplaces and APIs. As the IoT landscape continues evolving, the need for developers to build innovative applications comes into focus. Platforms must be furnished with full sets of tools and resources, including Software Development Kits (SDKs) and Application Programming Interfaces (APIs), to ease the development process. Such ease of access spurs active developer communities that support rapid deployment of applications taking advantage of IoT capability. In addition, the availability of IoT marketplaces eases the distribution and monetization of such applications, benefiting developers and organizations. As explained in current studies, these features are important in rendering IoT platforms competitive and capable of coping with the evolving demands of users and the market. Through facilitating the creation of applications, platforms not only enhance their performance but also trigger the growth of innovative solutions within the field of IoT (Figure 5.5).

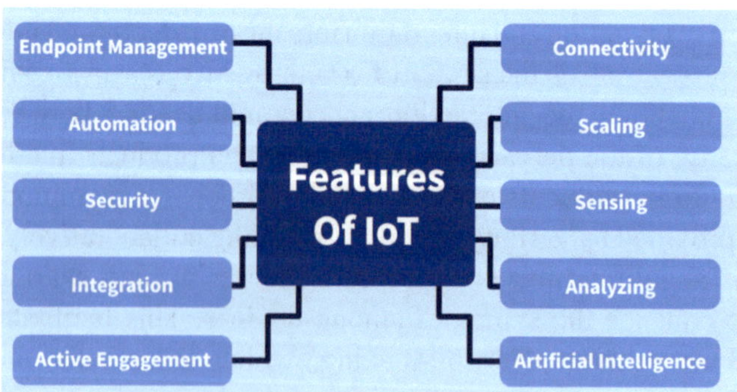

Figure 5.5. Key features of Internet of Things (IoT) platforms.

Source: https://www.interviewbit.com/blog/wp-content/uploads/2021/11/Image-2-12-1024x554.png.

5.1.3. Examples of Popular Internet of Things (IoT) Platforms

One of the leading IoT platforms is Amazon Web Services (AWS), which stands out due to its comprehensive collection of cloud-based computer resources fully tailored to IoT application use cases. AWS IoT Core enables devices to easily and securely connect to the cloud, and it is simple to handle large scales of connected devices. The platform includes device management, data processing, and analytics functionalities needed in developing scalable IoT solutions. Also, having machine learning services available on AWS allows developers to add advanced features such as predictive maintenance and anomaly detection to enhance operational effectiveness in numerous sectors. It is not only process-effective but also promotes innovation since one is able to prototype applications quickly that function on massive volumes of data generated by IoT devices. While the rollout of IoT devices keeps growing, AWS remains in the lead due to its well-established infrastructure and vast set of tools. Microsoft Azure is a core platform in the IoT ecosystem, which provides a dynamic solution for creating and sustaining IoT applications. Azure IoT Hub offers secure and reliable communication between the cloud and devices, while Azure IoT Central simplifies development through a low-code development interface and integrated analytics and monitoring capabilities. Moreover, Azures cloud capabilities optimize scalability, allowing small startups and big enterprises to deploy their IoT solutions without putting a strain on their infrastructure. The robust synergy among Azure and other Microsoft products, such as Power BI for analytics visualization and Azure Machine Learning for more advanced analytics, creates a solid ecosystem for businesses seeking to optimally utilize IoT data. The integration provides significant value to decision-making capability, allowing organizations to respond quickly to business needs and operational problems. Google Cloud IoT is another top platform that supports device integration to the cloud with an end-to-end set of services. The platform offers a strong foundation consisting of Google Cloud Pub/Sub for real-time messaging, Cloud Functions for event-driven development, and BigQuery for querying large datasets. These technologies support the ingestion, processing, and analysis of IoT data securely and at scale. Furthermore, Google Cloud's focus on incorporating artificial intelligence features allows the users to derive actionable insights from their data, paving the way for data-driven initiatives. With organizations desiring to embrace IoT solutions increasingly, Google Cloud's commitment to continuous evolution and interface with its machine learning and data analytics tools makes it the go-to choice among businesses and developers. As a group, these IoT platforms embody the shift of cloud integration into the IoT space to establish innovative applications for augmented user experiences and enhanced operational efficiency.

5.2. CLOUD COMPUTING FUNDAMENTALS

Cloud computing is a fundamental infrastructure for the Internet of Things (IoT) as it provides elastic services and resources that facilitate data management and application deployment. The model offers varying service models such as Infrastructure as a Service (IaaS), Platform as a Service (PaaS), and Software as a Service (SaaS), which accommodate various organizational needs and promote flexibility. With all the vast data coming from IoT devices, conventional computing methods might not suffice. Cloud computing saves the day here by offering effective data storage, processing, and analytics capability in far-off data centers. With the emergence of IoT systems, incorporating cloud services is crucial to take advantage of data-driven insights that can facilitate operational and strategic decision-making. While cloud computing provides immense value, the impending scale of IoT devices necessitates enhancement over the conventional cloud model. Fog computing emerges as an urgent enhancement to the conventional cloud model, especially for latency-critical processing applications nearer to the data source. Distributed processing on the network edge by Fog reduces the burden on the centralized cloud infrastructure and eases handling the heterogeneity and ubiquity of IoT nodes.

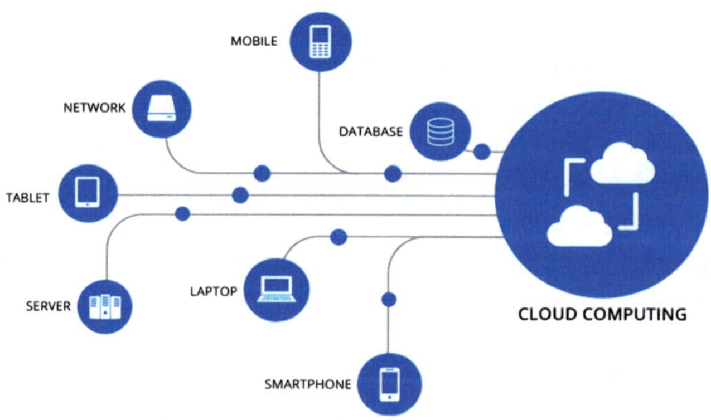

Figure 5.6. Cloud computing serves as a foundational infrastructure for the Internet of Things (IoT).

Source: https://media.licdn.com/dms/image/v2/D5612AQGIb671jsBShQ/article-cover_image-shrink_600_2000/article-cover_image-shrink_600_2000/0/1668088631564?e=2147483647&v=beta&t=1aGu-ayHgWzgBQmBUtqX7sy9y2wof9_2c2MBv9kb8-Y.

The Fog computing usage enables the processing and decision-making on data in real time, thus creating new opportunities for the uses of automation, such as health monitoring solutions capable of leveraging patient data obtained by wearables. As shown in recent studies, using Fog computing in conjunction with IoT applications can potentially introduce significant responsiveness and user experience improvements (Akrivopoulos et al., 2017). The convergence of IoT platforms and cloud computing underscores the shift from one-function programs to sophisticated, multi-function systems that create value for several stakeholders across industries. As companies desire to monetize the value of their IoT investments, cloud-based platforms that offer end-to-end data processing power is imperative. The platforms integrate different streams of data that emanate from sensor networks, enabling comprehensive analysis that drives innovation and operational excellence. Besides, Fog computing integration with cloud infrastructures allows organizations to effectively handle high amounts of data associated with IoT systems, developing a collaborative network that improves machine-to-machine (M2M) communication and heightened organizational objectives. Research indicates that the system architecture not only delivers sufficient interfaces to facilitate seamless interactions but significantly improves performance for practical cases (Akrivopoulos et al., 2017) (Figure 5.6).

5.2.1. Definition of Cloud Computing

In the modern era, cloud computing has evolved into a foundation paradigm that redefines the management of applications and data. In essence, cloud computing refers to the provision of computing services—such as software programs, processing capacity, and storage—over the internet. This model supports an on-demand character of resources such that users are able to access and manage services from nearly anywhere, minimizing the need for immense local infrastructure. This flexibility is particularly meaningful in the world of the Internet of Things (IoT), where sensors by the tens of thousands generate enormous amounts of data that must be processed, stored, and analyzed quickly. By virtue of cloud computing, IoT platforms can scale effortlessly, accommodating the growing demands of networked devices without unnecessary waste of resources and with enhanced functional performance. Moreover, the synergy between IoT platforms and cloud computing is not only synergistic in nature but also revolutionary in the case of smart city initiatives. With cities increasingly employing Information and Communication Technologies (ICT) to improve urban services and the quality of life among citizens, cloud computing comes in as a major player. Cloud computing makes it possible to develop software platforms that support varied applications loaded in smart spaces ranging from traffic control to green buildings. An analysis of ongoing projects reveals that this integration can be classified as a series of enabling technologies, alongside Cyber-Physical Systems and Big Data, significantly Cloud Computing (Chaves et al., 2017). This complex approach in leveraging cloud services ensures that smart city infrastructures can expand as new challenges emerge while leveraging the potential of networked IoT devices. The term Cloud of Things (CoT) also represents the intricate interdependency of IoT and cloud computing, which can solve the humongous amount of issues introduced by the explosive growth of interconnected devices. When billions of things produce humongous amounts of data, problems in data management, aggregation, and storage are increasingly prominent as well (Farahbakhsh et al., 2017). In this setup, cloud computing provides virtually limitless abilities to process the amount and type of data coming from IoT devices. Middleware deployment comes into focus in this setup by allowing free communication between heterogeneous devices with varied interfaces and architectures. By bridging the gap between cloud services and IoT devices, middleware plays a critical role in making IoT platforms more efficient and reliable, leading to smarter, more connected environments (Figure 5.7).

Figure 5.7. The cloud computing.

Source: https://www.forbes.com/sites/bernardmarr/2020/11/02/the-5-biggest-cloud-computing-trends-in-2021/

5.2.2. Types of Cloud Services (Infrastructure as a Service (IaaS), Platform as a Service (PaaS), and Software as a Service (SaaS))

Cloud computing has revolutionized the way enterprises deploy and host their infrastructure and applications, particularly in IoT environments. The three primary cloud service models—compute (IaaS), platform (PaaS), and software (SaaS)—are all built with distinct advantages to cater to different business needs. IaaS provides virtualized computer resources over the internet, allowing organizations to dynamically scale their infrastructure. This works particularly well for IoT implementations, where computational power might be needed more or less based on data generated by a number of devices that are in communication. With the use of IaaS, businesses can ensure that they have resources available to support their IoT applications without the costs of maintaining physical hardware. PaaS, on the other hand, offers an environment where developers can develop, test, and deploy applications without worrying about managing the underlying infrastructure. The service is key to companies that desire to innovate around their IoT ecosystems because it delivers pre-configured tools and services to accelerate development cycles. With PaaS, developers are at liberty to focus on creating and refining IoT applications that co-exist with cloud services.

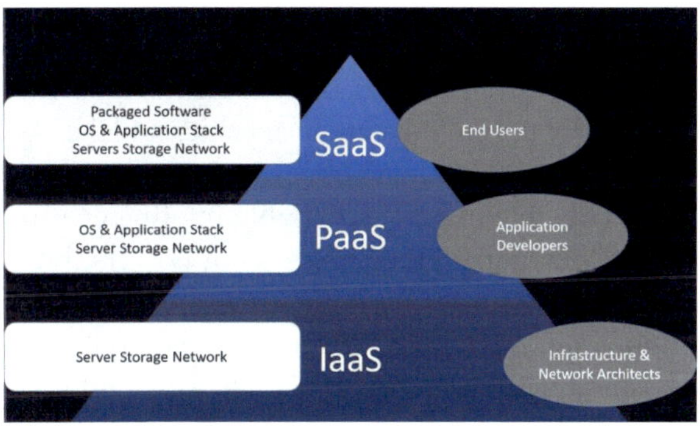

Figure 5.8. Types of cloud services (Infrastructure as a Service (IaaS), Platform as a Service (PaaS), and Software as a Service (SaaS)).

Source: https://www.researchgate.net/publication/349103776/figure/fig3/AS:988608836149248@1612714122299/Cloud-Service-Models-IaaS-PaaS-and-SaaS-Noor-et-al-2018-Infrastructure-as-service.jpg.

This is particularly relevant in the context of coupled simulations as outlined in recent research, which indicates the need for simulation management over heterogeneous infrastructures for creating more advanced, integrated IoT solutions (Clement et al., 2017). Therefore, PaaS facilitates fast experimentation and adaptation, essential in the rapidly evolving technology landscape. Lastly, SaaS delivers software applications over the web on a subscription basis, releasing organizations from the burden of installation and upkeep. With IoT platforms combined, SaaS applications can enable real-time data processing and analysis, allowing business organizations to derive useful insights out of their IoT data. This model also promotes the idea of serverless computing, in which applications can be deployed by developers without needing to manage servers, thereby streamlining processes to be more operationally efficient (Baldini et al., 2017). As businesses embrace IoT increasingly, the integration of SaaS offerings with cloud computing is necessary to drive innovation and meet consumer demands. Thus, the collaboration between IaaS, PaaS, and SaaS directly influences the development and functioning of IoT platforms, paving the way for more interconnected and intelligent systems (Figure 5.8).

5.2.3. Benefits of Cloud Computing for Internet of Things (IoT)

The fusion of IoT with cloud computing provides innumerable benefits that significantly enhance efficiency and flexibility in operations. One of the most significant benefits is the ability to handle massive amounts of data offered by innumerable connected devices. With the rise of IoT devices growing at an increasingly higher rate, so does the amount of generated data. Cloud computing offers virtually unlimited storage capacity and the ability to manage, consolidate, and analyze this data in a cost-effective manner. This is critical as organizations want to be able to gain actionable intelligence from information that is accessible. By leveraging cloud computing powers, businesses can use powerful analytics tools to process real-time data, thereby improving decision-making and building an intelligent operating system for smart buildings, homes, and cities (Farahbakhsh et al., 2017a). Also, cloud computing plays a vital role in addressing the heterogeneity problem of IoT devices, a challenge that poses relevant challenges in order to provide smooth integration and communication among devices. IoT systems consist of devices with disparate interfaces, operating systems, and architectures, so they need a solid platform for easy communication. Cloud computing delivers this since it allows for the deployment of middleware solutions that serve as middlemen between devices and applications. This middleware enhances compatibility and communication channels so that different IoT devices can operate together within a huge system. This interoperation is of the utmost importance in optimizing the performance of IoT platforms since it enables various devices to share data and collaborate in real time, thus achieving maximum value when added together (Farahbakhsh et al., 2017b). Moreover, the flexibility and scalability of cloud computing are central benefits that facilitate IoT deployments. The cloud allows companies to scale their IoT infrastructure quickly when and where the need arises, either through the addition of new devices, storage capacity, or processing. This scale-as-you-go approach keeps companies from investing in physical infrastructure that may lie dormant. Cloud platforms also facilitate easy addition of new services and analytics features, allowing for innovation and quick deployment of new features. Thus, companies are able to respond promptly to the evolving needs of the market and the demands of consumers through the establishment of a competitive edge in the fluctuating technology landscape with constant enhancement of IoT capabilities (Farahbakhsh et al., 2017a) (Figure 5.9).

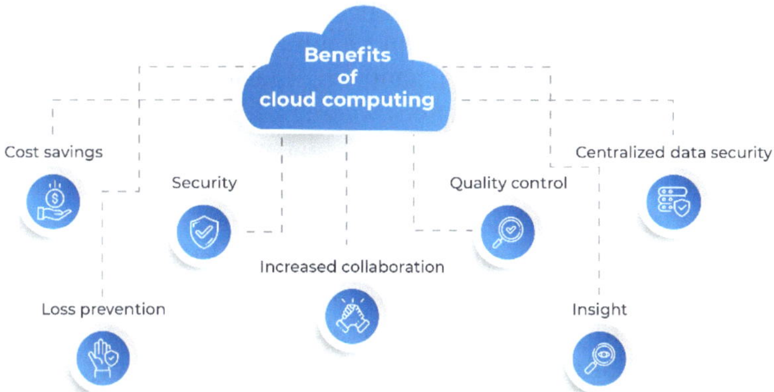

Figure 5.9. Benefits of cloud computing for Internet of Things (IoT).

Source: https://api.edoxi.com/assets/studyhub/Benefits_of_Cloud_Computing.webp.

5.3. INTEGRATION OF INTERNET OF THINGS (IOT) PLATFORMS WITH CLOUD SERVICES

The integration of IoT platforms with cloud computing is now a key solution to enhance the performance and efficiency of modern technology systems. The integration allows for data storage and processing of large amounts of data from IoT devices, which operate within real-time environments. Through the use of cloud computing capabilities, such platforms can attain scalability, allowing companies to handle changing levels of demand without having to invest in a large amount of on-premises infrastructure. Additionally, cloud offerings also provide for sophisticated analytics and machine learning functions, thus helping organizations distill useful insights from the streams of big data produced by IoT networks. The combination of IoT and cloud computing not only optimizes the efficiency of operations but also leads to creative usage in various sectors, further contributing to the overall creation of smart technologies. The addition of fog computing further enhances the overlap of IoT platforms with cloud services through addressing specific data latency and bandwidth constraints problems. Unlike typical cloud infrastructures, fog computing manages data closer to the source, a requirement for applications that require immediate response. For instance, with medical monitoring systems, being able to have a fog gateway allows critical patient details, such as ECG traces, to be securely shared with physicians without introducing extra latency. Not only does this structure improve response time in times of crisis, but it also delivers a continuous stream of information between devices and cloud services. The transition to fog computing is a significant IoT application transformation, particularly in situations where timely access to information is of prime importance, thereby reinforcing the inter-connectedness of IoT with cloud integration as a transformative force across sectors such as healthcare and smart grids, where speed and reliability are the top considerations. Smart Grids (SGs) are also a depiction of the synergetic coexistence between IoT and cloud computing, exemplifying their combined potential in revolutionizing energy usage and distribution. By integrating IoT devices such as smart meters and sensors into SG infrastructure, cloud services enable two-way energy flow, real-time analysis of power usage across different industries, and monitoring. Collectively, these technologies offer greater automation and surveillance, both improving the reliability of the grid and minimizing energy wastage. As indicated in current research, SGs are significantly augmented by cloud-based IoT platforms that support richer network functions that optimize the generation, transmission, and distribution of power. This not only leads to increased efficiency but also to the resolution of pressing issues such as climate change and resource utilization, illuminating a clear path for future innovations in the area of energy systems and beyond (Akrivopoulos et al., 2017; Copeland et al., 2019).

5.3.1. Mechanisms for Integration

The integration of Internet of Things (IoT) platforms with cloud computing is key to enhancing overall system function and efficiency. To facilitate this integration, several mechanisms have been designed that provide support for the complex interactions among heterogeneous devices, applications, and services. Middleware solutions, for instance, offer a basic layer that interfaces disparate IoT components and applications and facilitates interoperability and smooth communication. Middleware abstracts the fundamental technological complexities and thus supports the easier management of devices and data flows. Such a strategy is particularly important in enhancing the usability and readiness of IoT solutions since it caters to a vast population of stakeholders, ranging from application developers and device providers, and is hence a cornerstone in realizing the full potential of synergistic IoT ecosystems (Mazhelis et al., 2016).

Another key mechanism for integration includes the design and deployment of cloud-based services facilitating the IoT infrastructure. The integration of cloud computing and IoT, commonly referred to as the Cloud of Things (CoT), has significant advantages in the area of data handling, aggregation, and storage, particularly in a landscape where there are vast volumes of data generated by connected things. CoT avails near-infinite resources that ensure scalability of IoT applications by addressing challenges related to heterogeneity of connected things. Facilitating this convergence are middleware products that are specifically made to facilitate interaction across devices with varied interfaces, architectures, and operating systems.

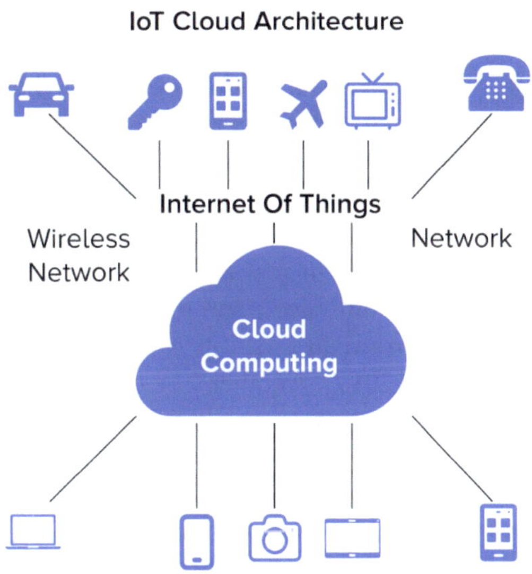

Figure 5.10. Mechanisms for integration.

Source: https://miro.medium.com/v2/resize:fit:1400/1*44gAqwB-JhxzTSMnvHh-W3g.jpeg.

Thus, the middleware technologies are a significant boost in surmounting integration challenges in the IoT realm (Farahbakhsh et al., 2017a). Lastly, the evaluation of integration mechanisms in IoT platforms is necessary to identify gaps and areas for improvement. A perceptive gap analysis, as identified in new studies, can reveal the specific needs and expectations of diverse actors in the IoT system. Ownership of data, security, and privacy implications should be considered to ensure integration mechanisms enhance not only interoperability but also users' interests. The symbiotic dynamics of technological progress, as attested by findings of expert surveys conducted within the Finnish IoT program, underline the necessity of continuous replenishment and evolution of IoT platform capability. By plugging these gaps with designed integration efforts, stakeholders can provide a truly holistic and functioning IoT ecosystem, with smooth interconnectedness being the norm rather than the exception (Mazhelis et al., 2016) (Figure 5.10).

5.3.2. Challenges in Integration

There are many challenges involved in integrating Internet of Things (IoT) platforms and cloud infrastructure that must be overcome to realize the full potential of smart environments. One of the underlying issues is the absence of standardized protocols that support interoperability among a broad range of devices and platforms. The current situation is one where there are a vast number of proprietary systems, each with separate interfaces and communication protocols, which cannot be easily integrated. This lack of uniformity not only hampers data sharing among devices but also limits scalability since new technology may not easily fit into existing frameworks. Overcoming these interoperability issues is crucial in realizing a harmonized and effective IoT ecosystem, where devices communicate and function effectively and ultimately improve service delivery in urban cities in smart city use as highlighted by the work of Chaves et al. (2017). Another key challenge is in IoT and cloud integration data ownership and management. As more data is generated by numerous devices, the matter of data ownership becomes essential, bordering on security and privacy. Different stakeholders such as device manufacturers, app developers, and users have diverse interests, and it's challenging to cope with data governance. Furthermore, the heterogeneity of IoT devices calls for robust data handling and sharing capabilities in a bid to form informative information out of the collected data. According to Mazhelis et al. (2016), current platforms do not possess adequate solutions to solve these multi-aspect data ownership problems, and that is hindering the design of secure and effective user and business environments. Lastly, the integration of IoT platforms and cloud services needs to have a comprehensive approach to providing adequate support to application developers. Several of the existing IoT middleware products are weighed against their ease of use and developer-friendliness, but they lack when it comes to providing end-to-end ecosystems and optimized marketplaces for apps. Closing these gaps is imperative because developers need the same tools that enable them to develop applications quickly while ensuring the strength and security of the solutions they produce. Gap analysis in Mazhelis et al. (2016) shows that without special development support, most IoT platforms will not be able to meet the rising demands of the ecosystem and will therefore arrest innovation and proper deployment of integrated IoT solutions in real-world applications.

5.3.3. Case Studies of Successful Integrations

The extensive use of IoT devices has necessitated innovative ways in which these technologies can be efficiently integrated with cloud services. Successful implementations of fog computing are highlighted by case studies that demonstrate the potential of fog computing as a solution for ensuring proper data management. Decentralizing cloud services and bringing computational resources close to end-users is the mechanism through which fog computing operates in addressing common latency and bandwidth problems. For instance, in smart city deployments, the application of fog computing entails real-time data processing and decreases the response time for traffic management and public safety applications significantly. The integration of fog and cloud paradigms not only guarantees enhanced system performance but also leads to opportunities towards better device interoperability, emphasizing the importance of a robust integration framework in the IoT environment (Carpio et al., 2019). The Cloud of Things (CoT) is another important case study that integrates cloud computing and IoT services to offer scalable smart environments. CoT offers virtually unlimited cloud resources, allowing massive data generation and effective aggregation and storage of data. Cloud services enable the addition of richer functions in various applications, such as smart homes and buildings, where various devices must seamlessly communicate. Most important in this effort is the addition of middleware that serves as a bridge between heterogeneous devices with different interfaces and operating systems. By offering a trustworthy communication platform, middleware plays a key role in effectively integrating IoT devices into cloud infrastructures, ultimately enabling the evolution of holistic smart environments that are capable of responding to new challenges (Farahbakhsh et al., 2017b). When these case studies are analyzed, it is possible to identify the issues experienced during the design and implementation stages of IoT and cloud integrations, especially with regard to heterogeneity and data management. Effective integrations depend on a strong sense of the particular requirements of the environment that they will be deployed within. For example, how communication protocols are chosen, as some of the investigations of other integration frameworks have reported, is key to establishing the performance of systems. As the IoT ecosystem matures, the need to create flexible and interoperable solutions becomes paramount to building effective system architectures. Future research and implementation practices should therefore be aimed at enhancing middleware technologies and communication protocols to provide better methods of integration, ultimately promoting the overall performance of IoT platforms in cloud environments (Carpio et al., 2019).

5.4. DATA MANAGEMENT IN INTERNET OF THINGS (IOT) AND CLOUD INTEGRATION

The intersection of data management strategies in IoT and cloud solutions lies at the center of increasing operational effectiveness and innovation. With the increasing spread of the IoT, heterogeneous data sources become more and more difficult to manage, putting additional pressure on the necessity for robust integration measures. Services that span several areas of application for IoT are faced with challenges due to heterogeneity of devices and systems employing varying communication protocols and techniques. Such fragmentation hinders service developers from creating homogeneous data-driven applications. To resolve such issues, cloud-based platforms have emerged in the form of central frameworks, which offer seamless integration of the resource-constrained IoT devices. These platforms not only facilitate data handling but also assist developers in building and deploying services more effectively, ultimately leading to better user experiences and operational intelligence (Demeester et al., 2015). In the evolving landscape of industrial settings, data handling from IoT devices is especially challenging and promising. In the past, industrial control systems were closed systems operating on proprietary protocols that limited connectivity to broader enterprise management systems. The shift towards increased connectivity has come with the possibility of optimized production processes through the exploitation of data captured from sensors and actuators. The new model calls for new data management schemes that walk the line between ease of access and security of sensitive information. Secondly, the merging of IoT and industrial systems heightens the need for robust cyber security policies as greater interconnectivity creates new vulnerabilities (Johnson et al., 2016). The dilemma highlights the critical role played by effective data management structures in delivering integration while guarding against potential threats and risks (Johnson et al., 2016). Also, the development of consistent data management strategies for IoT and cloud integration opens up the potential for enhanced synchronization and coordination among various platforms. Through service developers tapping onto advancements in cloud computing, they are subsequently well-placed to address the complex requirements of processing streams of data from a varied group of IoT devices. Effective data management systems not only collect data from numerous sources into one central repository but also improved data quality and reliability, essential for sound decision-making. Through an overarching strategy to data management, organisations can unlock the full potential of IoT data—turning raw, isolated information into usable information. Thus, the cloud is a building block to store massive amounts of data and facilitate intelligent analytics that drive innovation and improve service delivery (Demeester et al., 2015) (Figure 5.11).

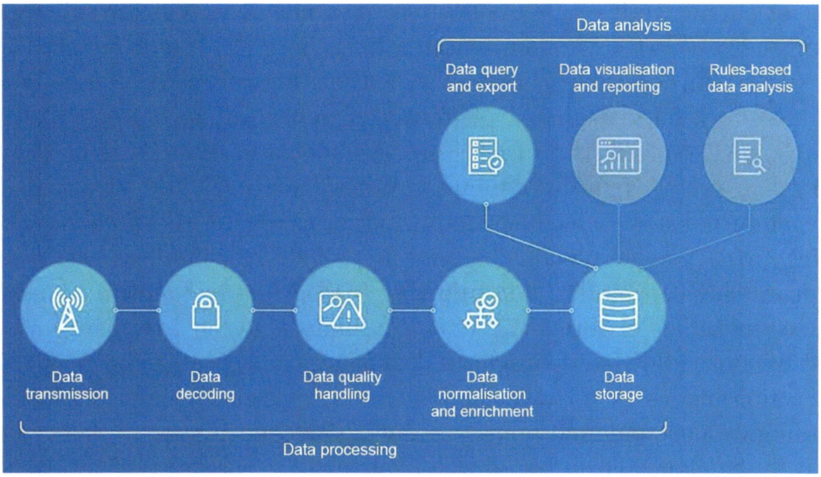

Figure 5.11. Data management in Internet of Things (IoT) and cloud integration.

Source: https://iot.electronicsforu.com/wp-content/uploads/2020/01/1-1.jpg.

5.4.1. Data Collection and Storage Methods

Data collection and storage are fundamental elements when it comes to IoT platforms and their amalgamation with cloud computing. With more devices becoming connected, they generate huge amounts of data for which strong mechanisms for aggregation and processing are needed. Conventional storage solutions tend to be inadequate in dealing with the exponential increase and diversity of IoT data. Here, cloud storage has been a savior as a prime solution. Cloud storage not only offers the scalability to keep up with the huge growth in data, but also enhances accessibility so that users can retrieve information from nearly any place. Such a cloud-based strategy can greatly enhance data management operations by reducing the burden on local infrastructure and providing advanced data analytics functionalities, which are imperative for real-time decision-making in smart spaces (Farahbakhsh et al., 2017a). Besides, middleware integration is the way forward towards ensuring data collection challenges in IoT settings are surmounted. Because IoT devices may employ different interfaces and architectures, one platform must integrate different types of data in a seamless manner. Middleware serves as a mediator that facilitates easy communication between devices and applications seamlessly, thereby ensuring data is efficiently collected and translated for storage. By implementing middleware solutions, it is possible for developers to raise the interoperability of IoT devices so they can function more easily in conjunction under one framework of data collection. It is not only vital for solving the heterogeneity of IoT devices but also for designing a stable architecture where data validity and consistency are primary issues. Effective middleware solutions are the bedrock for scalable IoT platform design to easily incorporate evolving technology (Farahbakhsh et al., 2017b). The advent of the "Cloud of Things" (CoT) also provides further stimulus towards the need for advanced data aggregation and storage methods optimized for use in IoT (Farahbakhsh et al., 2017a). CoT leverages cloud computing to provide near-limitless data storage and processing capabilities, thus allowing IoT devices to handle a lot of data without impacting performance. Interoperability between IoT platforms and cloud solutions allows self-contained data aggregation that can be analyzed to yield actionable insights. However, this convergence also incurs complications regarding data security and privacy as individuals' details are exchanged and stored in the cloud. Therefore, developing robust security and compliance is essential in order to mitigate risks while considering the innovative potential of CoT. With considerations for secure and viable storage techniques, IoT platforms have the potential to live up to its full capabilities in powering smart ecosystems and high-end uses (Farahbakhsh et al., 2017b).

5.4.2. Data Processing and Analytics

Convergence of data processing and analytics on IoT platforms and integration with the cloud is increasingly important as organizations strive to leverage the incredible volume of data generated by networked devices. Effective data processing capabilities enable such platforms to gather and analyze data from varied sources, from sensors and actuators, and ultimately transform raw data into actionable insight. For instance, IoT middleware solutions are emerging to facilitate sensor and internet connection seamlessly, enhancing the capability of end-users and application developers. Through the comparison of different platforms, it is apparent that support for heterogeneous sensing technologies is critical in delineating the effectiveness of data processing and data sharing features, especially in extracting maximum value from the data gathered across the IoT ecosystem (Mazhelis et al., 2016). Additionally, the importance of data ownership and its impact on security as well as privacy cannot be neglected when considering IoT analytics. As companies become increasingly reliant on cloud integration for data storage and processing, it is important that these issues are addressed. The collection, processing, and then sharing of sensitive information must be accomplished with vigorous security protocols in place to preserve user privacy as well as regulatory compliance. Sites that do not take these factors into consideration risk reduced adoption and distrust among users. Therefore, a comprehensive gap analysis of current IoT solutions demonstrates that the majority of them continue to encounter these challenges, pointing to the significance of data governance frameworks and user-centered designs prioritizing security (Mazhelis et al., 2016). Finally, in seeking cities and organizations to be more intelligent and efficient, the employment of data analytics becomes inevitable in the provision of information for decision-making. Smart City platforms, for instance, leverage data from multiple sources such as transportation, energy consumption, and public services to enhance the quality of life in cities. But for the maximum potential of such platforms to be realized, functional and non-functional requirements must be well understood. A recent survey of Smart City projects highlighted the integration of enabling technologies like Big Data and Cloud Computing as key to augmenting data analytics potential. As the elevated open research issues are explored, stakeholders will be in a position to create next-generation platforms that will enable enhanced data processing and analytics, leading to smarter and more responsive cities (Figure 5.12).

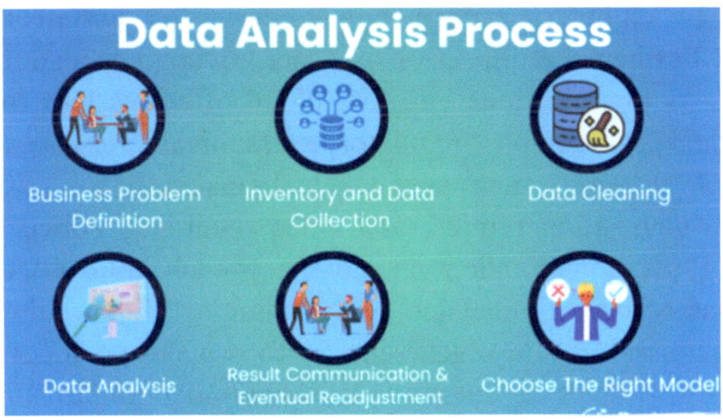

Figure 5.12. The intersection of data processing and analytics.

Source: **n**https://daxg39y63pxwu.cloudfront.net/images/blog/data-analysis-process/Data_Analytics_Process.png.

5.4.3. Security and Privacy Concerns

As far as IoT platforms and cloud integration are concerned, security and privacy have emerged as paramount concerns that affect users trust as well as the very viability of these technologies. Consolidating hundreds of devices into an integrated ecosystem offers tremendous opportunities but also vulnerabilities that can be exploited by nefarious forces. As data is always being collected and transferred on the networks, the risk of interception and illegal access increases exponentially. With the higher exposure, stronger encryption techniques and secure communication protocols must be implemented to safeguard sensitive information. In addition, privacy concerns are evoked as consumers become increasingly aware of the ubiquitous data collection culture within IoT systems, typified by a lack of openness as to how their information is handled and shared. Among the important aspects of privacy and security is the heterogeneity of systems and devices that constitute IoT ecosystems. Various devices may run on different interfaces, operating systems, and architectures, which provides a challenge to implement standardized security practices. As pointed out in the Cloud of Things (CoT) literature, middleware solutions become central to addressing the challenge of standing up to such heterogeneous devices in supporting communication as well as addressing the various security requirements of diverse devices. The absence of standardized protocols has a tendency to introduce vulnerabilities to be attacked and therefore, and subsequently, raise concerns about data confidentiality and integrity. Hence, the development and design of CoT platforms should take into consideration strong security solutions that solve device heterogeneity and the threats of interactions between heterogeneous components of IoT systems (Farahbakhsh et al., 2017a). Further, the impact of IoT data governance also merits careful consideration. As organizations integrate cloud offerings into IoT platforms, they have to deal with complex regulatory and ethical landscapes surrounding data privacy legislation, such as GDPR. Legislation prescribes stringent consent rules, data storage rules, and rules regarding notice of breaches, compelling organizations to adopt an active approach in their security efforts. The problem is not just technical deployments but instilling a culture of privacy insight and accountability in organizations. To prevent potential implications of data breaches, businesses must invest in continued employee training and awareness programs, as well as employing sophisticated security products. Ultimately, solving all the complicated security and privacy matters is essential for the successful implementation of IoT platforms and their cloud integration, thus avoiding distrust among users and system instability (Farahbakhsh et al., 2017b).

5.5. SCALABILITY AND FLEXIBILITY OF INTERNET OF THINGS (IOT) SOLUTIONS

Scalability of IoT solutions is one of the main features that enables systems to scale with rising demands without loss of performance and reliability. As the number of connected devices is expanding at lightning speed, the necessity for dependable architecture able to manage huge volumes of data becomes crucial.

With the advent of cloud computing comes the Cloud of Things (CoT) paradigm that offers virtually unlimited cloud services explicitly tailored to various IoT applications. This approach eliminates the traditional data processing, storage, and management limitations, hence enhancing IoT platforms' scalability. Moreover, the CoT facilitates deploying resources on demand, allowing organizations to manage computational resources effectively and respond to user demand changes rapidly. Scalability of IoT solutions, thus, not only benefits technological innovation but also business competitiveness in handling the complexities of a rapidly evolving digital environment (Farahbakhsh et al., 2017a). Flexibility of IoT solutions is also essential as it enables effortless adaptation to various application needs and shifting technology environments. The inherent heterogeneity of interconnected devices is a major problem since devices operate on varied interfaces, protocols, and architectures. Middleware is what comes to the rescue for this, offering an intermediary layer for ensuring communication between disparate IoT components in a secure fashion. Middleware facilitates interoperability, enabling devices to communicate effectively regardless of their underlying platforms. By eliminating the complexities that are associated with device management, middleware enhances the flexibility of IoT systems to support various functions such as data aggregation, reporting, and real-time processing. With this, organizations can create IoT solutions that are flexible and customized to fit specific operational needs, thereby enhancing innovation and efficiency in processes (Farahbakhsh et al., 2017b). In short, the interplay between scalability and flexibility is essential to the feasibility of IoT solutions, particularly under a cloud-integrated paradigm. The ability to scale and adapt to new challenges as IoT systems grow will be what makes them truly viable in practical deployments. The adoption of CoT principles allows organizations to take advantage of the enormous computational power of cloud infrastructures without relinquishing agility in the face of mounting technological challenges. In addition, the use of hardened middleware solutions will enable IoT platforms to handle multiple devices and applications' demands without any inconvenience. Putting flexibility and scalability at the top of the agenda places organizations in a leadership position to thrive in an increasingly connected world, tapping the full potential of IoT technologies in various sectors, including smart cities, industrial automation, and health care (Farahbakhsh et al., 2017a, 2017b).

5.5.1. Importance of Scalability in Internet of Things (IoT) Applications

Scalability in IoT applications is most critical, particularly in light of the ever-changing environments that these applications function in. The more devices that are connected, the systems must be capable of supporting various loads without hampering performance or functionality. This scalability is not only the incorporation of devices and users without interruption but also dealing with data generated by means of these interactions. Proper scalability enables IoT platforms to deliver assistance for smart city initiatives, where vast quantities of data collected through several sources need to be processed and analyzed in a timely manner for enhancing public services as well as citizens' standard of living (Chaves et al., 2017).

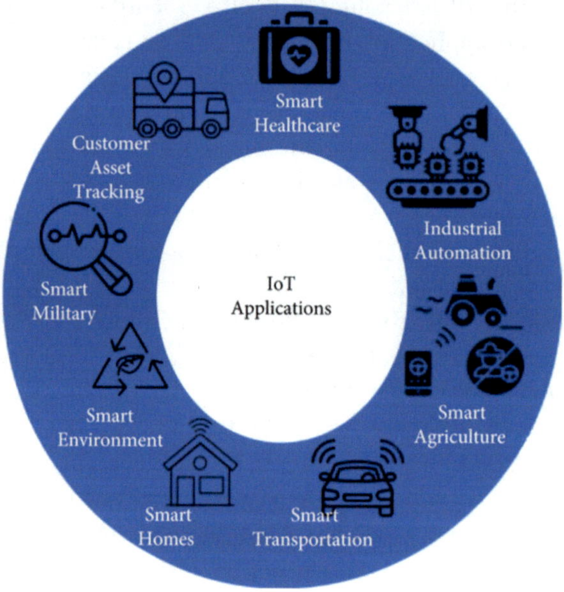

Figure 5.13. Importance of scalability in Internet of Things (IoT) applications.

Source: https://onlinelibrary.wiley.com/cms/asset/3b16943f-3a43-4870-8554-788dc923c016/jece7716956-fig-0001-m.jpg.

Thus, the success of IoT deployment is dependent on platforms with strong scalability attributes to deal with both current and future loads. In addition, scalability directly contributes to the operational adaptability of IoT applications. In rapidly evolving technology landscapes, where new devices and services are continuously being added with frequency, the ability of an IoT platform to scale and change is most crucial. Cloud of Things (CoT) strategy combines cloud computing and IoT to provide nearly unlimited scalability through virtual resources that are shareable as needed. This integration allows companies to manage big data efficiently, as well as address data aggregation and storage concerns resulting from an ever-increasing volume of data generated by IoT (Farahbakhsh et al., 2017a). Consequently, scalable architecture design and implementation make consolidation of various IoT functions easier, minimize latency, and maximize cloud resources to the best for performance. Finally, addressing scalability challenges in IoT applications matters for encouraging innovation and their deployment on various industries. Middleware solutions play a vital role in this regard, serving as middlemen that ensure various devices with varying protocols are able to communicate efficiently within scalable systems. Apart from realizing interoperability, such solutions also enable support of the varied needs of disparate applications like homes and smart cities. Since cities strive to be smarter and more efficient, platforms must enable the spread of data and device complexity and be reliable with ease of integration (Chaves et al., 2017). Continuous innovation in scalable IoT platforms is not just a technical requirement but also a key path to release the full potential of connected devices to redefine daily life and governance (Figure 5.13).

5.5.2. Role of Cloud Services in Enhancing Scalability

The scaling nature of cloud services is even more pronounced in IoT platforms, where unexpected increases in connected devices suggest the necessity of flexible and versatile infrastructure. On-premises systems are generally unable to cope with variable data workloads-generated changing demands. On the contrary, cloud computing offers unprecedented scalability such that organizations can quickly scale resources up or down in real-time depending on the demand of their IoT applications. This scalability not only enables instant resource allocation but also cost-effective management since users only get charged for what they consume. Accordingly, cloud services allow IoT platforms to handle vast data streams and devices in an efficient way, thus facilitating the development of innovative applications and services that can be easily adjusted according to changing user needs and environmental conditions, ultimately enhancing operational efficiency and service quality in the smart city ecosystem (Chaves et al., 2017). Secondly, the integration of cloud computing into IoT ecosystems offers a synergistic benefit that largely contributes to the scalability of smart technology. With devices becoming ubiquitous, they generate enormous volumes of data that need to be processed and analyzed almost in real-time in a bid to reach meaningful conclusions. Cloud platforms facilitate this through the offer of high-level data processing and management functionality, which is instrumental to the functioning of high-end IoT applications. The integration enables more effective interaction between heterogeneous devices, thereby the formation of what is known as the Cloud of Things (CoT). Cloud services address issues with different device architectures and operating systems through the application of middleware solutions that connect devices and platforms, enhancing the overall user experience. This integration is required to address the threat of the massive number of connected devices and to make cities the smart, efficient cities they can be (Farahbakhsh et al., 2017a). Finally, with the emergence of advanced technologies, the role of cloud services to make the scalability of IoT platforms more critical will increase day by day. The necessity for smarter cities demands environments that can support heterogeneous collections of applications from traffic management to environmental monitoring. Cloud computing, aside from providing an elastic infrastructure to run these applications, also fosters innovation through the available tools and services. Furthermore, as businesses move to leverage big data analytics within these spaces, cloud services may assist in storing and analyzing humongous datasets accumulated from millions of IoT devices. Beyond scaling operations, this combination of cloud and IoT also allows cities to leverage data-driven insights for more informed decision-making. As such, the symbiotic relationship between cloud services and IoT platforms is determinant to the attainment of smart city solutions that are fully integrated and scalable (Chaves et al., 2017).

5.5.3. Examples of Scalable Internet of Things (IoT) Solutions

The development of smart city initiatives is one such scalable IoT solution that captures the power of technology in projecting and expanding city life. With the convergence of various information and communication technologies (ICT), it is possible for smart platforms for the city to be embraced that facilitate improved services and quality of urban living. These platforms specifically utilize cyber-physical systems, big data, and cloud computing as key components. For instance, by leveraging live data analytics of IoT devices such as sensors and cameras, city administrators can control traffic flow, save energy, and render public places more secure. However, as outlined in the state-of-the-art software platforms for Smart Cities overview, significant scientific and technological challenges need to be addressed in order to enable the widespread application of such scalable solutions in cities (Chaves et al., 2017).

Furthermore, scalable IoT solutions over into the industrial use case in which manufacturing industries heavily depend on IoT platforms to automate operations. The convergence of cloud computing with IoT allows industries to achieve real-time monitoring and predictive maintenance, hence decreasing downtime and expense. Intelligent systems gather data from sensors embedded within machinery and transmit it to cloud-based platforms for analysis. The resultant implications allow organizations to anticipate equipment failures and make decisions. Based on current research, advances in cloud infrastructure allow for multi-provider services, which can offer fault tolerance and reliability and, thereby, allow IoT applications to scale in manufacturing environments (Buyya et al., 2017). This shift not only supports increased efficiency but also allows for more intelligent manufacturing practices. Other than urban and industrial applications, scalable IoT solutions also play an important role in agriculture, transforming traditional farming to technology-driven, precision farming. Farmers use IoT devices such as weather sensors and soil moisture sensors to obtain real-time information on the health of crops as well as environmental conditions.

This data is then analyzed in cloud infrastructure so that farmers can adjust irrigation schedules, use fertilizers cost-effectively, and initiate pest control measures based on current data. These systems not only increase the yield but also promote sustainable farming by minimizing wastage of resources. Applications of IoT in agriculture show how scalability can be facilitated by cloud-based systems, particularly as an increasing number of farmers adopt these technologies as a result of global food requirements and environmental factors, subsequently producing more smart agricultural systems (Figure 5.14).

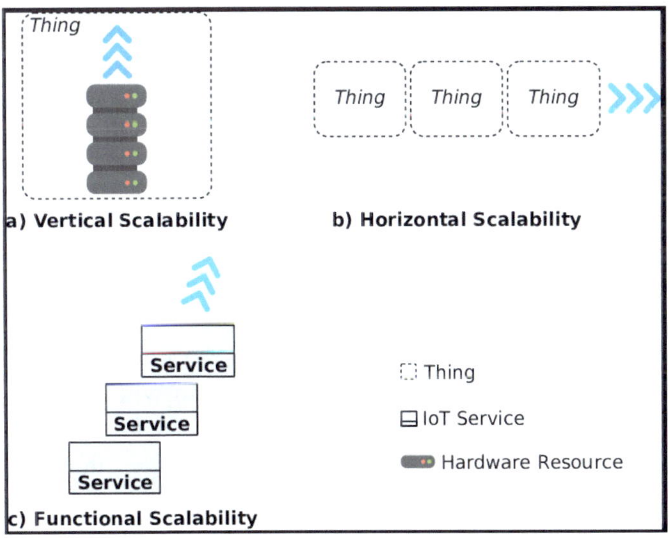

Figure 5.14. Scalability of Internet of Things (IoT) systems.

Source: https://www.researchgate.net/profile/Damian-Arellanes/publication/339892185/figure/fig3/AS:874111110762497@1585415737712/Scalability-of-IoT-systems.png.

5.6. FUTURE TRENDS IN INTERNET OF THINGS (IOT) AND CLOUD INTEGRATION

The convergence of IoT and cloud computing will redefine the digital interaction landscape in various industries. As smart devices increase, the demand for robust data management and analytics capability intensifies. The transformation calls for premium cloud integration solutions that are equipped to handle the volume of data generated by IoT devices effectively. By facilitating in real-time processing and analysis, cloud platforms allow companies to derive useful insights and, as such, enhance decision-making. Second, the convergence of IoT with cloud infrastructures is poised to provide enhanced operational efficiency, and companies will be able to streamline processes and save costs. All the same, effective realization of these advantages hinges on addressing current technological hindrances, and that is why innovation around IoT platforms and cloud architectures matters. Fixing these problems will ultimately pave the way for more sophisticated, large-scale, and smart adaptive solutions in cities and beyond (Chaves et al., 2017). IoT and cloud integration trends on the horizon suggest shifting towards more decentralized models of computing, enabling greater flexibility and durability in system design. As conventional single-provider cloud environments mature, multi-cloud and hybrid cloud approaches are becoming more popular; these designs enable better resource allocation and redundancy.

The decentralized method not only improves service availability but also permits transparent data transfer across various cloud providers, promoting interoperability among a wide range of IoT devices. This transition is needed in empowering new use cases, such as smart city initiatives, based on the integration of many technology platforms to deliver sophisticated services and improve quality of life among people. In addition, the advent of edge computing in which data processing occurs closer to the origin will assist in advancing cloud capabilities, further streamlining operations and reducing latency in IoT use cases (Buyya et al., 2017). As the future emerges, the use of artificial intelligence (AI) for IoT and cloud integration cannot be overestimated. AI technologies continue to gain deeper penetration into IoT platforms to facilitate automating decision-making and improving the delivery of services. Integration paves the way for self-learning systems that respond to diverse operating conditions and user choice, opening up the possibility for a more efficient and personalized experience. Moreover, AI-powered analytics can enhance predictive maintenance such that companies can anticipate equipment failures before they happen, something of particular importance for industrial usage. The combinations of IoT, cloud computing, and AI not only bring up operation capacities but also provide new business and research opportunities.

By embracing such innovations, organizations can create wiser infrastructures that can address emerging challenges and therefore lead the way in technological innovation during the digital age (Chaves et al., 2017).

5.6.1. Emerging Technologies Impacting Internet of Things (IoT)

The rapid evolution of emerging technologies has significantly altered the face of the IoT, leading to more sophisticated capabilities as well as effortless integration into cloud platforms. One of the greatest revolutions is sensor technology, the basis of IoT applications. Advances in sensor systems have enabled the collection and aggregation of vast amounts of real-time data, creating new business models and services. This has spurred the development of robust infrastructures required for the implementation of IoT solutions. When new companies enter the market, they technologically utilize these advanced sensor systems to offer differentiated products that are suitable for consumers, as revealed by available literature. By establishing a dense system of intelligent devices, the IoT market is at the crossroads of a revolution transformation, which can aggregate efficiency and connectivity in various industries (Bremner et al., 2016). The intersection of IoT platforms and cloud computing has a core role of facilitating data processing and management from hundreds of devices. In all this, the emergence of IoT middleware solutions has been a primary enabler in ensuring end-to-end connection between sensors, actuators, and the cloud. These middleware platforms need to cater to the diverse needs of stakeholders like device manufacturers, application developers, and end users. Both proprietary and open source IoT platform analysis gives a multi-dimensioned image where ease of use and readiness play a direct role in determining adoption levels. The ongoing gap analysis reveals some major gaps in current solutions, particularly in terms of data processing capability and the type of support available for application developers. As the IoT ecosystem keeps evolving, filling such gaps is crucial for platforms to ensure integration with cloud offerings is improved and widespread penetration can be realized (Mazhelis et al., 2016). Security and privacy challenges form another critical aspect which gets directly influenced by new technologies in the IoT space. With gadgets getting dispersed and data sharing happening everywhere, the importance of robust security frameworks cannot be stressed further. New technologies will have to address the challenges of data ownership and privacy implications on users. With the implementation of secure controls, such as encryption methods and access schemes, IoT platforms can generate more consumer trust in their products. Additionally, full-fledged IoT marketplaces have the potential to empower application developers and enhance the completeness of the ecosystem. With the growing focus on security, these platforms will not only grow user trust but also promote the sustainable growth of IoT solutions. Therefore, the integration of robust security features in IoT platforms is essential to build a secure arena to cater to future developments (Mazhelis et al., 2016) (Figure 5.15).

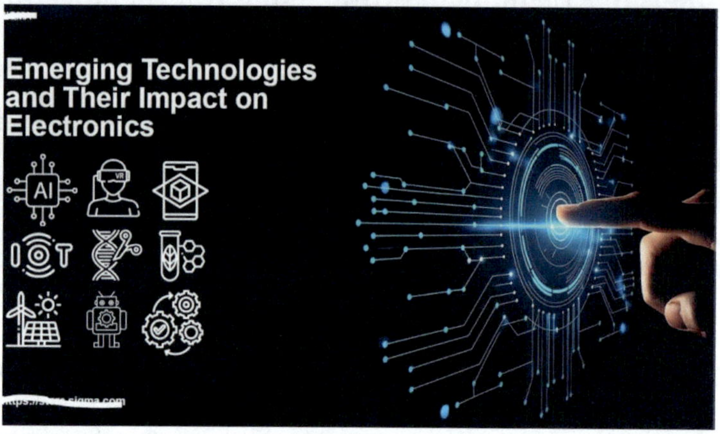

Figure 5.15. Emerging technologies impacting Internet of Things (IoT).

Source: https://blog.siqma.com/wp-content/uploads/2024/12/Emerging-Technologies-1024x564.jpg.

5.6.2. Predictions for the Future of Internet of Things (IoT) Platforms

The future for IoT platforms is tremendous with technology as it keeps on improving. With the help of advanced machine learning and cloud services, these platforms will be better equipped to manage huge volumes of data in real-time. For instance, home automation apps will use sophisticated architectures to facilitate the interaction of devices, including microcontrollers from Arm, with cloud providers like Amazon Web Services. The quality of interaction can potentially transform user experience by using improved functionality and real-time analytics. IoT projects that optimize communication bandwidth and speed will further come up with new uses, thus expanding the scope of appeal of IoT solutions to residential customers and businesses as well, thus paving the floodgates for mass adoption of smart home technologies (Cochran et al., 2018). As cities transform into being smarter, IoT platforms will lead the development of urban infrastructure. Smart city initiatives will rely on these platforms to enhance provision of services, thus elevating the citizens' quality of life. Through the consolidation of the enabling technologies such as cyber-physical systems, big data, and cloud computing, city governments can enhance efficiency in operations as well as decision-making based on data. The issues the platforms currently have to deal with must be overcome through sustained investment in research and development as well as collaboration across stakeholders. Emerging software platforms that enable various IoT applications will be paramount to future real-time responsiveness and flexibility in cities, more solidifying the roles of IoT platforms in future urban planning (Chaves et al., 2017). In the future, the convergence of IoT platforms with security, sustainability, and privacy will be paramount to breaking through public skepticism and regulatory challenges. Future growth will be towards establishing robust cybersecurity measures to ensure important data shared through these platforms. IoT platforms will also need to embrace green culture as the world is confronted with environmental challenges. With the use of power-efficient protocols and the inclusion of renewable energy sources, platforms will be able to minimize their footprint. Features encouraging greater privacy for users and better protection for data will also be a main emphasis, resulting in higher consumer trust. With such challenges being more dominant, IoT platforms will not only determine technological progress but also align with social responsibility as well as environmental responsibility, thus ensuring a sustainable journey for the industry.

5.6.3. Potential Challenges and Opportunities

The interfacing of Internet of Things (IoT) platforms with cloud computing presents a two-dimensional scenario that is full of opportunities and challenges. One of the greatest challenges is due to device and technology heterogeneity in the IoT system. IoT platforms must accommodate a vast number of heterogeneous actuating and sensing technologies, which can complicate the connection and interoperability of data. As revealed in gap analysis of current IoT platforms, the platforms are rated largely on how prepared and consumable they are for stakeholders like device providers and consumers (Mazhelis et al., 2016). Such rating calls for platforms to bridge technological divides and enable devices to communicate with each other without difficulties while guaranteeing data transfer integrity and security. Failure to address this problem appropriately may slow user adoption and the overall success of IoT implementations. Apart from the technical problem, data ownership and privacy are some of the largest issues confronting IoT and cloud integration. With increasingly more data being collected, transmitted, and processed in such networked systems, questions arise regarding who owns the data and how it is protected. As big data analytics continues to gain traction, organizations have more than ever before opportunities to learn from the vast amounts of data generated by the

IoT devices and, subsequently, transform their operational strategies (Perera et al., 2018). However, the looming threat of security risks and misuse of data is yet to disappear. The implications of poor data governance can potentially lead to user and stakeholder trust breaches, regulatory interest that can serve as a growth inhibitor, and hence, establishing solid frameworks for solving ownership and data privacy is essential in the development of a growth-friendly environment in the IoT space. Despite the issues highlighted, the convergence of IoT platforms and cloud capabilities provides tremendous potential for innovation and efficiency. Companies can leverage the technology of big data to access and analyze sensor data for insights of value, paving the way for enhanced operational effectiveness and personalized customer experience. The convergence of these technologies enables companies to explore new markets and streamline existing operations significantly. Besides, the development of isolated IoT marketplaces provides development enablement to application developers, ensuring rich ecosystems flourish (Mazhelis et al., 2016). As the IoT landscape continues to shift, bridging existing gaps while seizing future opportunities can help organizations forge models that not only stay abreast of current trends but also anticipate future requirements in the ever-evolving paradigm of technology.

SUMMARY

In Summary, integration of IoT platforms with cloud computing is essential in addressing the complexity induced by pervasive connectivity and data generation. With increasingly more IoT devices propagating across various disciplines, their diversity is turning out to be a huge challenge for service developers to leverage these technologies. Developers can mitigate the problems with heterogeneous connectivity options and communication protocols in the IoT context through the utilization of cloud-based methods. The proposed evaluation of a cloud platform enhances the effective integration of low-resource IoT devices with greater efficiency and seamless deployment of services over multiple application fields, as witnessed in the research findings that testified to the platforms ability to cause minimal impact on device operation while attaining maximum integration benefit (Demeester et al., 2015). Furthermore, the CoT is a crucial step in IoT evolution through scalable cloud services that enable the operation of large-scale IoT platforms. The massive data volumes generated by connected devices become difficult to manage and demand sophisticated solutions, and the facilitator of this becomes cloud computing. Efficient management, aggregation, and storage solutions are paramount for empowering the rapidly growing IoT ecosystem. As highlighted in recent studies, middleware technologies guarantee that the issue caused by device heterogeneity is resolved by enabling interoperability, which smoothens interactions among heterogeneous systems (Farahbakhsh et al., 2017a). Finally, the interoperability between IoT platforms and cloud computing provides an environment that fosters innovation, resulting in smarter homes, cities, and industries. Mitigating the pressing issues of device heterogeneity and data management using effective cloud integration strategies, in turn, will not only augment operational efficiency but also provide avenues for the development of new services. With the era of the Internet of Things continuously progressing, putting great emphasis on integrating cloud services will become crucial for researchers and practitioners seeking to excel in this complex field. The convergence brought about by interaction among middleware technologies and cloud characteristics will automatically chart the way forward for IoT applications, making them appropriate to address the needs of a highly connected world (Demeester et al., 2015; Farahbakhsh et al., 2017a).

Some key issues arising in current discussions regarding IoT platforms and their compatibility with cloud environments point to the evolving characteristics of the Internet of Things. Middleware products, which enable connectivity between sensors and actuators, play a vital part in ensuring seamless data exchange within IoT ecosystems. Open-source or proprietary platforms, these cater to a broad spectrum of user bases, including end-users and application developers down to device manufacturers. An analysis of current IoT solutions points out shortcomings in usability and readiness for a heterogeneous environment, specifically ownership of data, privacy, and overall support for developers. This shortcoming analysis highlights the necessity for platforms to mature and become adaptive in an attempt to satisfy the specific needs of the contemporary complex IoT world, as outlined in Mazhelis et al. (2016). Coupled with connectivity, the transition of legacy power infrastructures to Smart Grids signifies the implementation of IoT technologies in real-world applications. Smart Grids integrate IoT devices, such as sensors and smart meters, to facilitate two-way communication of information between the consumers and service providers. This delivers enhanced monitoring, analysis, and management of various segments of energy systems, which is critical to address inefficiencies and increasing energy requirements. The application of IoT in Smart Grid infrastructure is the ultimate example of seamless communication between devices and cloud services, enabling timely data exchange and processing. The acknowledgment of the sophistication of such applications opens up advantages and drawbacks in utilizing IoT-based solutions, as they were established in Copeland et al. (2019). In general, the analysis presents an urgent need for further research and discussion of IoT platforms and their integration into the cloud. In a bid to properly address the gaps thus realized, stakeholders must accord high priority to establishing platforms that facilitate interoperability, security, and data handling. Policy proposals based on surveys and expert evaluation center on a robust reinforcement of the existing infrastructure while also addressing the constraints confronted by developers and end-users alike. Moreover, creating a complete IoT ecosystem, including specialized marketplaces and back-end systems, can greatly increase the efficiency and accessibility of IoT applications. And as the conversation on IoT platforms goes on, embracing these strategic recommendations will be crucial to propelling innovation for even greater utility in future integrations.

REFERENCES

1. Ahmed, S., Atzori, L., Hankin, B., Basco-Carrera, L., Bastin, L., Beven, K., Beven, K., et al. (2019). Models of everywhere revisited: A technological perspective. https://core.ac.uk/download/322489447.pdf (accessed on 2nd October 2025).
2. Akrivopoulos, O., Amaxilatis, D., Chatzigiannakis, I., Tselios, C., et al. (2017). Enabling stream processing for people-centric IoT based on the fog computing paradigm. https://core.ac.uk/download/158328004.pdf (accessed on 2nd October 2025).
3. Akrivopoulos, O., Antoniou, A., Chatzigiannakis, I., Tselios, C., et al. (2017). On the deployment of healthcare applications over fog computing infrastructure. https://core.ac.uk/download/98343526.pdf (accessed on 2nd October 2025).
4. Baldini, I., Castro, P., Chang, K., Cheng, P., et al. (2017). Serverless computing: Current trends and open problems. http://arxiv.org/abs/1706.03178 (accessed on 2nd October 2025).
5. Bremner, D. (2016). Analyzing the IoT ecosystem: The barriers to commercial traction. https://core.ac.uk/download/42370075.pdf (accessed on 2nd October 2025).
6. Buyya, R., & Varghese, B. (2017). Next generation cloud computing: New trends and research directions. https://core.ac.uk/download/96664300.pdf (accessed on 2nd October 2025).
7. Carpio, F., Dizdarevic, J., Jukan, A., Masip-Bruin, X., et al. (2019). A survey of communication protocols for Internet of Things and related challenges of fog and cloud computing integration. https://core.ac.uk/download/187266039.pdf (accessed on 2nd October 2025).
8. Chaves, A. P., Gerosa, M. A., Kon, F., Milojicic, D., et al. (2017). Software platforms for smart cities: Concepts, requirements, challenges, and a unified reference architecture. http://arxiv.org/abs/1609.08089 (accessed on 2nd October 2025).
9. Clement, S. J., Davies, J., McKee, D. W., Ouyang, H., et al. (2017). The Internet of Simulation, a specialization of the Internet of Things with simulation and workflow as a service (SIM/WFaaS). https://core.ac.uk/download/74236601.pdf (accessed on 2nd October 2025).
10. Cochran, S., Heidari, H., Martin, S., Tousimojarad, A., et al. (2018). Arm Mbed – AWS IoT system integration [Open access]. https://core.ac.uk/download/151170100.pdf (accessed on 2nd October 2025).
11. Copeland, R., Crespi, N., Rehmani, M. H., Saleem, Y., et al. (2019).

Internet of Things-aided smart grid: Technologies, architectures, applications, prototypes, and future research directions. http://arxiv.org/abs/1704.08977 (accessed on 2nd October 2025).

12. Davidsson, P., Heyer, C., Mihailescu, R. C., Spalazzese, R., et al. (2017). A role-based approach for orchestrating emergent configurations in the Internet of Things. http://arxiv.org/abs/1809.09870 (accessed on 2nd October 2025).
13. Demeester, P., Hoebeke, J., Moerman, I., Van den Abeele, F., et al. (2015). Integration of heterogeneous devices and communication models via the cloud in the constrained Internet of Things. https://core.ac.uk/download/55746971.pdf (accessed on 2nd October 2025).
14. Farahbakhsh, R., Farahzadia, A., Rezazadeh, J., Shams, P., et al. (2017). Middleware technologies for Cloud of Things–a survey. http://arxiv.org/abs/1705.00387 (accessed on 2nd October 2025).
15. Farahbakhsh, R., Farahzadia, A., Rezazadeh, J., Shams, P., et al. (2017). Middleware technologies for Cloud of Things–a survey. http://arxiv.org/abs/1705.00387 (accessed on 2nd October 2025).
16. Fischer, M., Meiling, S., Purnomo, D., Schmidt, A., et al. (2018). MONICA in Hamburg: Towards large-scale IoT deployments in a smart city. http://arxiv.org/abs/1803.06854 (accessed on 2nd October 2025).
17. Johnson, C. (2016). Securing the participation of safety-critical SCADA systems in the industrial Internet of Things. https://core.ac.uk/download/46664188.pdf (accessed on 2nd October 2025).
18. Mazhelis, O., Mineraud, J., Su, X., Tarkoma, S., et al. (2016). A gap analysis of Internet-of-Things platforms. http://arxiv.org/abs/1502.01181 (accessed on 2nd October 2025).
19. Perera, C., Chen, F., Suciu, G., Hashem, I. A. T., Xu, L. D., Belgaum, M. R., Sun, Y., et al. (2018). Challenges of Internet of Things and big data integration. http://arxiv.org/abs/1806.08953 (accessed on 2nd October 2025).

CHAPTER 6

DATA PROCESSING AND ANALYTICS

Contents

INTRODUCTION

In today's digital age, data processing and data analytics have become imperative to organizations as well as researchers. With the accelerative nature of data creation, having the ability to extract meaningful information from extensive datasets becomes a strategic necessity. Social media sites, in particular, have become goldmines of information, giving organizations unusual opportunities to make sense of consumer behavior, societal patterns, and market forces. This data usage shift is a step back from traditional data analytics methodologies towards more advanced, multimodal ones that handle a range of data forms, including text, image, and video. By embracing these innovations, organizations can better manage the intricacies of today's information landscapes. Multimodal learning systems are significant components in maximizing the efficiency of social media space data processing. By virtue of their use of advanced neural embedding models, analysts can uncover, categorize, and retrieve user profiles from diverse user interactions across multi-platforms. An analysis of such systems is technical in nature but also indicates an intimate understanding of the behavior of online community users. By analyzing interactions within platforms such as Twitter and forums for a particular ideology, data practitioners can produce extensive insights. As observed in the literature, the experiment's interaction with artificial actors reveals the potential of such systems to imitate user behavior and measure the quality of multimodal representations, paving the way for a more serious analysis of user interaction patterns (Salton et al., 2019). Finally, as firms increasingly rely on data analytics, designing customized tools and services becomes priceless. With the advent of big data technologies, social data mining has become a significant market segment, and organizations are starting to develop proprietary solutions or send work out to specialist companies. Such technologies as cloud computing services demonstrate the unobtrusive integration of analytics applications into organizational processes so that practitioners are able to carry out advanced queries at once and receive real-time results. The creation of services like twearch is an excellent illustration of how companies are able to tap into the power of social media analytics as a service, reducing costs substantially while enhancing capabilities. This shift not only allows organizations to make data-driven decisions but is also reflective of the widening paradigm of data processing and analysis for the contemporary digital era (Ayguadé Parra et al., 2014).

Definition of Data Processing

Processing data is an orderly series of activities consisting of collecting, manipulating, and processing data in order to receive meaningful information. With organizations continuing to generate ample data, the necessity for efficient methods of processing data is imperative. Data processing entails processes such as collection, storage, retrieval, and conversion. Each process plays a distinct role in the entire process of data life cycle, enabling companies to convert raw data into informative information. This is necessary since businesses attempt to cleanse their data, just like oil is treated to yield valuable products. Improved data complexity and volume have further resulted in higher demand for new models, such as the ones seen in data-driven scientific discovery, well-deserved recognition of data to generate great insights and drive decision-making (Amen et al., 2015). The advent of Big Data has made it imperative for organizations to develop sophisticated data processing techniques and technologies that are capable of handling real-time analysis of data. The ability to process large volumes of data in a short time enables organizations to make decisions based on real-time data analysis. New models for real-time analytics reflect this requirement, with data processing models that optimize the speed of data processing without sacrificing accuracy. For instance, advancements in cloud computing have introduced enhanced capacities for real-time analysis, thereby equipping organizations with the capacity to meaningfully make sense of big datasets.

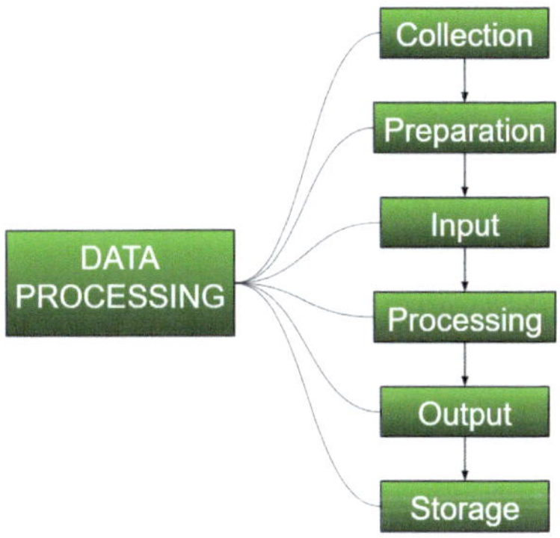

Figure 6.1. Data processing.

Source: https://media.geeksforgeeks.org/wp-content/uploads/data-processing.png.

This use of advanced technologies into data processing systems highlights ongoing developments in the field and its reliance on advanced methodologies to remain competitive in a data-driven economy (Heiskanen et al., 2019). Furthermore, the term data processing now needs to encompass the intricacies that have been added by various streaming data processing systems (SDPSs). The need for scalable and efficient stream analysis has propelled the development of a number of open-source systems like Apache Storm, Apache Spark, and Apache Flink with their respective different performance characteristics. There is a need to know about the latency and throughput of the platforms so that their efficiency can be compared in real-world situations. Evaluating SDPSs allows organizations to choose appropriate technologies for their specific data processing needs, especially for those applications that necessitate real-time performance. Sharper definition of data processing then further includes the special methodologies and tools that allow efficient management of streaming data, thus making enhanced data analytics and decision-making abilities (Heiskanen et al., 2019) (Figure 6.1).

Importance of Data Analytics in Modern Business

In today's business environment, the relevance of data analytics has grown as businesses seek to leverage the power of information towards strategic gain. With more emphasis on using data to inform decisions that can make businesses more competitive and operationally efficient, organizations are being pushed more to make data-driven decisions. By leveraging huge volumes of data, businesses are able to uncover information that shapes marketing, enhances customer interaction, and maximizes resource utilization. Besides that, analysis of data enables trends and patterns to be seen which, if dealt with beforehand, can bring in the highest revenue and growth. Implementation of data-driven techniques not only enables businesses to prevent risks but also enables them to react swiftly to the forces of shifting business environments, thereby solidifying their footing in the ever-changing business world. Among the most significant elements of the significance of data analytics is its contribution towards creating innovative solutions, particularly in urban planning and infrastructure management. As cities become smart cities, the application of big data becomes more prominent in addressing urban issues. For instance, new analytical techniques employ big data and computing power to enhance different industries, such as transportation and energy management. These intelligent systems have the capability to analyze real-time input data to optimize traffic flow and reduce congestion, improving the quality of life of the people. Such pragmatic uses illustrate the value conversion of data analytics, eventually contributing to sustainable development and efficient city management. As explored in existing research, established architectures like the Hierarchical Spatial-Temporal State Machine (HSTSM) can underpin this pathbreaking work by effectively modeling urban dynamics' complexity (Doctor et al., 2020). Data visualization is also a very significant component of data analytics because it enables the stakeholders to participate meaningfully and intuitively with data. By applying data visualization tools, businesses are able to unveil complex data in an open graphical presentation so that decision-makers can readily point out patterns and correlations. With the clearness comes increased understanding and capacity to develop an environment where timely decisions can be made based on data. Good data visualization, aside from aesthetics, facilitates sense-making processes by allowing users to navigate the data while emphasizing primary findings. Thus, the organizations that implement data visualization not only provide information for all but also facilitate collaboration among teams. The integration of compact visual representation and advanced analytics forms a robust foundation for the modern firms to respond effectively to uncertainties as well as drive innovations with success (Bikakis et al., 2018).

6.1. TYPES OF DATA PROCESSING

Data processing incorporates various methods to transform raw information into meaningful information, which can be utilized for decision-making and data analysis. Batch processing is one of the popular types in which massive amounts of data are collected and processed in batches at set time intervals. The process is most appropriate where real-time processing is irrelevant, allowing organizations to handle massive volumes of data without excessive computational burdens. However, freshness-related issues can arise as batch processing tends to cause delays that compromise timely insights. With the requirement for more responsive outputs to data on the rise, other forms of processing have taken notice and stressed the need for techniques that can accommodate quick, real-time analytics regardless of scalability-efficiency compromises. By contrast, real-time processing or stream processing responds to data as it streams in, enabling immediate insights and action. This is essential in industries like finance and healthcare, where real-time decision-making from live feeds of data can have critical effects on outcomes. Real-time processing enables the use of complex algorithms that deal with data streams in real time, enabling organizations to exploit insights to optimize operations. As mentioned in recent research, the union of big data analytics and intelligent systems provides more capabilities, particularly in areas such as intelligent transportation for intelligent cities. To be effective in real-time processing, there has to be high-end architecture and technology, not only emphasizing the capability to handle enormous amounts of data inflows but also the expertise to derive actionable insights promptly (Doctor et al., 2020). Finally, a combination of both batch processing and real-time processing has also been a feasible choice for modern data environments. By leveraging the best aspects of both, organizations can maximize efficiency and accuracy in data analysis. The approach allows organizations to thoroughly analyze historical data through batch processing and at the same time also leverage real-time data for real-time operating decisions. The worth of such a hybrid approach can be realized in fields like energy management, in which predictive analytics is precious to maximize the utilization of resources and provide stability to the grid. In addition, as computational techniques like artificial intelligence continue to advance, they again enable hybrid processing methods. Accordingly, the marriage of big data analytics and innovative processing techniques adapts to the ever-changing landscape of data utilization, emphasizing its relevance to contemporary analysis frameworks (Gharehpetian et al., 2019) (Figure 6.2).

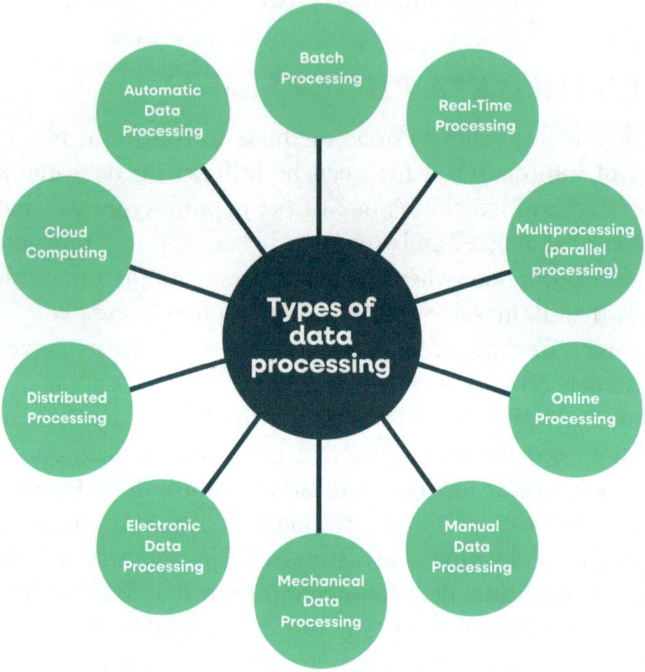

Figure 6.2. The types of data-processing.

Source: https://images.ctfassets.net/dkgr2j75jrom/3SmuSeYhoh05YNaHjlSKdO/2194fe092d6941a155543fed54d36c33/types-of-data-processing.jpg?w=1679&h=1397&fl=progressive&q=50&fm=jpg&bg=transparent.

6.1.1. Batch Processing

Batch processing is a fundamental mechanism in the framework of data processing and analytics in the context of handling gigantic datasets common to Big Data configurations. This technique involves the collection and processing of large volumes of data within fixed periods, contrary to real time. The strength of batch processing lies in its efficiency to handle resource consumption effectively, leading to low operation costs and enhanced throughput. This technique supports advanced analytics, such as model calibration in statistical analysis where optimal setting of parameters is most important. However, the process is time-consuming, with several configurations run from scratch for each one-of-a-kind dataset, which can disrupt timely access to insights supporting informed decision-making, pointing to the need to enhance conventional batch practices to meet current applications (Qin et al., 2014). Although useful, batch processing has its weaknesses because it seldom addresses the demand for real-time analysis in today's data-driven world. Organizations are ever-more reliant on real-time understanding, which demands a more active data handling process. Innovative solutions such as speculative parameter testing assist in mitigating some of the inherent shortcomings of batch processing through parallel evaluation of diverse setups. The innovative use of advanced parallel multi-query processing accelerates and simplifies identifying the best solution by data scientists.

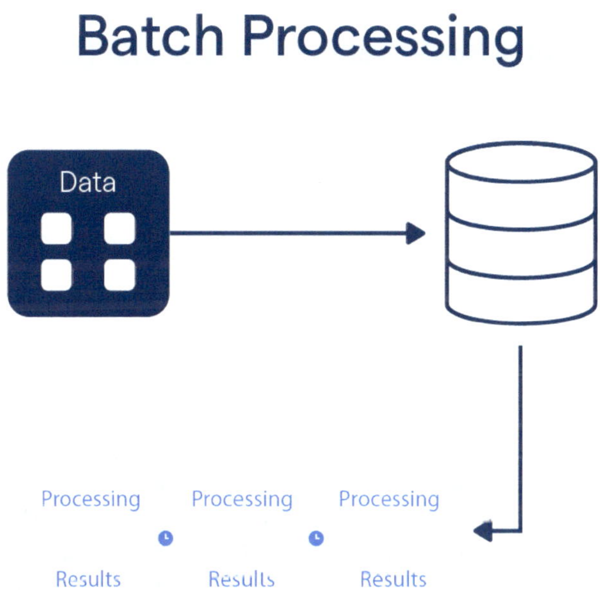

Figure 6.3. Batch processing.

Source: https://cdn.botpenguin.com/assets/website/Batch_Processing_2b9e11d5bd.png.

Their use in systems like GLADE PF-OLA is a demonstration of the potential to harness batch processing to facilitate Big Data analytics requirements through timely analysis while still harnessing the benefits of batch operations (Qin et al., 2014). Real-time Big Data analytics offers a contrast opposite to batch processing, with immediate return feedback on data inputs. But where batch and real-time analytics overlap is a fertile research ground to try to take the best of both worlds. Optimizing batch operations and combining them with real-time analysis allows companies to unlock value hidden in big data, exposing raw data to become actionable insight. The future of data processing thus rests in a hybrid model that combines the reliability and efficiency of batch processing with the responsive adaptability of real-time analytics. As these paradigms come closer to merging, the capability to discover deep insights and enable data-driven scientific discovery becomes increasingly accessible, allowing organizations to more effectively contextualize and take advantage of their data assets (Figure 6.3).

6.1.2. Real-Time Processing

Real-time processing is now an essential component of data analytics because of the need to glean instant insights from fast-flowing data streams. The emergence of Internet of Things (IoT) technologies further accelerated this demand since widespread sensors produce volumes of data at hitherto unseen speeds. Traditional data processing methods, in most instances tied up with batch processing, are not capable of handling such floods. Consequently, methods like Complex Event Processing (CEP) have emerged to carry out real-time analysis of data streams. Specifically, CEP is tasked to embed higher-level semantics of knowledge and execute queries smoothly across temporal dimensions, which underscores the necessity of frameworks that marry real-time analytics with persistent history datasets. This integration is crucial for enabling streams of data with diverse disciplinary attributes in order to enhance the analytical systems' insight generation capability (Prasanna et al., 2016). The most significant advancement in real-time processing paradigms is epitomized by the development of the Knowledge-infused CEP (X-CEP) model. By integrating domain-sensitive query schema along with temporal operators into CEP, this paradigm enhances analytical capability over both real-time and prolonged data streams. X-CEP addresses two primary gaps in existing models: querying based on higher-level knowledge semantics and support for cost-effective evaluation across past, real-time, and predictive timelines. The latter is particularly valuable in real applications, such as the Smart Power Grids example, where complex data interdependences must be searched for optimal performance. This new model of query is first put into action as a prototype, SCEPter. It not only validates the theoretical design but also exhibits improved execution efficiency with query optimizations of the designed high-volume data environments, thus highly optimizing the analytics workflow (Prasanna et al., 2016). Moreover, important real-time processing advances are encapsulated in the employment of Spark-MPI, which combines the functionality of the Spark data platform with the high-performance messaging interface (MPI). This integration responds to the growing demand for sophisticated processing systems that can meet the growing complexity regarding the three Vs of big data: volume, velocity, and variety. The resilient distributed datasets (RDDs) offered by Spark enable a decoupling of algorithmic processing from data sources, thereby broadening the spectrum of data-intensive analytics from basic SQL queries to more sophisticated machine learning operations and graph processing. The extension of this space by Spark-MPI is a revolutionary step towards real-time analysis, particularly in research configurations in experimental contexts such as ptychographic and tomographic reconstruction pipelines. This versatile method underscores the importance of robust and malleable frameworks in handling contemporary data challenges (Figure 6.4).

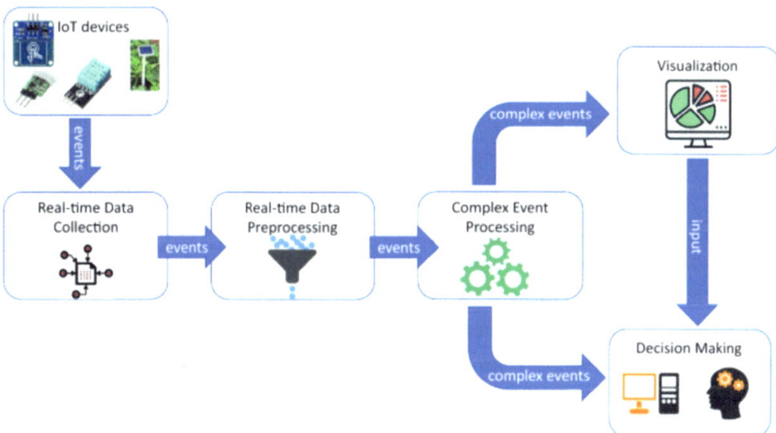

Figure 6.4. Real-time processing.

Source: https://media.springernature.com/lw685/springer-static/image/chp%3A10.1007%2F978-3-030-71069-9_11/MediaObjects/503914_1_En_11_Fig1_HTML.png.

6.1.3. Online Processing

In data analysis and processing, online processing is a crucial method of processing and handling huge amounts of data in real-time. Unlike batch processing, which operates on accumulated data, online processing allows instant input, processing, and output of the data with little delay. This is particularly important for applications requiring instant decision-making according to recent data, such as financial transactions, customer support, and real-time fraud prevention. Organizations can enhance their responsiveness to shifting data patterns and end-user requirements through online processing, with a further competitive advantage in dynamic situations. One of the inherent strengths of online processing is its ability to support intricate model calibration, particularly when dealing with Big Data applications. Data scientists find model calibration to be a difficult task, and they have to find the optimal model parameters from an enormous universe of configurations. Techniques such as speculative parameter testing and online aggregation have been developed to make this easier so that several configurations are tested together, with their performance being judged incrementally. Online processing in this capacity is essential because it not only enhances the process efficiency of the calibration but also enables dynamic learning from data distributions, further improving model accuracy and performance under real-world usage (Qin et al., 2014). Furthermore, inclusion of processing via the internet in data analysis extends to the field of accounting information systems (AIS), where inclusion of new technology is key to reconfiguring the curriculum to accommodate new technologies. Changing to the incorporation of big data and data analysis in AIS accommodates rising recognition of the need for real-time processing of financial data. As accounting evolved with the implementation of enterprise-wide systems and IT audits, online processing has become essential to process transactions efficiently and check for fraud. Emphasis on instantaneous availability of information and analytics not only revolutionizes the training model but also equips the next generation of accountants with the skills to mitigate the complexity that has arisen due to technology ramp-up and data proliferation (Larkin et al., 2020) (Figure 6.5).

Figure 6.5. The online data processing services.

Source: https://5.imimg.com/data5/SELLER/Default/2023/1/OH/PW/DA/90995205/online-data-processing-services-500x500.jpg.

6.2. DATA COLLECTION METHODS

Data collection methods are one part of the overall analytics and data handling framework, the first steps necessary for organizations to convert raw information into actionable intelligence. Various methods exist, including surveys, interviews, observational research, and automated data collection, each applicable to specific research queries or business objectives. The application of mixed-method approaches has the tendency to render data collected fuller by maximizing qualitative and quantitative data. However, selection of methods must also consider the ethical approach because the collection of personal data brings great risks when privacy and confidentiality are concerned. The overlap between the achievement of quality data and respect for individuals' rights is always a gigantic concern, hence the need to embrace stringent data collection procedures. With the evolving digital landscape, the necessity for advanced means of collecting data is increasingly apparent. To illustrate, the advent of Internet of Things (IoT) technology allows for the collection of data in real-time from a broad range of sources such as machinery and sensors monitoring activity and behavior.

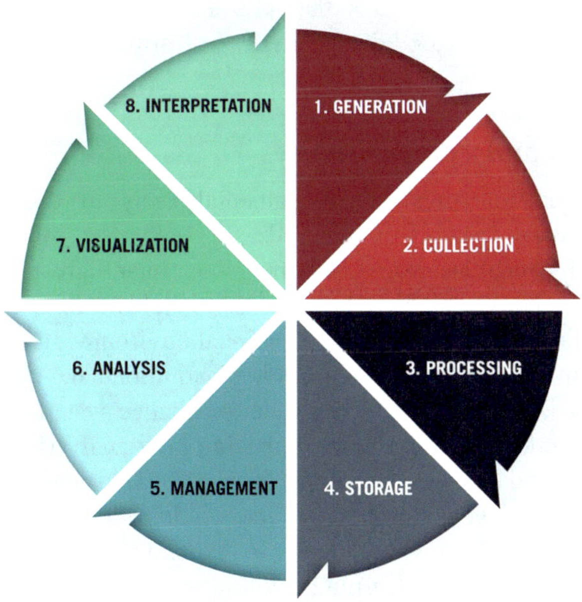

Figure 6.6. Data collection methods.

Source: https://online.hbs.edu/PublishingImages/HBS-Cycle-Chart-v2.jpg.

This advancement is a movement away from laborious and often time-intensive means of collecting data toward automated systems for data capable of handling vast quantities of information with unprecedented speed. The shift is especially relevant in industries such as utilities, where massive data collection both creates opportunities and challenges. Data deluge tends to lead to rich data and weak information, thus making innovative analytical techniques indispensable to distill meaningful conclusions from the torrents of data produced in day-to-day activities (Koronis et al., 2018). In educational institutions, learning analytics is another key aspect of data gathering structures, which seek to improve students' performance by monitoring learners' degrees of engagement and accomplishment. However, the practice has potential ethical concerns, most notably the unintended disclosure of individual details during data processing. To counter such risks, researchers advocate for the use of de-identification techniques that make it possible to analyze learning activities without compromising the privacy of students (Ebner et al., 2016). By anonymizing data, learning institutions are able to enhance analytic capability while safeguarding sensitive data, and thus facilitate a more ethical application of data-gathering tools. This dual focus on efficiency and ethics validates the necessity of thoughtful measures that preserve individual privacy without sacrificing the benefits of good data analytics in learning environments (Figure 6.6).

6.2.1. Surveys and Questionnaires

Surveys and questionnaires are fundamental tools in data analysis and processing, offering systematic methods for gathering information. Their systematic design enables researchers to gather quantitative and qualitative information effectively, which can be analyzed to yield meaningful conclusions. This approach is most important in educational environments, where gauging student engagement is important in informing institutional development. Recent research emphasizes the ability of Big Data to enhance systems for monitoring and analyzing students' engagement, proposing how analysis tools encourage decision-making and enhance students' experiences (Cao et al., 2013). Through questionnaires and surveys, teachers are more able to identify trends, monitor program outcomes, and act with planning to respond to students' requirements, thus creating a data culture in learning environments. Survey and questionnaire design must consider the very purposes for which they are to be used, and this involves careful question planning, order, and general design. Open-ended questions can provide qualitative richness, while closed-ended questions are more appropriate for quantitative measurement. Iterative refinement renders the tools more reliable and valid as gathered data become representative of the targeted population's view. Furthermore, the use of technology, for instance, web-based survey tools, has significantly enhanced the data-gathering processes and facilitated increased accessibility to various populations (Austen et al., 2019). Such technologies have transformed traditional methodologies, making it possible to achieve increased participation and instant analysis of data, with the efficiency of the research process as a whole enhanced. Finally, the implementation of surveys and questionnaires in data processing systems is meant to emphasize the necessity of engaging stakeholders actively in research. As emphasized through education programs, empowering students to engage with data not only enhances a feeling of control but also results in more holistic, advanced interpretations of findings (Austen et al., 2019). By engaging students as active subjects and not passive ones, institutions are able to build a more inclusive research culture that accounts for diverse voices. This supports the wider purpose of leveraging data to drive better educational outcomes, so insights obtained through surveys and questionnaires inform actionable recommendations that genuinely capture student needs and ambitions (Figure 6.7).

Figure 6.7. Surveys and questionnaires.

Source: https://cdn.prod.website-files.com/64022de562115a8189fe542a/6403ef254011213ade81b20b_The%20Difference%20Between%20Survey%20and%20Questionnaire.jpg.

6.2.2. Web Scraping

Web scraping has emerged as an important technique in data analysis and processing through which researchers and businesses can collect enormous amounts of data from websites at will. The technique consists of extracting structured data from unstructured web pages and transforming it into a usable form that can be analyzed. It is more in focus now because it offers the ability to scrape current data, helping institutions observe trends, monitor consumer behavior, and gain competitive intelligence. Being a methodology, web scraping poses unique challenges and possibilities; it not only requires technical expertise but also ethical use of data and user privacy considerations. These considerations are significant in deciding that web scraping processes significantly contribute to the fields of social research and market research while adhering to legal requirements and ethical guidelines. Integration of web scraping as a component of social research goes a long way in asserting its potential to revolutionize traditional data gathering paradigms. With the utilization of this instrument, scholars can utilize live data streams to power more dynamic and timely exploration of emergent issues relating to contemporary society. For instance, social media data scraping on websites like Twitter can provide an insight into the views of the masses on significant issues, e.g., economic policies.

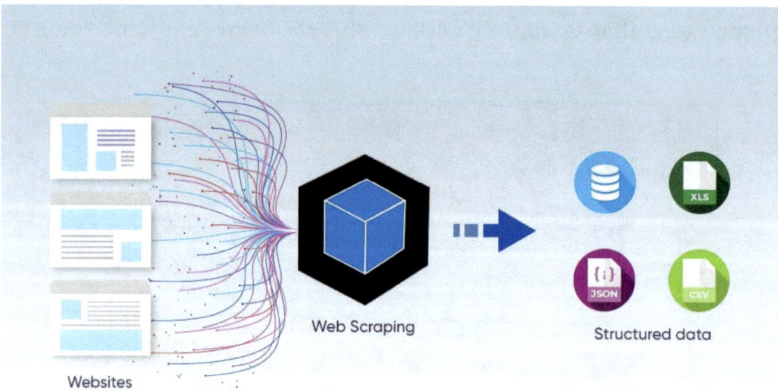

Figure 6.8. Web scraping has emerged as a pivotal technique in the realm of data processing and analytics.

Source: https://blog.skyvern.com/content/images/2024/12/Creative-1.jpg.

Not only does it facilitate the formation of access to vast amounts of information but also enables researchers to access pre-established structures and categorizations which are already present in digital environments. As proposed in recent research, advantages of employing scraped data include the capacity to analyze ongoing conversations and trends, creating a stronger context for understanding social events and emerging issues, as suggested by Back et al. (2013). Moreover, web scraping implications transcend academic research; they significantly affect business analytics and user-centric applications. With the immense demand for tailored job search experience, for example, companies are now spending more on advanced web scraping methods to enhance their websites. Scraping job postings information and studying significant attributes, companies can tailor search engines to deliver better outcomes to job seekers and employers. This data-driven approach promotes more, and more interlinked, interactions among recruiters and applicants and enables better matches on the basis of specialized skill sets and company features. The structure outlined in current studies outlines a new data-driven jobs search engine that exploits such scraping techniques to automate job searching processes, pointing toward a tendency for more customized user experiences within the data space, as highlighted in Gao et al. (2017) (Figure 6.8).

6.2.3. Sensor Data Collection

The rise of sensor technology brought with it a new age in data gathering, particularly in the context of the Internet of Things (IoT). With hundreds of sensor devices, organizations can gather enormous amounts of data sets that contain in-depth information on phenomena. This has proven to be incredibly beneficial in fields such as healthcare, manufacturing, and urban planning, where there is a constant need for environmental and operational condition monitoring. For instance, wearables sensors are able to capture real-time health information, making possible timely intervention and enhancing patient treatment. With the omnipresence of the sensors, the data generated doubles exponentially, with consequential data processing and analytical techniques becoming paramount to extract meaningful information from such massive data. Real-time analytics become a critical capability within the field of sensor data collection, by means of which organizations are able to rapidly detect and respond to events while they are occurring. The ability to analyze real-time streaming sensor data enables improved understanding of normal and abnormal patterns, which is critical for predictive maintenance and anomaly detection use cases. Advanced machine learning algorithms, like the ones used to detect falls from wearable sensors, are a demonstration of the capability of real-time analytics. By leveraging next-generation technologies like deep neural networks, analysts can provide high precision rates for identifying primary events. This capability facilitates improved decision-making procedures and supports automatic alerting systems, transforming sensor information into a precious asset for organizations that want to refine their activities and improve safety outcomes (Isah et al., 2018). While challenges posed by handling and processing huge volumes of sensor data are numerous, they also present great opportunities for innovation. The Big Data concept emphasizes the capacity to gain useful insights from the large and diverse data generated through the use of sensor technology. With increasingly more companies applying data-driven decision-making, sophisticated analysis frameworks are now necessary. An example of this is a system based on the cloud, which can offer real-time processing of data and, with it, improve analytics and data management systems' effectiveness. By processing raw sensor data to uncover hidden value, organizations can better facilitate not only operational improvement but also strategic vision. Improvements allow for a new generation of data analytics tapping the power of sensor data to drive decision and innovation across markets (Amen et al., 2015) (Figure 6.9).

Figure 6.9. The surge in sensor technology has revolutionized data collection.

Source: https://www.advancedtech.com/wp-content/uploads/2024/01/Reliability-360-Engineer_Industrial-Technology_939x604-1200x675.jpg.

6.3. DATA CLEANING AND PREPARATION

Data cleaning and preparation are the foundation of the data processing and analytics process, having a direct effect on the quality of insights derived from data sets. The raw data collected from data sources are prone to errors, inconsistencies, and duplications, which can have a significant impact on analysis. By identifying and repairing these errors, organizations are capable of improving data integrity and credibility. This exercise is particularly important in commercial environments, where the creation of decisions on the basis of erroneous data has hard financial implications. As the data quality management workshop highlighted, making this kind of awareness a prerequisite for the implementation of a data-analytical way of thinking to organizational work is crucial (Wilbik et al., 2018). Thus, a formal approach to data cleaning not only enhances the quality of datasets but also promotes culture of data-driven decision-making. Secondly, the tremendous volume of data currently being generated present unique challenges that need rigorous cleaning and preparation. With 90% of data available that has been created in the past two years, very few of them are actually processed through analysis (CIACCIO et al., 2016). This means that much of the data collected is not being realized due to inadequate cleaning processes that withhold detailed analysis. Pre-positioning data in its clean state prior to working on it can unlock its actual value potential; calling it the oil of the 21st century signifies its value in driving insights into various domains.

Figure 6.10. Data cleaning and preparation serve as foundational steps in the data processing and analytics pipeline.

Source: https://miro.medium.com/v2/resize:fit:1400/1*dkXUqK368gwrvL3Rqdnr tQ.png.

Cleaned data enhances analytical capacity to identify trends, predict analytics, and informed business decision-making, ultimately transforming the way organizations operate within competitive landscapes. Finally, the process applied in cleaning and preparing data must have a tailored strategy, depending on the nature of the dataset and the intended analytical objectives. Processes such as removing duplicates, standardization of data formats, and handling missing values are crucial to preparing data for analysis. Further, participation of domain experts at this phase can provide additional information to specific data quality issues in the context of industry standards. An organized data preparation process also involves rigorous test processes to ensure that the cleaned datasets meet quality thresholds determined beforehand. This aligned purpose not only enables accuracy in analytical outcomes but also sets a solid platform for subsequent data governance procedures that can adapt to future data concerns, as suggested during the eSCF workshop discussions (Wilbik et al., 2018) (Figure 6.10).

6.3.1. Identifying and Handling Missing Data

Missing data and their handling represent a crucial element in data processing and analysis, particularly with respect to its strong impact on the integrity of research results. Missing data may be due to various sources like human errors, technical malfunctioning, or even flaws in data collection mechanisms. For each case of missing data, there is uncertainty, eventually impairing analytical findings and hampering decision-making. Thus, strict identification methods, such as exploratory data analysis, are needed to identify the patterns and mechanisms of the absences. According to challenges in big data analytics, useful information recovered from large datasets requires comprehensive insight into missing data to maintain analyses accurate and reliable, yielding results in conformity with the objectives of asset management operations in various industries (Koronis et al., 2018).

Figure 6.11. Identifying and handling missing data.

Source: https://dataladder.com/wp-content/uploads/2020/10/7.jpg.

Missing data requires a multi-faceted approach that is tailored based on its nature and volume in a dataset. Some protocols that are followed include imputation procedures, where an estimate is made for missing data based on data that does exist, and deletion procedures, where incomplete records are removed from analysis. Each of the procedures has advantages and disadvantages, and the need for careful thought before application arises. Imputation, for instance, can maintain useful information but may lead to bias if not carried out correctly. As has been argued in the case of frameworks to handle data quality, greater consciousness of uncertainty due to missing data can lead organizations to employ more effective methods for handling such situations (Wilbik et al., 2018). Thus, the decision regarding the technique is based on an accurate knowledge of the specific situation under which the missing data arises and how it will impact the analysis. Finally, incorporating solid mechanisms for detecting and handling missing data into the overall data processing and analytics framework is key to furthering accurate and actionable insights. Organizations, in all industries, must deal with massive amounts of data while attempting to keep data integrity and quality intact. With maturing big data and advanced analytics technologies, an active approach towards handling missing data can maximize data-driven decision-making efficiency. This active approach goes beyond routine data handling operations and also guides long-term data governance and usage strategies. Thus, effective handling of missing data not only enhances the quality of analysis now but also lays the groundwork for smarter, data-driven activities in the future (Koronis et al., 2018) (Figure 6.11).

6.3.2. Data Transformation Techniques

Data transformation techniques are a significant part of data processing and analysis because they are responsible for preparing raw data for analysis. These processes involve converting data from its original form to a structured and useful form that permits intelligent analysis. It is not merely a type conversion; it includes enhancing data quality by cleansing, normalizing, and aggregating. These processes ensure the data are accurate, consistent, and reliable, which is essential in decision-making. Emphasis on change has grown with the latest developments in information technology. As brought out in the study of accounting information systems, curriculum reform has encompassed significant topics such as big data and data analytics, referencing existing trends in data transformation approaches and their impacts on accounting practice and education (Larkin et al., 2020). Also, data transformation techniques lead to more sophisticated methods such as dimensionality reduction and feature engineering. Dimensionality reduction reduces the variables accounted for, simplifying models and removing opportunities for overfitting. Feature engineering, however, involves creating new features from available data for improved model performance. These conversions allow analysts to derive more significant patterns from big data, which can be particularly important in industries such as finance and healthcare where the insights gained from data can contribute to improved decision-making and process efficiencies. Data manipulation and conversion capacities determine the success of analytical outputs, and thus the increasing need for professionals who possess strong data conversion abilities (Anderson et al., 2019). With more reliance on data-driven approaches, the choice of effective data transformation techniques is central to deciding the outcome of analytics projects. Combining hierarchical models, such as cascades, yields a good model for query times reduction and efficient handling of large volumes of data. Through techniques such as resolution scaling and color-depth reduction, which are applied to such models, analysts are able to optimize content extraction workflows. This not only reduces processing time but also accommodates the processing of large datasets, paving the way for real-time analytics applications. The transformation of data transformation practices and technology represents a pivotal shift in how enterprises leverage data, bringing into prominence the urgency to constantly adapt in light of shifting analytical environments (Anderson et al., 2019).

6.3.3. Importance of Data Quality

The need for quality data in data processing and analysis cannot be emphasized enough, particularly with growing reliance on data-driven decisions from organizations. Quality data provides the foundation for good insights and well-educated strategies, directly impacting an organizations performance and operational efficiency. When data integrity is compromised through inaccuracies, incompleteness, or inconsistency, consequences are catastrophic. These shortcomings will lead to flawed analyses, suboptimal decision-making, and ultimately impair the organizations ability to compete effectively in the marketplace. As determined through academic literature, ensuring data quality and data management processes are of prime importance for leveraging the full potential of analytics and big data methods in numerous industries, including corporate and government environments (Guirguis et al., 2020). Furthermore, the association of data quality with organizational capabilities emphasizes the need for an up-to-date data governance framework. Data governance is a good means of possessing policies, procedures, and technologies to ensure that data is accurate, consistent, and accessible across organizational silos. When organizations prioritize data quality, they improve decision-making processes as well as maximize overall performance in line with strategic objectives predetermined. Research has shown that there is a high correlation between good data management and improved decision-making capacity, complementing data quality as the top driver to organizational achievement (Guirguis et al., 2020). Based on such evidence, organizations must invest in systems and training that bring about a culture of data, thereby empowering employees across levels to appreciate data quality. In addition to enhancing performance and decision-making, data quality plays an important role in addressing contemporary issues presented by technology growth at a rapid pace, particularly with regard to smart city planning and big data use. The combination of computational intelligence techniques with reliable data analytics is the keystone in transforming urban areas into smart cities capable of addressing multidimensional socioeconomic issues, such as transportation, utilization of resources, and sustainable living.

Effective data quality ensures that findings derived from data inform policy decision-making and operational interventions in these smart cities. While cities increasingly use data infrastructures as a resource to solve problems, the implications of data quality extend from organizational performance to wider societal development and the effectiveness of city living. Consequently, prioritizing data quality would seem to be an optimal theme in the debate about data processing and analysis (Figure 6.12).

Figure 6.12. The significance of data quality within the domain of data processing and analytics.

Source: https://cdn.prod.website-files.com/6108e07db6795265f203a636/6286b502f13e7755c6dcf5eb_What%20is%20Data%20Quality_.jpeg.

6.4. DATA ANALYSIS TECHNIQUES

Data analysis techniques play a key role in transforming raw data into meaningful information, thereby enabling informed decision-making in many areas. A key technique is visual analytics, an approach that blends computational algorithms with interactive visual interfaces. The technique allows analysts to sift through large datasets and detect trends, patterns, and relationships that might be missed under normal analysis. Visual analytics, not only enhancing comprehension of complex data structures, but also facilitating collaborative discovery, whereby various stakeholders are able to access the data in parallel, has been highlighted through recent advances in visual analytics techniques as playing an important role within the framework of movement data analysis, where visualized data is able to play an important role towards comprehending the moving dynamics in varied scenarios, as highlighted in Andrienko et al. (2012). Apart from visual analytics, machine learning has emerged as a central skill in data analysis, particularly in the management of real-time data streams. By bringing together supervised, unsupervised, and reinforcement learning algorithms, analysts are able to discover insights from big data sets more efficiently and effectively. By using predictive analytics techniques, organizations are able to forecast future trends from historical data so that they can act ahead of time rather than in reaction.

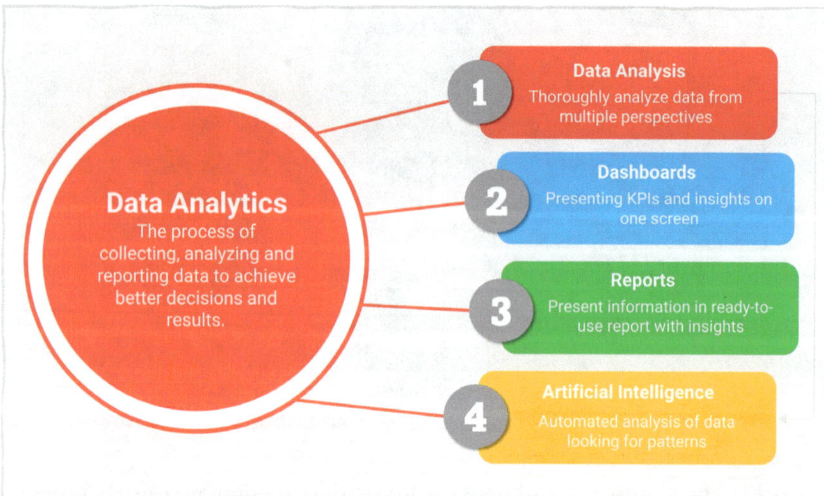

Figure 6.13. Data analysis techniques.

Source: https://www.passionned.com/wp/wp-content/uploads/data-analysis-sectors.png?v=0827.

Furthermore, diagnostic analytics provides the appropriate frameworks to identify anomalies and deviant patterns in data, enhancing the analytical level and facilitating on-time effective interventions. The capability of machine learning methods to process and examine complex datasets has proven to be worth its weight in gold in several sectors including health, finance, and public policy, as discussed in Abualigah et al. (2025). Additionally, the unique variations of analytics—descriptive, diagnostic, predictive, and prescriptive—

cumulatively provide an end-to-end understanding of information. Descriptive analytics is the backbone in that it provides an overview of the past information, while diagnostic analytics goes further to identify why something happened in the past. Predictive analytics projects what will happen in the future based on patterns, making a connection between the past behavior and future expectations. Finally, prescriptive analytics not only recommends action based on predictive forecasts but also applies optimization techniques in order to present the most optimal solution. This systematic approach to data analysis presents firms with a strong arsenal of tools with which to meet the complexities of today's data landscapes, with the assurance that decision makers will be positioned to better meet challenges with confidence and precision. Through the application of these diverse data analysis techniques, organizations can optimize their planning for strategy and their operational effectiveness, and therefore create a competitive edge (Figure 6.13).

6.4.1. Descriptive Analytics

Descriptive analytics is a foundation component in the broader range of data processing and analysis that allows organizations to analyze historical data and learn from past performance. By applying various statistical techniques, descriptive analytics allows companies to summarize large data sets into useful aggregates that point out industry patterns, trends, and anomalies over a period. Such information is paramount to decision-makers as it paves the way for insights into business competence and consumer behavior. Moreover, companies that prioritize the use of descriptive analytics position themselves for achieving data-driven insights that propel organizational performance. As elaborated in current research, effective use of business analytics can significantly differentiate top-performing companies from competitors, reiterating the imperative of building an effective analytical approach and organizational practices that ensure data-driven decision-making within the business environment (Cao et al., 2017). The integration of descriptive analytics within a firm's decision-making process is based on various factors like the technological landscape and data culture in the organization. In managing the complexities of big data, firms would have the need for advanced analytics frameworks. It is here that descriptive analytics is of use, allowing companies not just to consolidate data but to analyze it in a systematic way to identify major insights to guide business strategy. For example, industries seeking mass customization and optimizing production processes are highly reliant on predictive and descriptive analyses to anticipate customer demands and refine supply chain dynamics. The challenge, however, is implementing effective smart analytics tools that would be capable of unleashing big data potential while facilitating a seamless integration into existing processes, as noted in current research spanning the synergy of analytics, cloud computing, and industrial applications (Chen et al., 2016). In addition, descriptive analytics has gone beyond the general data summarization; it has far-reaching implications in predicting the future directions and influence on strategic initiatives. Through the systematic classification of past data, organizations are able to prepare reports that point out major performance indicators necessary for ascertaining accomplishment. Such empirical evidence prepares the ground for knowledge-based decision-making, allowing organizations to make proactive adjustments to strategies. Additionally, as companies invest in enhancing their analytical abilities, they must consider how descriptive analytics can be combined with other types of analytics, including predictive analytics, to better understand the dynamics of the market. Lastly, descriptive analytics not only provides insight into historical performance but also stands as a key stepping stone on the road to achieving more mature, unified means of processing data that results in continued improvement and competitive market advantage within an increasingly data-reliant world (Cao et al., 2017) (Figure 6.14).

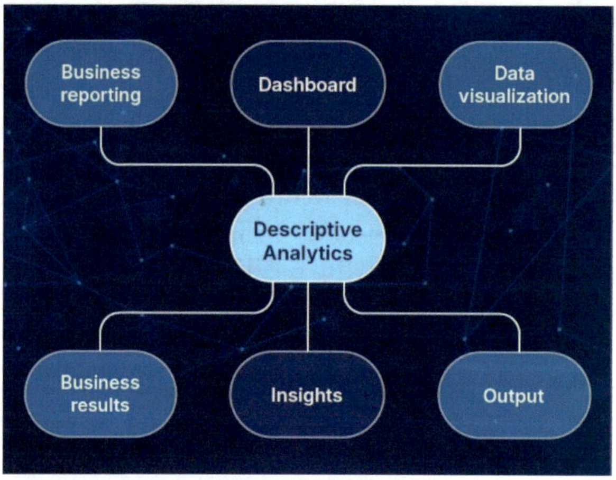

Figure 6.14. Descriptive analytics.

Source: https://www.softwebsolutions.com/wp-content/uploads/2023/09/What-is-descriptive-analytics-1200x900.webp.

6.4.2. Predictive Analytics

In the field of data analysis and data processing, predictive analytics is an innovative tool that helps organizations make data-driven decisions. Employing statistical algorithms and machine learning techniques, predictive analytics analyzes historical data to forecast future events and behavior patterns. The process engages vast amounts of data, commonly termed Big Data, that facilitates the generation of actionable insights. It not only aids businesses in anticipating customer demand but also enhances operational efficiency through the identification of potential market trends. As predictive models become more accurate, accuracy and relevance are made a standard from finance to medicine, underlining the need to have robust data processing systems in place to support such advanced analysis. The applicability of predictive analytics also extends to city planning, where the application of effective utilization of data can enhance the management of cities. Urban planners and policymakers can leverage predictive analytics through the strength of Big Data to address essential issues in traffic congestion and environmental sustainability. Emerging technologies like the microservice-based architecture underlined in recent studies allow real-time context-aware responses with predictive streaming processing of data. Such a structure not only enhances the reliability of predictions but also solves complicated event processing concepts, allowing cities to respond dynamically to various conditions. Developing such frameworks proves the strength of big data and predictive analytics in creating resilient and intelligent urban environments, ultimately leading to citizens' improved quality of life (Puig et al., 2019). Also, the union of predictive analytics and computational intelligence gives rise to new applications tailored towards addressing actual issues, particularly for smart city initiatives. As demonstrated through research into intelligent transport systems, predictive analytics can enhance traffic safety and flow, demonstrating the real-world applications of such advanced analytical approaches. Deployment of Big Data, however, comes with inherent challenges like privacy, commercialization, and policy-making issues. Consequently, the formulation of total strategies that encompass data protection and proper use is critical to ensure the advantages of predictive analytics are achieved without compromising citizen trust. As predictive analytics develops in the general background of data processing and analytics, these issues will need to be addressed to unlock its full potential (Doctor et al., 2020).

6.4.3. Prescriptive Analytics

Prescriptive analytics is a strong development in the processing and analytics space, where information provides organizations specific suggestions that inform decisions. It's distinct from basic descriptive or predictive models in that it provides instead recommendations based on data. By using advanced algorithms and optimization methods, prescriptive analytics examines potential outcomes of various actions and thereby enables organizations to optimize strategy across a variety of activities, from supply chain management to customer engagement. According to research, companies that can effectively make the most out of prescriptive analytics perform better than their competitors, which proves the importance of having a well-designed analytical strategy that integrates such capabilities as a part of day-to-day business operations (Cao et al., 2017). Prescriptive analytics involves a formalized approach of processing data with emphasis on data quality and usefulness of data. It is comparable to standard frameworks such as SEMMA and CRISP-DM, which operate data-driven projects by providing formalized phases of sampling data, exploratory data, and modeling. Prescriptive analytics is faced with particular challenges in making accurate business cases and preserving data granularity consistent with organizational needs. These challenges are significant to address because they may impede the efficacy of analytical knowledge and the effective deployment of prescriptive recommendations. Moreover, organizations must create the right skill sets among personnel to interpret and implement such advanced analytics solutions effectively (Ivanov et al., 2017). While creating prescriptive analytics, organizations must also consider the operational and strategic implications of their data initiatives. As outlined, developing an end-to-end analytics platform is not merely a matter of technological advancement; it also entails balancing business goals and analytical capabilities. Such a holistic approach enables organizations to manage the complexity involved with big data projects better, such as storage limits and security concerns for data. Moreover, while organizations incorporate prescriptive analytics into decision-making, continuous deployment of models becomes essential in order to maintain pace with evolving market situations. By incorporating prescriptive analytics, business firms are likely to make key decisions that drive long-term business growth and company success, thus reflecting the significance of analytics in data business (Cao et al., 2017).

6.5. TOOLS AND TECHNOLOGIES FOR DATA PROCESSING

Data processing is greatly impacted by various tools and technologies designed to process lots of information efficiently. Within the field of big data analysis, these tools fall into three broad categories: storage, processing, and visualization. Hadoop and Apache Spark are examples of devices that have emerged as front-runners in processing, whereby companies can deal with big data by decomposing computational processes into little pieces and presenting them to groups of machines for execution. In addition, technological developments in cloud have transformed the storage capacity to include scalable and elastic solutions that are essential for the handling of increased data requirements. By integrating results of current research, it is easy to understand the sophisticated functionalities these tools utilize to improve data analytics and current developments that keep shaping this practice (Roy et al., 2019). Besides, including security measures in data processing programs becomes increasingly important as organizations utilize cloud computing solutions for storing secret data. With the transition towards cloud technology, there has been an increase in concerns regarding data exposure, necessitating stronger security practices. New methods, such as advancements in cryptography methods, are imperative in their function to ensure data safety on non-trustworthy servers. The proposed Computing on Masked Data (CMD) tool is a good example of the trend since it harmoniously unites database technologies and cryptography, enabling mathematical calculation to be performed securely with the cloud platform offloading time-consuming calculation. Such efforts not only promote data integrity but also deliver effortless analytics to diverse sets of users, ultimately putting organizations in the best position possible for data-based decision making in the data age (Gadepally et al., 2015). In addition to security and processing features, visualization tools are a critical component of the data processing environment. They transform complex datasets into comprehensible formats, allowing stakeholders to derive effective actionable insights. Tools such as Tableau and Power BI have gained traction due to their ability to transform raw data into visual narratives that enhance understanding and engagement. Adequate visualization not only facilitates the communication of findings to technical stakeholders but is also useful for exploratory data analysis as it reveals trends, patterns, and relationships that might otherwise remain obscure. The close integration of visualization tools with the broader system of data processing is a sign of an integrated analytics strategy in which disparate components work together to generate insightful conclusions and inform strategic pursuits (Roy et al., 2019).

6.5.1. Software Solutions (e.g., Excel, R, and Python)

In today's data-driven world, software solutions such as Excel, R, and Python have emerged as essential enablers for data processing and analysis. Excel, widely used in academia and professional setups, serves as an introduction to data analysis with its user-friendly interface and versatile capabilities. Excel can be used by users for anything from simple data input to complex statistical analysis with in-built functionality, pivot tables, and add-ins. Merging the software with multiple data sources makes it even more handy, allowing analysts to manipulate and inspect data efficiently. In addition, the common knowledge of Excel across various professional groups makes it a basic competence for future data analysts and thus a necessary tool in data processing and analytics. Although Excel provides a solid basis, more advanced software packages such as R and Python are increasingly used to manage high-level data sets and apply advanced statistical techniques. R, in particular, is intended for statistical computing and has a lavish set of packages that contribute to its strength in data visualization and special analyses. The area is conducive to the increasing need for analysts who can efficiently operate within large sets of data. According to research, the increasing sophistication of data analytics work calls for experience with tools such as R and Python, where features such as machine learning, data mining, and advanced statistical modeling can be effectively leveraged (Dou et al., 2017). Hence, the use of these programming languages in the toolkit of data analysts must be undertaken to achieve better data-driven outcomes. Applicability of software to data processing transcends technical ability, as their utilization can have direct impacts on industry outcomes. R and Python, for instance, are reputed to be efficient and precise for data analytics, a fact substantiated by research confirming correlations between software proficiency and improved performance measures (Coto et al., 2016). Besides, the tools also enable data analysts to prototype and deploy data-driven applications to impact decision-making across industries. As pressures rise for organizations to use the power of data for competitive advantage, the curriculum in education must also become more focused on the skills employers demand. By developing a solid understanding of data processing tools such as Excel, R, and Python, schools can prepare students to cope with the shifting trends of data processing and analytics.

6.5.2. Database Management Systems (e.g., SQL and NoSQL)

The advent of Database Management Systems (DBMS) has revolutionized the way organizations store and process data, particularly in Data Processing and Analytics. Amongst them, SQL (Structured Query Language) databases have formed the foundation, offering structured models for data storage, organization, and retrieval. SQL databases are best suited for scenarios where data relational schema and integrity are the priority, and therefore, suited for transactional applications that require precise data manipulation. They offer a relational model with strict schemas, enabling complex querying and data relationships. However, with the emergence of big data, traditional SQL systems are not able to perform and scale up well, especially in the case of unstructured data types. Thus, the new paradigm of data necessitates the implementation of alternate solutions. As a result of limitations in SQL databases, NoSQL (Not Only SQL) systems have gained immense popularity, particularly for applications involving large data. Unlike their relational equivalent, NoSQL databases are designed to be highly performant and scalable with flexible schemas and tailored data access patterns. Their flexibility allows them to optimize both structured and unstructured data and hence is a viable option for heterogeneous applications.

As is identified in current studies, the transition to NoSQL is precipitated by the growing trend of big data, generally characterized by their size, velocity, and variety. NoSQL databases such as MongoDB, Cassandra, and the graph database Neo4j are best positioned to manage big data from systems such as the Internet of Things (IoT), which produces vast amounts of real-time data from sensor networks (Küçükkeçeci et al., 2017). Processing and querying big data produced by multimedia sensor networks need advanced database mechanisms that ensure efficient querying and retrieval of meaningful information. One of the interesting approaches is to utilize the capabilities of both SQL and NoSQL systems, creating a hybrid setup to ensure best performance and analytics. Combining conventional relational schemes with NoSQL databases is able to make them more efficient, especially when handling analytical loads (Abelló Gamazo et al., 2016). This approach allows companies to harness the power of built-in structured querying in SQL while also gaining from the performance benefits offered by NoSQL databases. These advances are necessary because the need for real-time data analysis becomes absolutely crucial across a number of fields, ranging from smart cities to surveillance systems, to stretch the boundaries of what database administration can achieve in the field of data processing and analysis.

6.5.3. Cloud Computing Platforms

Cloud computing platforms have contributed immensely to transforming the landscape of data processing and analytics. Cloud services provide elastic resources that are custom-designed to cater to the growing demands of data-driven entities, thereby facilitating large datasets' collection, storage, and computation. One of the prominent aspects of cloud services is their ability to harness distributed computing for computational efficiency. Incorporation of intelligent, automated resource management techniques is core in achieving maximum utilization of such systems. In line with Buyya et al. (2012), effective management techniques, particularly autonomic provisioning of resources, play a significant role in providing service dependability alongside cost-orientation in Software as a Service (SaaS) implementations hosted on cloud infrastructures. Therefore, the confluence of innovative management practices and analytics capabilities is vital to organizations that aim to make their data processing process lean. Cloud infrastructure dynamics are constantly changing, a stark departure from traditional models contained in single-provider data centers. New-generation cloud platforms are gravitating toward decentralized architectures that aggregate resources from various providers. This paradigm supports more flexibility and resource availability, which is crucial in data-intensive computing environments where processing and analysis of data should be highly interactive.

As described in Buyya et al. (2017), this revolution demands new architectures of computing that are designed to accommodate the emerging issues of connectivity, service delivery, and autonomous learning systems. Shattering the traditional infrastructure frameworks, organizations have managed to leverage cloud computing platforms for the creation of sophisticated analytics solutions that enable the diverse needs of today's data processing. Ultimately, the union of cloud computing platforms and data analytics is transforming the way firms leverage data for business advantages. Dynamic resource scaling combined with advanced management tactics enables organizations to derive insights from big datasets while enabling operational efficiency. As cloud providers drive innovation in this field, it is probable that attention will remain focused on enhancing cloud service interoperability and intelligence. Such development not only addresses present requirement but actually looks to the future needs of data analytics, adopting a forward-looking approach to information handling. The evolution discussed by Buyya et al. (2012, 2017) illuminates the trajectory of cloud computing as a crucial component for businesses to maintain pace with the challenges of staying competitive during an age dominated by data.

6.6. APPLICATIONS OF DATA ANALYTICS

Data analytics has emerged as an invaluable resource in a majority of industries, resulting in a paradigm shift in how businesses manage and utilize information. Within business environments, analytics is at the core of supporting decision-making through enhanced understanding of customer behavior and market patterns. Take IBM Watson Analytics as an illustration of how advanced analytics platforms can provide organizations with relevant facts from large volumes of data. It not only enhances information handling capacity, but also enables improved customer service, ultimately resulting in informed decision-making that brings about business success. The integration of such analytics systems into traditional business intelligence strategies illustrates the significant change in data-driven decision-making processes, illustrating the demand for organizations to keep pace with these innovations in order to stay competitive in the market (Jiang et al., 2017). In the public sector, data analytics is increasingly being used to assist with addressing the current urban issues, particularly in the building of smart cities. With the application of Big Data, urban planners will be well positioned to take knowledgeable decisions regarding infrastructure and the deployment of resources, hence enhancing the quality of life. The integration of computational intelligence approaches with Big Data analytics is the foundation for addressing some of the real-world problems that comprise traffic regulation and public safety. An exemplar case study is focused on smart city intelligent transportation systems, where data analytics facilitates simulation of complicated patterns of usage and mobility. Not only does this advance urban planning but also fosters a sustainable urban ecosystem through the effective utilization of resources. Therefore, the convergence of data analytics and smart city initiatives highlights the impactful role of data in determining urban environments in the modern era (Doctor et al., 2020). Moreover, data analytics uses extend beyond organization or city lines, affecting broader economic activity. As more sectors become dependent on data-driven reactions, analytics itself becomes an important component in enabling innovation and optimizing operations. For example, firms across diverse sectors utilize predictive analytics for trend projection and risk mitigation to enable them to respond before actually responding to market developments. This strategic approach is vital in maintaining competitiveness in a progressively changing economy, whereby an ability to predict and adapt can make or break organizations. The broader relevance of this reliance on data analytics supports its ability to remodel, not only to improve operational efficiency but also to drive economic growth and social progress across sectors. As a result, the integration and analysis of data analytics become vital to organizations wishing to manage the complexities of contemporary markets (Jiang et al., 2017).

6.6.1. Business Intelligence (BI)

The integration of BI into organizational systems offers a key means of harnessing vast amounts of data to feed decision-making processes. As corporations rely more and more on data analytics, BI software technologies facilitate the process of raw data being transformed into actionable intelligence. This is a process that integrates many techniques like data mining, reporting, and online analytical processing that are critical in tracking trends, optimizing operations, and predicting customer behaviors. By using BI, organizations can achieve a decision-making culture that is data driven, thereby gaining a competitive edge in dynamic markets. As argued in studies, effective utilization of BI is essential in handling sophisticated issues through enabling organizations to access complementary information with ease. This underscores the need not only to adopt BI tools but, more significantly, to embed them strategically into the operations framework for the purpose of maximizing their potential contribution to overall performance. The Business Intelligence implications extend much further than technology adoption; they are rooted in the way organizations go about building analytics capabilities. Experiences have proved that variance in performance of top- and bottom-ranked companies can be explained frequently due to their applications of Business Analytics (BA). Design alignment of BA with the organizational environment has an eminent influence on decision-making effectiveness and overall performance outcomes. Companies that prioritize evidence-based strategies and make a good analytical foundation outperform their peers. Conversely, laggards in this space have difficulty applying data to strategic ends. Thus, maintaining a culture that supports analytical ability not only facilitates immediate business objectives but also provides the basis for long-term organizational ability and resilience, as documented in contemporary research on BA practices in industries (Cao et al., 2017). Additionally, the changing profile of smart cities demonstrates Business Intelligence's city-changing capabilities in urban development and planning. The dynamics between Big Data and computational intelligence technologies highlight how BI has the potential to transform the functioning of the public sector and interact with citizens. For example, in intelligent transportation systems, BI technologies can monitor real-time traffic patterns to maximize traffic flow and safety. Through advanced analytics, cities are able to optimize their operational efficiencies, which result in better living standards for citizens. But this integration also comes with its own set of issues, including maintaining data privacy and creating proper policies to harness and secure the data properly. As research shows, collaborative deployment of BI in addressing modern urban challenges not only enhances business efficiency but also raises relevant concerns regarding commercialization and ethical use of information in modern society (Doctor et al., 2020).

6.6.2. Healthcare Analytics

The use of healthcare analytics by modern healthcare systems has transformed the utilization of data to enhance patient outcomes, enhance operational efficiency, and reduce costs. Such paradigm shift is primarily due to the vast quantities of electronic health information that are being systematically collected and processed today. With advanced analytics, healthcare professionals can extract useful information from such data, which contributes to well-informed decision-making processes. The new techniques that healthcare analytics offers span from predictive modeling, trend analysis, to personalized medicine strategies. Most significantly, analytics applies to a broad area of applications including healthcare informatics and bioinformatics, whose aggregate impact is to establish an improved and responsive healthcare system. The coming together of these subjects further highlights the key role of analytics in improving healthcare provision, thereby defining such key problems as patient safety and cost management as crucial for sustaining healthcare systems of the present time (Isewon et al., 2015). The arrival of sensor technologies and the availability of wearable devices have played an important part in the development of healthcare analytics, making it possible to monitor patients' health conditions in real-time on an ongoing basis. Remote health monitoring systems use data gathered from a variety of devices—smartwatches to advanced medical devices—to measure vital signs, identify discrepancies, and alert healthcare professionals to any issues in a timely manner. These broad health interventions enable providers to address patients earlier on, potentially preventing severe illness and hospitalization. Moreover, by enabling patients and providers to freely communicate with each other, these technologies individualize care and enhance patient activation. The capability has implications that extend beyond the care of individual patients; they also have the capacity to guide public health management by aggregating data in order to identify trends and make recommendations for policy. This alignment of analytics and technology is at the heart of developing a well-being and efficient healthcare landscape in the future (Hosseinpour et al., 2019). With developing healthcare analytics, a number of challenges and opportunities will need to be continually researched and developed. Data security, ownership, and privacy are paramount concerns, especially with patient information increasingly available in an Internet of Things and networked world. Framing strong data fusion and analysis frameworks is crucial to maintaining ethical practices while benefiting from the insights provided by big data. In addition, artificial intelligence and machine learning have the potential to transform healthcare analytics by enhancing the accuracy and specificity of predictive modeling. There has to be an interdisciplinary effort on the part of technologists, healthcare professionals, and policymakers to overcome the inherent complexities of this domain. By forging alignments and concentrating on shared goals, healthcare analytics has a shining future, with the ability to make significant inputs towards enhanced means of giving and getting care (Hosseinpour et al., 2019; Isewon et al., 2015).

6.6.3. Social Media Analytics

Growing relevance of social media analytics cannot be overstated in today's data-driven world. Social media platforms generate massive quantities of user-created content, which, when analyzed properly, provide organizations with deep insights into the behaviors and tendencies of customers. This information can inform marketing campaigns, simplify customer interactions, and enhance user experiences. Employing advanced analyzing tools and techniques, firms are capable of segmenting audiences based on engagement and sentiment, allowing for more targeted communication. Equipped with the power of social media analytics, companies can change direction in real-time according to changes in consumer behavior and optimize their overall performance in the market. Also, the role of location analytics within social media analytics is really significant. As noted in recent reports, location analytics refers to the statistical examination of geographic or spatial data related to user locations, which delivers insights into visit patterns and interest. This analysis assists in understanding various demographics better, ultimately assisting in marketing decision-making. For instance, through the analysis of social media data of travelers, marketers are able to know which sites are most visited and tailor tour packages according to the target audience. The importance of integrating location analytics into social media analysis cannot be overstated, as it helps business organizations fine-tune their outreach programs more effectively and design more efficient marketing strategies, as seen in the case of Australian outbound tourists' model (Gammack et al., 2019a, 2019b). Further, the use of new pattern data mining techniques and image processing within social media analytics is a significant breakthrough in the field. This new approach facilitates the processing of dense data content in order to derive useful insights that were previously impossible with conventional methods. As organizations employ such sophisticated methodologies, they can identify behavioral patterns and participation trends that could inform their strategy. The architecture model envisioned in the research not only improves data processing capabilities but also suggests management decision support. This adaptability in methods attests to its potentiality for use across various locations and populations, and thereby, it solidifies its position as an essential tool for data processing and analytics professionals (Gammack et al., 2019a, 2019b).

SUMMARY

In Summary, data processing and analytics is characterized by its rapid evolution, which is sustained through the rapid advancement of technologies and increasing emphasis on efficient data management. As this essay has shown, such approaches as KniMet capture the integration of data processing steps into a unified workflow. The emergence of tools that accumulate processes such as feature filtering and normalization captures effectively the necessity for practical solutions for fields such as metabolomics, where complexity operates against effective analysis (Atzori et al., 2018). These technologies not only simplify the workflow but also enhance accuracy and dependability in results, making it more critical for research and application across various fields. More importantly, implications of shifting data processing methodologies extend beyond direct technological benefits. As big data expands and increasingly permeates industries, the need for data analytics professionals becomes increasingly urgent. The findings of accounting information systems research point to a paradigm in education, as programs react by integrating topics of data analytics and enterprise systems (Larkin et al., 2020). Such changes reflect an acknowledgment of the overarching imperative to equip professionals for an environment dominated by information. Hence, comprehensive training in theoretical and practical areas of data processing is essential for budding professionals. In unifying the subject matter covered in this essay, it is clear that the evolution of data processing techniques and pedagogical frameworks are symphony forces that propel academia and industry ahead. The issues described in the various data-intensive industries signal the imperative for continually advancing methods for handling data that leverage new technologies in addition to pedagogical developments. While businesses struggle to harness the power of data analytics, it is essential that they invest not only in cutting-edge tools but also in developing a workforce proficient in utilizing these tools. In this manner, stakeholders in various industries can stay ahead in the current data-driven era, making better decisions and driving innovation.

Data processing and data analysis are among the most important aspects in processing large amounts of information today. As organizations generate and process increasing amounts of information, the efficiency tools and techniques are essential. Overall strategies focus on maximizing data acquisition, enhanced data quality, and employing sophisticated analytical techniques to derive insights. For instance, techniques like ExplainIt! have been designed to support root-cause analysis in advanced environments, ordering causal hypotheses and representing intricate structures of data. Such characteristics assist operators to identify issues of performance promptly and repair them effectively, and prove the efficiency and effectiveness of systematic data processing in operating settings (Benjamini et al., 2019). When companies continue growing, adopting robust data processing models is required for the sustenance of competitive advantages. In big data analysis, traditional methodologies such as SEMMA and CRISP-DM are the driving forces, but they are often found lacking while addressing the distinctive challenges posed by large-scale projects. Data-driven projects require a re-evaluation of the methodologies to adapt to existing technological advancements as well as organizational needs. Business analysts and data practitioners face many challenges including poorly defined project objectives, poor data quality, and scarce resources. These challenges emphasize the need for a different methodological approach that considers the character of big data environments in order to enable professionals to implement effective solutions for a host of business applications, from supply chain optimization processes to customer value enhancement (Ivanov et al., 2017). Recognizing and surmounting the challenges is critical to effective project outcomes. Furthermore, the use of data analytics in organizational structures needs to be addressed on a multi-dimensional level. Repeated model deployment, training, and implementation of best practices are important to successful deployment of data processing methods. As businesses increasingly rely on data to make strategic decisions, building an environment that fosters data literacy and cross-functional discussion is important. The new innovative methods are not only set to bridge gaps in existing models but to also prepare upcoming professionals with how they will approach the evolving nature of data analysis. Ultimately, how an organization is able to use its data depends on overcoming technical as well as business challenges, thus integrating data analytics into its operating plans (Ivanov et al., 2017).

REFERENCES

1. Abelló Gamazo, A., Herrero Otal, V., & Romero Moral, Ó. (2016). NOSQL design for analytical workloads: Variability matters. https://core.ac.uk/download/81580670.pdf (accessed on 2nd October 2025).
2. Abualigah, L. (2025). Enhancing real-time data analysis through advanced machine learning and data analytics algorithms. https://core.ac.uk/download/640243360.pdf (accessed on 2nd October 2025).
3. Amen, B., & Lu, J. (2015). Sketch of big data real-time analytics model. https://core.ac.uk/download/30732896.pdf (accessed on 2nd October 2025).
4. Anderson, M. R., Cafarella, M., Ros, G., Wenisch, T., et al. (2019). Physical representation-based predicate optimization for a visual analytics database. http://arxiv.org/abs/1806.04226 (accessed on 2nd October 2025).
5. Andrienko, G., & Andrienko, N. (2012). Visual analytics of movement: An overview of methods, tools and procedures. https://core.ac.uk/download/18294977.pdf (accessed on 2nd October 2025).
6. Atzori, L., Fjeldsted, J., Griffin, J. L., Hall, Z., et al. (2018). KniMet: A pipeline for the processing of chromatography-mass spectrometry metabolomics data. https://core.ac.uk/download/157857918.pdf (accessed on 2nd October 2025).
7. Austen, L., & Jones-Devitt, S. (2019). Guide to using evidence in higher education. https://core.ac.uk/download/323056077.pdf (accessed on 2nd October 2025).
8. Ayguadé Parra, E., Cea, D., Nin Guerrero, J., Torres Viñals, J., et al. (2014). Towards the cloudification of the social network's analytics. https://core.ac.uk/download/41777542.pdf (accessed on 2nd October 2025).
9. Back, L., Becker, H., Bollier, D., Bowker, G. C., Danowski, J. A., Duhem, P. M. M., Weltevrede, E., et al. (2013). Scraping the social? Issues in live social research. https://core.ac.uk/download/2794969.pdf (accessed on 2nd October 2025).
10. Benjamini, Y., Cohen, I., Jeyakumar, V., Pedregosa, F., Seth, A. K., Shimizu, S., Tenenbaum, J. B., et al. (2019). ExplainIt!–A declarative root-cause analysis engine for time series data (extended version). http://arxiv.org/abs/1903.08132 (accessed on 2nd October 2025).
11. Bikakis, N. (2018). Big data visualization tools. http://arxiv.org/abs/1801.08336 (accessed on 2nd October 2025).

12. Boubeta Puig, J., Caravaca Diosdado, J. A., Chávez de la O, F., García de Prado Fontela, J., et al. (2019). Real-time context-aware microservice architecture for predictive analytics and smart decision-making. https://core.ac.uk/download/288159008.pdf (accessed on 2nd October 2025).
13. Buyya, R., & Varghese, B. (2017). Next generation cloud computing: New trends and research directions. https://core.ac.uk/download/96664300.pdf (accessed on 2nd October 2025).
14. Buyya, R., Calheiros, R. N., & Li, X. (2012). Autonomic cloud computing: Open challenges and architectural elements. http://arxiv.org/abs/1209.3356 (accessed on 2nd October 2025).
15. Cao, G., & Duan, Y. (2017). How do top- and bottom-performing companies differ in using business analytics? https://core.ac.uk/download/130975349.pdf (accessed on 2nd October 2025).
16. Cao, G., Duan, Y., Ong, V. K., Woolley, M., et al. (2013). Big data in higher education: Action research on managing student engagement with business intelligence. https://core.ac.uk/download/29821720.pdf (accessed on 2nd October 2025).
17. Chaudhary, A., Cowan, M., Hanwell, M., Jourdain, S., et al. (2018). Building near-real-time processing pipelines with the Spark-MPI platform. http://arxiv.org/abs/1805.04886 (accessed on 2nd October 2025).
18. Chen, Y., Flores Saldivar, A., Goh, C. S. F., Li, Y., et al. (2016). Identifying smart design attributes for Industry 4.0 customization using a clustering genetic algorithm. https://core.ac.uk/download/74230589.pdf (accessed on 2nd October 2025).
19. Coto, J. A., & Wilder, C. (2016). An exploratory study of patient falls. https://core.ac.uk/download/144553583.pdf (accessed on 2nd October 2025).
20. Di Ciaccio, A., & Giorgi, G. M. (2016). Statistics in the big data era. https://core.ac.uk/download/98335907.pdf (accessed on 2nd October 2025).
21. Doctor, F., Iqbal, R., Mahmud, S., More, B., et al. (2020). Big data analytics: Computational intelligence techniques and application areas. https://core.ac.uk/download/228156656.pdf (accessed on 2nd October 2025).
22. Dou, M., Li, Y., Ozgur, C., Rogers, R., et al. (2017). Selection of statistical software for solving big data problems for teaching. https://core.ac.uk/download/216809009.pdf (accessed on 2nd October 2025).
23. Ebner, M., & Khalil, M. (2016). De-identification in learning analytics.

https://core.ac.uk/download/53025461.pdf (accessed on 2nd October 2025).

24. Gadepally, V., Hancock, B., Kaiser, B., Kepner, J., et al. (2015). Computing on masked data to improve the security of big data. http://arxiv.org/abs/1504.01287 (accessed on 2nd October 2025).
25. Gammack, J. G., Miah, S. J., & Vu, H. Q. (2019). A location analytics method for the utilization of geotagged photos in travel marketing decision-making. https://core.ac.uk/download/223241640.pdf (accessed on 2nd October 2025).
26. Gao, H., Jin, Y., Muthyala, R., Qin, Y., et al. (2017). Data-driven job search engine using skills and company attribute filters. http://arxiv.org/abs/1709.05050 (accessed on 2nd October 2025).
27. Gharehpetian, G. B., Marzband, M., Moradi, J., Nafisi, H., et al. (2019). Attributes of big data analytics for data-driven decision making in cyber-physical power systems. https://core.ac.uk/download/305121394.pdf (accessed on 2nd October 2025).
28. Guirguis, K. (2020). From big data to big performance–Exploring the potential of big data for enhancing public organizations' performance: A systematic literature review. https://core.ac.uk/download/304636108.pdf (accessed on 2nd October 2025).
29. Heiskanen, H., Karimov, J., Katsifodimos, A., Markl, V., et al. (2019). Benchmarking distributed stream data processing systems. http://arxiv.org/abs/1802.08496 (accessed on 2nd October 2025).
30. Hosseinpour, F., Ilic, A. S., Jakobik, A., Jarynowski, A., Krzyszton, M., Marks, M., Moldovan, D., et al. (2019). Medical data processing and analysis for remote health and activities monitoring. https://core.ac.uk/download/224993779.pdf (accessed on 2nd October 2025).
31. Isah, H., Mahfuz, S., Nicholls, P., Zulkernine, M., et al. (2018). Detecting irregular patterns in IoT streaming data for fall detection. http://arxiv.org/abs/1811.06672 (accessed on 2nd October 2025).
32. Isewon, I., Obembe, O. O., Oyelade, O. J., Soyemi, J., et al. (2015). Bioinformatics, healthcare informatics and analytics: An imperative for improved healthcare system. https://core.ac.uk/download/83592967.pdf (accessed on 2nd October 2025).
33. Ivanov, P., & Zlatev, V. (2017). Applied business analytics approach to IT projects–Methodological framework. https://open.bu.edu/bitstream/2144/29281/3/CSECS%202017-proceedings_ivanov-zlatev.pdf (accessed on 2nd October 2025).

34. Jiang, F. (2017). Data analytics helps business decision making. https://core.ac.uk/download/228820531.pdf (accessed on 2nd October 2025).
35. Kanti Roy, M. (2019). A survey of big data tools and technologies. https://core.ac.uk/download/643705543.pdf (accessed on 2nd October 2025).
36. Koronis, A., Zhou, C., Tamilselvan, P., Khuntia, S. R., Khuntia, S. R., Khatri, V., & Peng, X. (2018). Smart asset management for electric utilities: Big data and future. http://arxiv.org/abs/1706.09711 (accessed on 2nd October 2025).
37. Küçükkeçeci, C., & Yazıcı, A. (2017). Big data model simulation on a graph database for surveillance in wireless multimedia sensor networks. http://arxiv.org/abs/1708.03878 (accessed on 2nd October 2025).
38. Larkin, T. P. (2020). The transformation of accounting information systems curriculum in the last decade. https://core.ac.uk/download/322959194.pdf (accessed on 2nd October 2025).
39. Prasanna, V., Simmhan, Y., & Zhou, Q. (2016). Knowledge-infused and consistent complex event processing over real-time and persistent streams. http://arxiv.org/abs/1611.00676 (accessed on 2nd October 2025).
40. Qin, C., & Rusu, F. (2014). Speculative approximations for tera scale analytics. http://arxiv.org/abs/1501.00255 (accessed on 2nd October 2025).
41. Salton, G., Lokoč, J., Lokoč, J., Zahálka, J., Zahálka, J., Barthel, K. U., Rossetto, L., et al. (2019). Interactive search and exploration in online discussion forums using multimodal embeddings. https://core.ac.uk/download/489773862.pdf (accessed on 2nd October 2025).
42. Wilbik, A. (2018). FSS++ workshop report: Handling uncertainty for data quality management. http://arxiv.org/abs/1810.02091 (accessed on 2nd October 2025).

CHAPTER 7

CASE STUDIES IN IOT-BASED SHM

Contents

INTRODUCTION

The advent of the Internet of Things (IoT) has revolutionally changed many industries, but most prominently so in structural health monitoring (SHM). IoT-centered SHM systems use sensor networks to track the integrity of structural components in real-time continuously, enhancing safety and maintenance strategies. These systems use a sensor network to track structural stress, vibrations, and environmental factors, which are delivered to central processing units for analysis. With the convergence of such technologies, engineers and scientists are able to automatically monitor the health of buildings, bridges, and other infrastructures ahead of time and identify potential issues before they result in catastrophic failures. This proactive response not only increases security but also improves maintenance work by allocating resources better and reducing overall costs in infrastructure management. With a variety of applications, IoT-based SHM has been highly beneficial in a range of applications. For instance, in urban areas, monitoring of public infrastructure under varying loads calls for advanced solutions. Internet of Medical Things (IoMT) model informs the design of similar monitoring systems with emphasis on distant data acquisition and analysis capabilities. In recent case studies, scientists designed portable devices for the assessment of structural conditions, drawing an analogy with chronic condition health monitoring systems, such as COPD, where real-time information continues to be central to management. Structural response monitoring by non-contact methods draw an analogy with the development of e-health platforms, heralding a broader scope of IoT applications. These advances guarantee not just the safety but also the resilience of infrastructure, adapting to emerging challenges imposed by environmental dynamics. However, the incorporation of IoT in SHM also raises fundamental questions of system complexity and user privacy. The traditional engineering practice was accompanied with a degree of control over design parameters that is being gradually undermined by cyber-physical systems (CPS). With autonomous data interchange among IoT devices, the stakeholders are at higher levels of privacy and security risks. As suggested, the transition to systems where parameters may be dynamically tuned introduces a degree of complexity that necessitates an examination of dominant design paradigms. Data integrity and user privacy protection in SHM systems require an interdisciplinary integration of technology, as evident in studies dealing with the convergence of mechatronics and cloud technologies. Through these challenges, researchers and practitioners can better create robust IoT-based SHM solutions that prioritize both operational efficiency and user privacy first.

Definition of Internet of Things (IoT) and Its Relevance to Structural Health Monitoring (SHM)

The IoT is the set of devices that are connected physically and exchange and add meaning to data via the internet. These technologies range from simple home appliances to complex industrial equipment, all connected to a common digital system. The relevance of IoT is particularly well exemplified by the domain of SHM, an application committed to continuous monitoring of structures such as bridges, buildings, and dams. Through the embedding of sensors within these structures, IoT technology enables real-time information to be recorded with regard to parameters ranging from strain to temperature, and vibration. This data flood provides proactive maintenance and risk management, increasing public safety and business efficiency. The use of IoT for SHM enables a shift from conventional, manual inspection to data-driven, automatic inspection. Such a shift is especially applicable for aging infrastructure, where conventional methods could prove insufficient. IoT allows SHM systems to carry out continuous monitoring and anomaly identification, lessening the inherent threats more probable with aged structures.

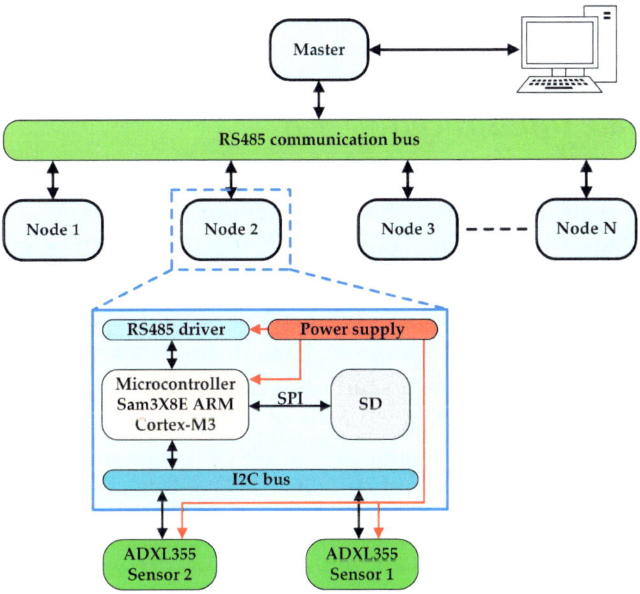

Figure 7.1. The Internet of Things (IoT) refers to the interconnected network of physical devices.

Source: https://www.mdpi.com/sensors/sensors-20-04908/article_deploy/html/images/sensors-20-04908-g002.png.

Moreover, the potential integration of IoT with emerging analytical technologies such as machine learning and artificial intelligence enhances predictive maintenance. The convergence of IoT devices and data analysis not only streamlines the monitoring process but also enhances more informed decision-making for stakeholders involved in infrastructure management. Apart from that, the impact of IoT on SHM is not merely about collecting simple data but also encompasses more extensive economic and social impacts. Using IoT platforms, stakeholders can attain more sustainable maintenance and construction operations, which in turn reduce the cost of operation and extend the lifespan of buildings. For instance, the implementation of a smart infrastructure management system can lead to reduced downtime, improved resource utilization, and increased defense against threats. As defined by existing research, IoT contribution to services sector development, in this case for SHM strategic management, is being progressively employed in both academic and applied applications (Britchenko et al., 2018). In addition, the utilization of IoT technologies helps reinforce a culture of safety and accountability, as well as ensuring compliance with rising societal needs for infrastructure integrity and safety standards (Figure 7.1).

Importance of Structural Health Monitoring (SHM) in Civil Engineering and Infrastructure

SHM in civil engineering and infrastructure cannot be overstated, particularly in relation to ensuring the longevity and safety of structures. As there is more urbanization and the development of sophisticated infrastructure, there is a need for reliable monitoring systems that is at the top priority. SHM makes it possible to make critical structural integrity evaluations constantly through the collection and analysis of data from sensors embedded in structures, which is vital when identifying potential failures in an early stage. It is not only vital for enhancing safety but also helps in optimal decision-making regarding maintenance and repairs, hence minimizing potential catastrophic failures and extending the life of infrastructure assets. With real-time data analysis integrated into it, SHM aligns with the contemporary requirements of smart cities and paves the way for the integration of intelligent systems into civil engineering structure monitoring. The use of IoT technology integrated with SHM systems enhances them by virtue of data-driven algorithms that process big amounts of data gathered using various sensors. The algorithms enable detectability as well as localizability of damage, with significant advantages in terms of cost reduction and operational efficiency. According to current literature, detection of damage in SHM can be carried out through two prominent methods: data-driven algorithms and model-based. Data-driven methods are said to typically prevail because they can provide rapid feedback through real-time analysis of sensor data, thus aiding rapid decision-making for maintenance action. Emphasis on data-driven methods indicates a significant direction toward increasingly automated, smart infrastructure in civil engineering that connects high-performance computing with the need for real-time monitoring of structural health (Cherta et al., 2020). In addition, the application of Building Information Modeling (BIM) is increasingly being incorporated with SHM, particularly to further enhance sustainability in the construction sector. Utilization of BIM in combination with

SHM allows data representation and management throughout the lifespan of a building from conceptual design to extended maintenance. The integrative thinking not only advantages architects and engineers in decision-making but also supports sustainable operation by enhancing resource use efficiency and energy saving. As suggested in existing bibliometric analyses of BIM research, focus varies across various themes like construction management and operations. This integration of BIM and SHM is a reflection of how infrastructure performance needs to be taken into account in an overall way, with focus brought to the advent of a new multidisciplinary partnership of professionals in the fields of architecture, engineering, and construction. Further research into the relationships is critical to advancing both disciplines (Aish et al., 2019) (Figure 7.2).

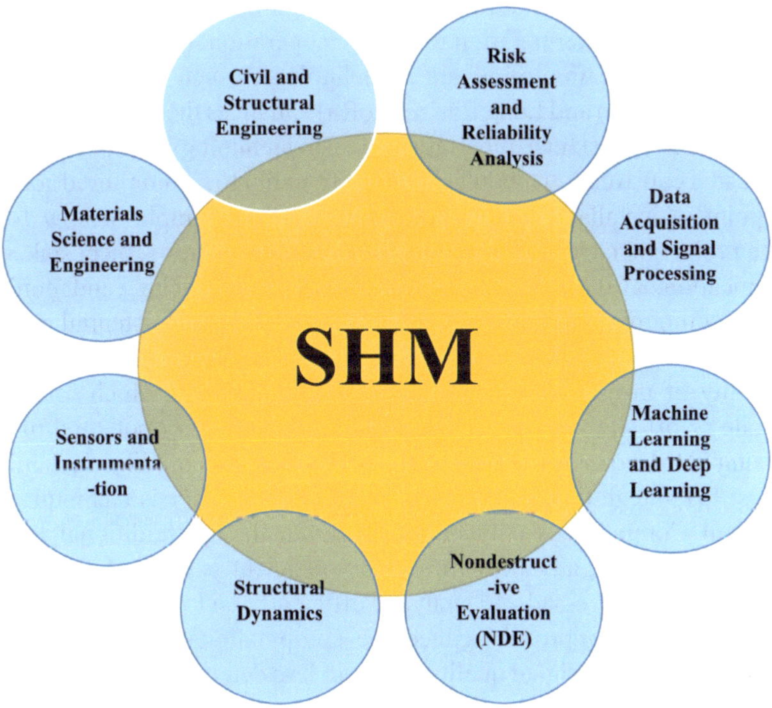

Figure 7.2. Importance of structural health monitoring (SHM) in civil engineering and infrastructure.

Source: https://media.springernature.com/lw1200/springer-static/image/art%3A10.1007%2Fs40430-025-05697-5/MediaObjects/40430_2025_5697_Fig1_HTML.png.

Overview of the Integration of Internet of Things (IoT) in Structural Health Monitoring (SHM) Systems

Integration of the IoT in SHM systems marks a significant milestone in the evolution of the civil engineering and infrastructure management sectors. IoT technologies enable real-time monitoring and data acquisition, enabling engineers to assess the structural integrity of buildings, bridges, and other critical infrastructure sufficiently. With a combination of sensors, such as accelerometers and strain gauges, stress, strain, and vibration information can be sampled continuously and transmitted. Real-time reliable data stream provides a good foundation for active maintenance actions and enhances decision-making. As seen in a number of case studies, the convergence of IoT with traditional SHM methods not only increases accuracy but also significantly reduces the duration of data analysis and interpretation to enhance operational efficiency and safety. Moreover, the application of advanced technologies such as Radio Frequency Identification (RFID) and Long Range (LoRa) enhances the operations of IoT in SHM systems. For instance, integration of such technologies with BIM enables prefabricated construction modules and functionality to be monitored accurately at the point of installation. This is a quintessential example of how IoT can enhance project management through effortless site laborer-project stakeholder communication. RFID technology allows easy location tracking and monitoring of critical components in real-time, minimizing delays and potential accidents. In addition, integrating cloud computing with IoT configurations makes remote accessibility of project information in real-time possible, which can be vital for on-site or off-site workers, indicating the revolutionary potential of IoT in optimizing SHM experiences (Liu et al., 2019). Besides simplifying monitoring activities, IoT integration in SHM has superior data analysis techniques, such as Principal Component Analysis (PCA). Even though traditional PCA can be memory-intensive, advances in memory-efficient parallel algorithms have enabled it to be compressed and analyzed efficiently on low-cost IoT gateways. Not only does this approach reduce the computational burden, but it also maintains high-quality signal quality, wherein less deterioration of data analysis quality is realized. The resulting system has the potential to provide significant speed improvement and efficiency, pointing towards the necessity of application of advanced analysis methods in IoT systems. Through a combination of the potential of IoT and best data handling methodologies, SHM systems are capable of increasing accuracy and reliability to a higher level, eventually resulting in more secure infrastructure management practice (Abdelgawad et al., 2019) (Figure 7.3).

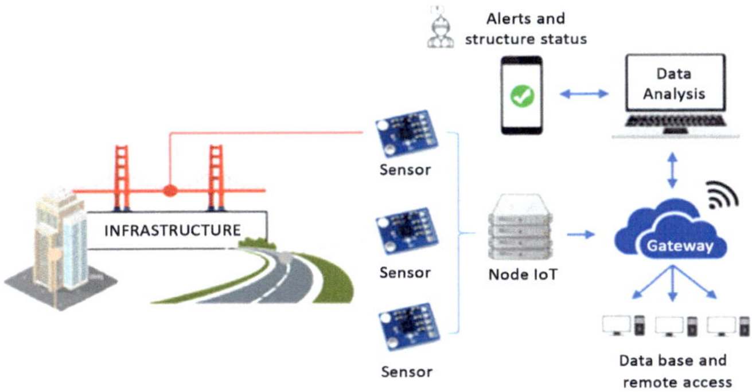

Figure 7.3. The integration of Internet of Things (IoT) in structural health monitoring (SHM) systems.

Source: https://www.iieta.org/sites/default/files/styles/inline_image/public/medias/2024-04/1_82.png?itok=cHe8MKIH.

7.1. HISTORICAL CONTEXT OF STRUCTURAL HEALTH MONITORING (SHM) TECHNOLOGIES

The history of SHM technologies is of direct association with changing engineering practices and materials science. Emerging in the late 20th century, SHM was aimed at enhancing structures' safety and longevity by online monitoring. SHM technologies in their initial stages were mainly dependent on traditional sensors and fixed monitoring systems that felt data sporadically. However, these methods were limited in their ability to provide real-time responses and dynamic analysis. With the development of digital technologies, including the introduction of the Internet of Things (IoT), the SHM industry has largely been transformed by enabling the transition towards more interactive, sophisticated monitoring models. With the development of these technologies, there was growing realization of their potential to synthesize various streams of data, creating opportunities for the ready interaction between monitoring structures and predictive maintenance methods (Ahmed et al., 2019). With the evolution of SHM technologies, the application of high-strength materials and manufacturing technologies progressively gained prominence. From conventional monitoring methods involving direct human observation to fully automatic systems based on advanced sensors, progress in SHM technologies has kept pace with developments in fields that have a direct bearing on it. Advances in low-cost, highly effective sensors have made possible wider applications of SHM systems across a variety of structures from bridges to buildings and dams. These developments have not just made monitoring cheaper but also prompted scholars and practitioners to think about the likely synergies between SHM and emerging technologies, such as big data analytics and machine learning. Such systems can now process massive amounts of data in real-time, making it possible for a more advanced understanding of the performance and integrity of structures that was not possible before (Ewart et al., 2019). While the historical background to SHM technologies shows key technological milestones, it also offers illumination of shifting paradigms of data use and decision-making in engineering. The "models of everywhere" concept, for instance, indicates a shift away from brush-stroke modeling approaches towards localized context-driven analysis that takes account of the unique properties of specific environments. This new paradigm is not only a methodological advancement but an epistemological shift in characterizing structural behaviors governed by multifaceted geographic and social factors. In this manner, SHM technologies are not only the answer to engineering challenges; they are also a revolutionary tool of evidence-based decision-making in infrastructure management. The ongoing dialogue between technology innovation and practical application underlines the need for creating evolving research agendas that attempt to overcome the persistent issues in the SHM field (Ahmed et al., 2019) (Figure 7.4).

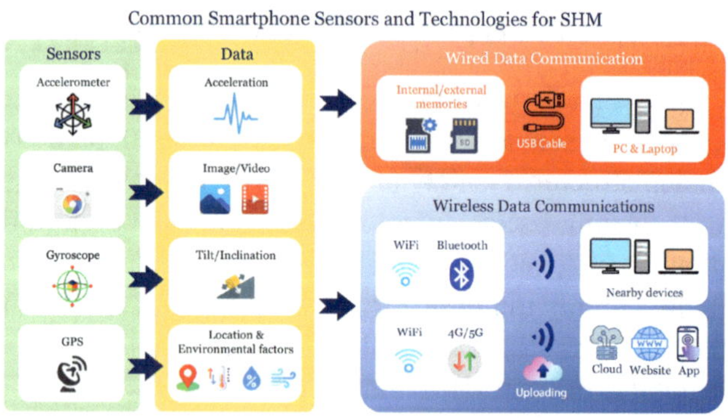

Figure 7.4. The historical context of Structural Health Monitoring (SHM) technologies.

Source: https://ars.els-cdn.com/content/image/1-s2.0-S0263224123012800-gr1.jpg.

7.1.1. Evolution of Traditional Structural Health Monitoring (SHM) Methods

The evolution of traditional SHM methods has undergone a revolutionary shift due to technological innovation as well as increased use of the Internet of Things (IoT). Initially, SHM greatly relied on static monitoring systems that used physical testing and visual inspection to estimate structural capacity. Such methods, while beneficial, were typically marred by their time-consumptive process as well as susceptibility to human-induced errors. The integration of sensors and data collection systems marked a milestone in monitoring operations. Ongoing data collection and analysis were made feasible through this development, improving timeliness and effectiveness of testing. Further, as industries began integrating digital technologies, the potential for real-time monitoring and remote access to data became a reality, thus rendering SHM systems more effective in various applications in engineering. As IoT technologies developed, the tasks of SHM systems extended from data acquisition to advanced analytics and predictive maintenance. With sensor networks complemented by advanced communication technologies, efficient transfer of data to cloud-based solutions was made, informing the stakeholders in real time about significant information on structural performance. For instance, utilization of Radio Frequency Identification (RFID) and Long Range (LoRa) technologies together with sensor networks has facilitated real-time performance tracking and monitoring of prefabricated components on site, as reported by corresponding research. Gathering data through such advanced methods not only supports richer decision-making but also assists in reducing costs as well as optimizing resource distribution. Subsequently, the climate of SHM evolved into an active approach to structural integrity management that allows timely intervention to maintain asset lives. Parallel to this, the adoption of IoT by the railway sectors is indicated by the transition away from conventional communications networks toward new, broadband-based networks. Industry-wide adoption of Internet of Things solutions has enabled railway infrastructure to leverage wireless sensor networks for predictive maintenance and real-time monitoring of assets. According to relevant research, the advancement of these technologies has paved the way for improved monitoring of the essential infrastructure, enhancing efficiency and safety in operation. This is part of a broader movement within a number of industries increasingly reliant on IoT-based solutions for structural monitoring. With the use of such technologies, leveraging connectivity, scalability, and insights based on data, traditional SHM techniques have evolved into dynamic systems capable of keeping up with the demands and challenges of modern engineering.

7.1.2. Transition from Manual Monitoring to Automated Systems

Transition from manual systems to automated systems is one of the largest transitions in the methodology of structural health monitoring (SHM) when applied to IoT applications. Traditional data collection methods conventionally utilized physical site visits and manual recording, which limited the speed and accuracy of data gathering. By using IoT devices, researchers and engineers can implement an embedded monitoring system that records real-time data in real-time, improving the responsiveness to any likely structural issue. This enhancement not only supports easier data collection but also reduces human error, making it possible to have more advanced data analysis features. As indicated in recent case studies, such automatic systems have considerably improved monitoring and control of structural health, allowing proactive maintenance interventions rather than reactive repair measures, thus prolonging the lifespan of assets and reducing operating costs. Automation times also allow the use of advanced data analysis techniques such as machine learning, which can derive meaningful information from complex datasets accumulated by IoT sensors. Machine learning has demonstrated potential for multiple applications, like fault diagnosis and detection, to enhance the overall performance of buildings through optimized operations and energy management. Though publication numbers for this topic increased immensely, as evidenced by systematic reviews of the literature, problems remain with model transferability and the justification of costs of these autonomous systems. The absence of a standardized method for applying machine learning across different buildings is questioned by its applicability and validity, necessitating continuous research and development in this field, which can bring about innovative solutions to enhancing efficiency and stability in structural monitoring (Hong et al., 2020). Transition issues to automated monitoring systems are typically further exacerbated by the lack of integration and harmonization across existing technology platforms. With organizations adopting the new solutions, the granularity level within microservices becomes central to the development of effective IoT architectures. According to recent studies, systematic mapping of microservices indicates that technical design decisions corresponding to desired value in such architectures is central to the transition. Development of a common microservitization framework can assist in designing flexible and resilient monitoring systems. This not only must prioritize short-term deployments but also long-term sustainability of SHM practice so that automated systems have the flexibility to evolve with new technologies and evolving industry demands (Andrawos et al., 2019). Thus, combining automation with architectural strategy has the capability to redefine the future of structural health monitoring effectively.

7.1.3. Role of Technology Advancements in Structural Health Monitoring (SHM) Development

The technology advances have seen the SHM field revolutionize as a discipline, particularly with the advent of IoT. Embedded sensing devices have evolved to be more sophisticated, not only performing passive data gathering but also proactive analyses. As stated, 'Advances in sensing and computing power are enabling the embedding of more computing power in small devices.' This is enabled as analysts can obtain data at high resolutions, hence the possibility to monitor structural integrity in real time. By utilizing networked devices within a construction or infrastructure project, engineers can have insights not previously possible, generating more informed decision-making.

Figure 7.5. Role of technology advancements in structural health monitoring (SHM) development.

Source: https://dt.mdx.ac.uk/wp-content/uploads/2022/08/london-digital-twin-research-centre.jpg.

Finally, the intersection of these technologies brings life to the SHM market, opening up opportunities for improved safety features and better maintenance processes. Also, the application of big data analytics in SHM is a demonstration of the way innovation in technology has developed the field. With the stream of data from IoT sensors increasing further, prospects of leveraging analytics to improve the performance of structures have been made available. The implications are extensive; higher analytical models facilitate better prediction of structural behavior, thereby minimizing risks and optimizing maintenance strategies. The controversy surrounding big data indicates that 'big data analytics has been identified as a key enabler for the IoT,' providing a feedback loop whereby real-time data not only informs present operational processes but also inputs into future predictive designs. This loop fosters a culture of continuous improvement and results in SHM being a dynamic and adaptive discipline which can react to new issues as they arise. Last but not least, the innovation in mobile technologies and awareness-based applications further enhanced the functionality of SHM systems. The introduction of smartphones and their surrounding ecosystems provided developers with opportunities to create innovative applications exclusively for monitoring and controlling structural health. Computers now are able to understand their environment, improve their contextual sensing by being more well-informed, and facilitate responsive data acquisition and analysis management. As repeated, 'the presence and awareness of the context provides the capability of being aware of physical environments.' The umbrella of the context enables SHM systems not only to operate independently but also to make well-informed, real-time decisions. Despite current challenges such as resource limitation within devices, continued research into these technologies promises to make SHM more resilient and powerful in safeguarding infrastructure (Figure 7.5).

7.2. KEY COMPONENTS OF STRUCTURAL HEALTH MONITORING (SHM)-BASED STRUCTURAL HEALTH MONITORING (SHM) SYSTEMS

The integration of IoT technology with SHM systems is a groundbreaking advancement in infrastructure management. At the core of such systems is a sensor system that takes real-time measurements of various structural parameters, such as strain, displacement, and vibration. Such sensors, commonly employed in integration with technologies like Radio Frequency Identification (RFID) and Long Range (LoRa) communication, allow real-time observation of building components and facilitate prompt intervention for structural safety assurance. For instance, integration of RFID and sensor networks within a Building Information Model (BIM) allows stakeholders to observe valuable project data remotely and facilitates accurate tracking of prefabricated components on the site. These advancements are indicative of the promising intersection of IoT and SHM whereby real-time data informs decision-making in construction projects by a significant degree (Liu et al., 2019). Data analysis plays an important role in enhancing the efficiency of IoT-based SHM systems. Voluminous data managed by sensor networks need to undergo advanced analysis techniques in order to derive meaningful patterns and trends for the identification of maintenance and management actions. One such data analysis method is Principal Component Analysis (PCA), which is data reduction with maintaining important information. Classical PCA algorithms are not free from computational inefficiencies, however, particularly when implemented on low-cost IoT gateways whose processing capabilities are not high. Research on memory-efficient parallel PCA implementations has exhibited significant improvements, achieving a 10x compression ratio while reducing memory usage without sacrificing data integrity. It enables IoT-based SHM systems to implement data analytics without trading off between reliability and performance (Abdelgawad et al., 2019). The application in real-time scenarios of IoT-based SHM systems showcases their effectiveness in various contexts. Case studies have depicted how such systems can enable it to monitor infrastructure such as bridges, buildings, and other essential structures more efficiently. Embedded sensor networks alongside cloud-based analytics enable maintenance personnel to repair issues on an active footing before they reach full-blown safety concerns. For example, real-time information provided by intelligent sensors can notify engineers of structural vulnerabilities so remediation can be undertaken quickly. Moreover, the incorporation of IoT technologies also facilitates more cooperation among the stakeholders of projects, improving communication and results of the projects. In brief, all key components of IoT-based SHM systems, including sensor networks, algorithms for data processing, and incorporation in the cloud, contribute towards a stronger and responsive structural health monitoring strategy, ensuring long life and safety of infrastructure (Liu et al., 2019).

7.2.1. Sensors and Data Acquisition Technologies

Technology in the form of sensors and data acquisition has significantly transformed the Structural Health Monitoring (SHM) field, as it facilitates real-time testing and the verification of integrity of built environments. The technology spans a broad range of devices from traditional strain gauges to advanced wireless systems that use Radio Frequency Identification (RFID) and Internet of Things (IoT) platforms. For instance, in prefabrication construction (PC) construction, the use of RFID and LoRa technology alongside sensor networks facilitates comprehensive gathering of data on the structural performance of components. As cited in several studies, such practices facilitate precise tracking of materials within sites, coupled with structural status monitoring during installation on a continuous basis. These advancements not only improve project performance but also provide valuable insights that allow industry experts to make sound decisions regarding current projects (Liu et al., 2019). Furthermore, the integration of cloud computing in data acquisition processes has revolutionized how project stakeholders access information and engage with real-time data. With cloud-based Building Information Modeling (BIM), experts can view project elements and be aware of their status remotely without needing to be onsite. This capability, particularly, comes in handy in managing and reducing the risks associated with malfunctioning elements or structural vulnerability. Insights gathered from sophisticated sensor networks can be rapidly processed and depicted, allowing on-time action and better overall project productivity. These systems have been demonstrated in real-world field applications in actual field environments to validate their accuracy and reliability in monitoring different elements of construction performance (Liu et al., 2019). Hence, the combination of sensors and cloud technology is one significant step in SHM practice. In medicine, particularly in the scenario of COPD management, sensors and data collection instruments are essential in remote patient care. The development of non-contact capacitive sensing devices integrated in smart vests illustrates the trend, as it facilitates real-time measurement of respiratory rate without the need for tedious setup. Preceded by the Internet of Medical Things (IoMT) model, the devices facilitate seamless data acquisition and transmission with the potential to grant more patient autonomy in rehabilitation. Studies have proved the efficacy of such devices and confirmed their functionality through comparative studies with traditional measurement benchmarks. The historic accuracy achieved in respiratory rate estimation endorses the potential of sensor-based technologies to revolutionize patient monitoring and eventually improve health outcomes among patients struggling with chronic illnesses. The applications of complex sensor systems across multiple fields, hence, accentuate their transformative impact on construction and healthcare sectors (Figure 7.6).

Figure 7.6. The advent of sensors and data acquisition technologies.

Source: https://cdn.adlinktech.com/webupd/en/upload/1200x630.jpg.

7.2.2. Communication Protocols and Data Transmission

Emergence of communication protocols has been a key factor in evolution of Internet of Things (IoT)-based Structural Health Monitoring (SHM) systems. The protocols have facilitated exchange among multitudes of devices, enabling smooth data transmission and processing across heterogeneous environments.

The key has been to optimize the protocols to support the unique conditions of SHM, where sensors are often utilized under high-risk and dynamic conditions. For instance, MAC protocols are essential to the planning of how equipment accesses common bandwidth. Protocols that have inefficiencies within them can lead to bottlenecks that interfere with data flow and potentially jeopardize the reliability of the SHM system. As per the recent study layout, re-rigging traditional communication protocols, most prominently ALOHA-type structures, would enable the data transmission to be resilient in the event of IoT-based SHM applications, enabling improved communication between numerous sensors that offer real-time structural integrity monitoring. Incorporating distributed queueing (DQ) algorithms is one area of future work in enhancing the IoT-based communication protocol. Such algorithms hold potential in the aspect of being able to surpass what is possible with conventional MAC protocols, especially during high levels of traffic or where the number of devices is more than expected. DQ algorithms work efficiently because they can manage data packets from an infinite number of devices, thus ensuring that data transmission remains efficient even when it is full.

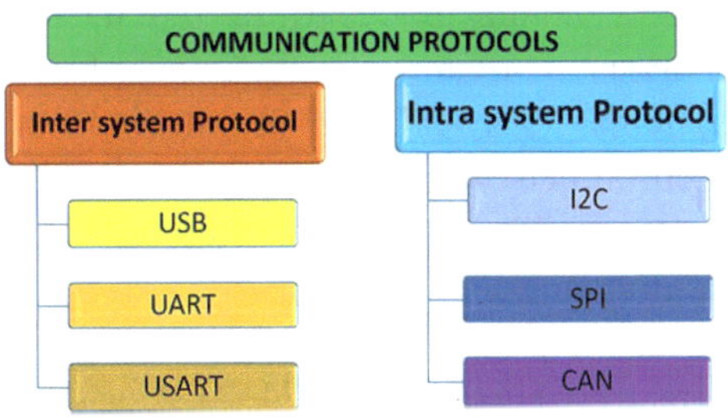

Figure 7.7. Communication protocols and data transmission.

Source: https://mexico.newark.com/wcsstore/ExtendedSitesCatalogAssetStore/cms/asset/images/common/technology/articles/2924710/2924710-communication-protocol-types.jpg.

As evidenced in varied research studies, these mechanisms can potentially make SHM systems perform significantly better, relying on timely information for accurate structural condition estimations. Application of DQ solutions not only provides efficient communication but also fosters resilience and agility in various operating environments encountered in IoT-based monitoring systems (Barbarov-Rostán et al., 2018). Also, the development of specialized e-Health platforms with integrated advanced communication protocols depicts the revolutionizing potential of IoT in applications such as healthcare and structural integrity monitoring. For instance, the application of a non-contact respiratory monitoring system using an IoT platform demonstrates the viability and versatility of utilizing future technologies for patient care and data analysis. Such an approach not only offers seamless data transmission but also remote monitoring, which improves the responsiveness of patients' conditions. Experimental confirmation of such systems highlighted the validity and accuracy of data collected, thus signifying the importance of improved communication protocols for overall system performance improvement. The findings support the linear correlation between strong communication protocols and the effectiveness of IoT-based projects, which is vital in healthcare monitoring and structural reliability tests (Barbarov-Rostán et al., 2018) (Figure 7.7).

7.2.3. Data Analytics and Processing Techniques

In the domain of IoT-based Structural Health Monitoring (SHM), data analysis and processing techniques play a core role in the transformation of raw sensor data into useful information. By using sophisticated algorithms and large sets of data, engineers can test the integrity of intricate system structures in real-time. Predictive modeling and machine learning-based data analysis techniques facilitate real-time analysis of streams of data gathered with sensors integrated into infrastructure systems. This real-time monitoring not only enhances the understanding of structural behavior under various conditions but also enables decision-making before possible catastrophic failure. The integration of these approaches highlights how important information can be employed to enhance safety protocols and operational efficiency in modern civil engineering. Fundamentally, the effectiveness of IoT-based SHM systems relies significantly on the chosen processing approaches. Preprocessing of data, i.e., filtering and normalization, is crucial in the production of quality data that can serve to give meaningful insights. Certain advanced computational methods, which have been explored for further enhancement of SHM capacity, are big data and computational intelligence-based approaches. For instance, Hierarchical Spatial-Temporal State Machine (HSTSM) offers one a biologically inspired modeling paradigm for understanding structural evolution dynamics over time. This approach not only aids in anomaly detection but also tremendously supports predictive maintenance planning, thus lessening risks of old infrastructure and structural failure, as mentioned in various case studies (Doctor et al., 2020). Besides, more advanced technologies like the combination of Radio Frequency Identification (RFID) and Long Range (LoRa) networks have emerged to support optimal data analytics in IoT-based SHM systems. These advancements provide for the real-time capture and processing of structural element information automatically, enhancing project management practice. Incorporating sensor networks monitoring performance metrics, such as stress and strain, these systems provide full visibility into construction integrity and communicate with cloud-based Building Information Modeling (BIM). This integration gives project teams remote access to critical information, thus facilitating collaboration and efficient decision-making throughout construction. These advanced data analytics techniques address modern problems in construction performance and offer gigantic benefits for future advancements in the industry (Liu et al., 2019) (Figure 7.8).

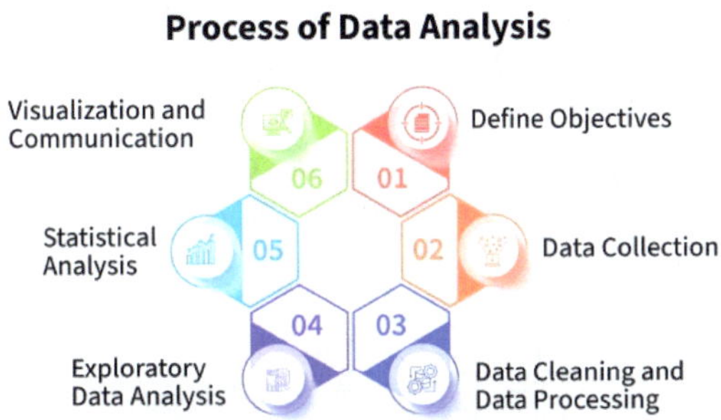

Figure 7.8. The process of data analysis.

*Source:*https://media.geeksforgeeks.org/wp-content/uploads/20250111153001887786/process-of-Data-Analysis.webp.

7.3. CASE STUDY: SMART BRIDGES

The development of Smart Bridges with Internet of Things (IoT) technologies is a groundbreaking advancement in the field of Structural Health Monitoring (SHM). These bridges have a number of sensors that constantly collect data on structure performance, traffic, and environmental conditions. All this data collection enables real-time monitoring, which allows for rapid assessment of a bridges integrity and safety. As more and more transportation systems rely on advanced infrastructure, the necessity for proactive maintenance has now become an imperative. Structural weaknesses identification before catastrophic failures are enabled by innovative IoT applications, not just guaranteeing such structures' lifespan but also public safety and resource use optimization. Smart Bridges thus reveal how technologies can transform infrastructure management by data-driven decisions and intervention at the appropriate moment. The operational efficiency of Smart Bridges is heavily dependent on data-rich algorithms that analyze the vast datasets that are created using embedded sensors. These algorithms possess the ability to achieve great breakthroughs in damage localization, identification, and prognosis, which is critical to ensure bridge safety in the long run. Data-driven methods, as compared to model-based methods, allow real-time assessment of the structural health based on real-time monitoring of the collected information. These algorithms are, however, computationally intensive due to their complexity and the processing of large amounts of data. To bridge such constraints, different optimized approaches have been put forward by researchers, such as memory-efficient algorithms for PCA. They improve data handling capability along with less computational burden, thereby making IoT-based SHM possible in resource-constrained settings, as can be ascertained from the implementation aspects elaborated in pertinent literature, e.g., Abdelgawad et al. (2019). Smart Bridges not only enhance the monitoring capability but also contribute to overall intelligent infrastructure development since they allow integration of advanced diagnostics and predictive analytics.

Figure 7.9. The emergence of smart bridges equipped with Internet of Things (IoT) technologies.

Source: https://buildipedia.com/media/k2/items/cache/1387325ab956c8c74bb3ab7914759e29_XL.jpg.

With the use of many different types of sensors, including accelerometers and strain gauges, more subtle details of structural behavior can be analyzed exhaustively. The sensor data aids in constant evaluation of structural health and directs maintenance planning. In addition, the knowledge established through these data-driven approaches can optimize traffic flow and reduce stress on the structures via reasoned design adjustments. Therefore, this transition towards intelligent infrastructures enabled by IoT technologies heralds a groundbreaking new era in civil engineering and urban planning, wherein a sustainable and resilient methodology for structure management is possible.

Scientific appraisal of these developments has emphasized the pivotal role of data-based methodologies to upgrading SHM applications, as reflected in Cherta et al. (2020) (Figure 7.9).

7.3.1. Overview of Internet of Things (IoT) Applications in Bridge Monitoring

Using IoT technologies for monitoring bridges has been a tremendous leap forward in structural health monitoring (SHM). With the ability to gather and analyze real-time data, IoT systems enable constant assessment of bridge conditions without a requirement for large-scale manual inspection. IoT systems harness a network of sensors like strain gauges, displacement sensors, and temperature sensors, which transfer data in wireless formats to centralized systems. For instance, by implementing Radio Frequency Identification (RFID) and Long Range (LoRa) technologies, as has been documented in literature, enables the accurate localization of prefabricated components at the point of installation and hence offers improved monitoring for structural health. These technologies do not just improve safety procedures but enable immediate action on detected anomalies, thereby enabling proactive maintenance rather than reactive. Implementation of wireless sensor networks (WSNs) in IoT systems adds another layer of effectiveness and efficiency in bridge monitoring. WSNs consist of energy-hungry sensors, which play a central part in the unhampered data flow. One of the major challenges comes from the protection of these networks, as incorporation of security measures like encryption adds additional energy requirements to sensors. But security lapses can potentially introduce vulnerabilities, such as denial of sleep attacks undermining sensor life. Rectifying these challenges is essential; studies have shown that giving top priority to energy efficiency schemes, such as duty cycling techniques, can optimize sensor performance while guaranteeing security. Consequently, successful deployment of such systems relies on a thorough balancing act between security and energy efficiency, noting the complexity of IoT deployments in bridge monitoring applications (Udoh et al., 2019). In addition, advances in cloud computing and Building Information Modeling (BIM) have revolutionized how bridge monitoring data is utilized. The possibility of consolidating data in a cloud-based BIM model enhances access and visualization for engineering teams. This convergence allows stakeholders to monitor structural performance in real-time, facilitating decision-making processes based on comprehensive data analysis. Data collection and visualization systems are capable of detecting likely issues before they become full-blown structural failures. For example, the effectiveness of cloud monitoring through cloud technology has been validated in real applications, both with enhanced reliability of data and significant developments in project performance (Liu et al., 2019). In conclusion, the integration of IoT technologies in bridge monitoring is extremely transforming the field of structural health monitoring, both enhancing safety and operational efficiency (Figure 7.10).

Figure 7.10. An overview of Internet of Things (IoT) applications in bridge monitoring.

Source: https://www.capetti.it/cms/img/articoli/big/52.jpg.

7.3.2. Benefits of Real-Time Data Collection and Analysis

In the Structural Health Monitoring (SHM) field, the implementation of real-time data collection and analysis has revolutionized the management and maintenance of infrastructure. Through the application of advanced technologies such as Internet of Things (IoT) sensors, information on various structural components can be received in real time, providing enhanced visibility of their current status. Through the provision of real-time data at their fingertips, immediate decision-making is facilitated, allowing engineers and maintenance staff to react rapidly to structural anomalies and dangers. As realized in prefabricated construction instances, a convergence of technologies like RFID and LoRa makes accurate tracking and monitoring of structural health easier, ensuring the entire process is more effective. By leveraging such aspects, stakeholders can ensure efficiency and safety throughout the life cycle of a structure, hence preventing the likelihood of high-impact failures (Liu et al., 2019). Real-time data gathering has benefits that extend beyond providing security in real time, since it results in cost-effectiveness and resource optimization in construction and urban management. As cities expand and resource allocation becomes an issue of highest priority, real-time analysis facilitates the monitoring and predictive repair of infrastructures. This adaptability becomes paramount in handling the ever-present issues of urban environments, especially as populations expand and tend to overstretch existing resources. In such cases, the IoT plays a pivotal role in that enormous data streams from a network of sensors are analyzed to determine usage patterns and trends in performance. Such insights lead to smart decision-making, as indicated by various Smart City initiatives utilizing IoT for sustainable resource management. Thus, real-time analysis of data becomes the pivot to enhancing urban living and reducing the adverse effects of urbanization (Akyildiz et al., 2013). Furthermore, cloud technologies' capability to give real-time data guarantees cooperation between stakeholders in a project, with better communication and transparency. Stakeholders are able to access and share data easily via cloud-based systems anywhere, thus guaranteeing information is not held in silos but cooperatively utilized. This holistic approach encourages forward-thinking action against building up structural issues, a responsive culture of safety and transparency throughout a projects life span. The feasibility of real-time monitoring, exhibited in prefabrication projects, identifies the potential for continuous improvement within the construction industry. By utilizing real-time data collection and analysis, professionals are able to enhance project performance, spur innovations, and utilize resources effectively, positioning themselves to mitigate present and future challenges (Liu et al., 2019) (Figure 7.11).

Figure 7.11. Benefits of real-time data collection and analysis.

Source: https://estuary.dev/static/f2bbee096a76ea47b9057d8838663360/6a1af/8d94ef_01_Real_Time_Data_What_Is_Real_Time_Data_92e6e5522e.jpg.

7.3.3. Challenges Faced in Implementing Internet of Things (IoT) Solutions

The implementation of IoT solutions, particularly in Structural Health Monitoring (SHM) systems, is also subject to a wide range of challenges that can hinder their effectiveness. One of the greatest challenges is the integration of different information and communication technologies (ICTs) needed to aggregate and dissect data. Effective smart city initiatives have shown that effective integration between various stakeholders is a significant component in overcoming such challenges. However, facilitating this type of cooperation is normally impeded by departmental silos, opposing priorities, and inaccurate resource allocation. Without appropriate management of such technical and organizational obstacles, it has the potential to result in inefficiencies and mismanagement of the SHM processes, hence defeating the potential benefits of IoT solutions in monitoring the building integrity in urban environments (Stahl et al., 2017). Another critical challenge encountered in the use of IoT solutions for SHM is the implementation of advanced data-driven algorithms for damage detection. Even though the algorithms offer the advantage of real-time processing, their use is challenging since it requires the use of high-performance processors due to the high computational requirements associated with analyzing vast amounts of sensor data. This not only raises issues regarding availability and cost of necessary technology but also issues regarding scalability of the systems and energy consumption. Furthermore, the performance of these algorithms is contingent upon the accuracy of models that are developed, and this can in turn generate discrepancies if the models are not sufficiently accurate in modeling the structures being monitored. The need for precise data and robust modeling is a primary hurdle that must be overcome by researchers as well as practitioners to enhance the reliability of SHM systems (Cherta et al., 2020). Finally, cybersecurity concerns are a primary challenge in IoT solutions for SHM implementation. As such systems tend to carry sensitive data on networks, they are exposed to inherent cyber risks that could compromise the integrity and reliability of structural monitoring. Networked interconnectedness of IoT devices also renders such threats greater inasmuch as a breach in one device could result in cascaded failures within a network. Adhering to regulations and standards for purposes of ensuring data privacy introduces additional complication. In response to these threats, organizations are going to have to invest in cybersecurity capabilities that can transform over time to accommodate new threats, a demand that will involve technical and financial investments. Overcoming these cybersecurity challenges is essential for establishing confidence in IoT-based SHM systems and their effective implementation in the defense of urban infrastructure.

7.4. CASE STUDY: BUILDING MONITORING SYSTEMS

In the field of Structural Health Monitoring (SHM), the implementation of IoT technologies entirely overhauls the techniques employed in building monitoring systems. An interesting case study makes this revolution apparent, demonstrating the advantages of employing enhanced sensor networks augmented with cloud computing. Using Radio Frequency Identification (RFID) and Long Range (LoRa) technologies, researchers were able to come up with an adaptive system that can automatically collect and analyze real-time information about prefabricated components on construction sites. The adaptive process not only facilitates ease in following structural pieces but also allow for monitoring of performance while being installed, hence guaranteeing structural stability and reducing the possibilities of future failure. The successful implementation of this method implies that properly designed building monitoring systems can potentially double industry norms of construction safety and efficiency, an indication of potential worth to the project delivery and stakeholder investment. Moreover, testing of this monitoring system in an actual prefabricated construction project was rigorous and persuasive. Seamless data transfer via LoRa technology provided confident insights into the structural performance of elements, demonstrating the possibility of calculating and pinpointing elements in real-time.

Utilizing a cloud-hosted Building Information Modeling (BIM) model allowed project teams to see and access critical information from mobile platforms, fostering collaborative decision-making and communication. This level of openness not only enhances the efficiency of project management but also empowers stakeholders with a dynamic oversight of construction processes. This case study represents the joining of IoT technologies and traditional construction practice as a trajectory towards improved project performance and decreased disruption in construction operations. Furthermore, as monitoring systems in building construction become increasingly reliant on data reduction, the difficulty then is to implement these methods within IoT platforms with limited capabilities. PCA is a key use in managing huge lists of sensor data, but its computationally intensive nature necessarily makes it impossible for use within real-time applications. Among the most promising developments in this field is the suggestion of a memory-efficient parallel version of the streaming History PCA algorithm with a considerable data processing efficiency and storage requirement boost. Achieving a compression factor of 10x with significant memory savings provides a good answer to embed SHM systems within low-cost IoT gateways. This algorithm runs well on boards like the Samsung ARTIK 710, demonstrating that improved analytics can be achieved without compromising data integrity. Thus, the continuous development of building monitoring systems illustrates an adherence to leveraging new technology to address long-standing challenges in construction administration and structural upkeep (Figure 7.12).

Figure 7.12. Building monitoring systems.

Source: https://miro.medium.com/1*E7e0My_DGgAIKPPv92vCnw.jpeg.

7.4.1. Use of Internet of Things (IoT) for Monitoring Structural Integrity in Buildings

The application of the Internet of Things (IoT) in structural health monitoring (SHM) revolutionized the manner in which buildings remain intact and secure. Through the incorporation of advanced sensor networks within the building, IoT facilitates real-time monitoring of vital parameters such as strain, temperature, and humidity. This proactive methodology allows for early detection of probable issues, e.g., material fatigue or fractures, that can significantly minimize the risk of failures with serious impacts. Moreover, the application of these technologies enables permanent data collection, thus providing building managers with critical information for maintenance and risk assessment. For instance, through the utilization of Radio Frequency Identification (RFID) and Long Range (LoRa) communication systems, construction teams can monitor precast components' structural behavior as they are being installed, thereby improving project execution and safety (Liu et al., 2019). Through the utilization of IoT solutions in SHM, not only is there an improvement in efficiency and safety but cost savings as well, particularly when considering building life cycles management. Sensors placed throughout a building can feed pertinent information to a cloud platform, employing a Building Information Modeling (BIM) system to provide a visual image of the building condition in real time. Through this pioneering integration, stakeholders are able to make decisions out of real-time data, thus reducing unnecessary expenditures on repair and maintenance. These technologies have been found to increase overall project performance by a significant margin while providing valuable information that can guide future building practice and shape design modifications. With the construction sector increasingly adopting such technologies, it will be able to achieve greater productivity and reduced wastage (Liu et al., 2019). But along with growing reliance on IoT infrastructure is growing demand for robust energy management and security features to safeguard integrity of data collected. Wireless sensor networks are typically battery-constrained and need to pay special attention to energy awareness in behavior and design. Energy-efficient protocol improvements, such as duty cycling, are essential to maximize performance while maintaining sensor life. In addition, the implementation of security measures to safeguard against potential attacks, i.e., denial of sleep attacks, should also be accorded proper significance so that sensors can function in an optimal manner for more extended periods. Ensuring a balance between security and energy needs within the IoT infrastructure is important to further enable the long-term support of SHM systems and the resilience of modern infrastructure.

7.4.2. Impact on Safety and Maintenance Practices

The integration of Internet of Things (IoT) technology in Structural Health Monitoring (SHM) has revolutionized safety protocols in the construction as well as the maintenance industry. With the deployment of advanced sensor systems, there is constant real-time monitoring of structural health, which reduces the risks involved in delayed inspection. In prefabrication (PC) schemes, IoT-based solutions allow monitoring and assessment of components with high accuracy during the installation process. Structural performance information can be collected and transferred effectively by employing Radio Frequency Identification (RFID) with strain sensors. Such computerized data collection not only enables timely maintenance actions but also aids in instilling a proactive safety management culture in organizations, as revealed by various case studies on technological innovation in construction projects (Liu et al., 2019). Moreover, the impact of IoT-based SHM also goes beyond temporary safety improvement to include long-term benefits in terms of maintenance practice. Maintenance organizations are increasingly being pushed to rationalize resource allocation without compromising safety and operational performance. IoT technologies facilitate these objectives through enabling stakeholders to achieve actionable information through data analysis. For instance, instant feedback on structural performance allows predictive maintenance, whereby possible issues are found early before turning into serious failures. This shift from reactive to predictive practices engenders a more sustainable approach to maintenance management, ultimately resulting in lower operation expenses and better asset life. Such methodologies are required to provide high levels of service, especially when confronted with challenging situations like winter maintenance operations that must be continuously fine-tuned to better adapt to safety and mobility (Shi et al., 2017). The application of IoT in SHM not only improves safety responses but also goes towards streamlining maintenance protocols in the industry. With enterprises adopting IoT solutions, data gathering and insights can be streamlined, enabling more consistent maintenance practices aligned with industry standards. With cloud-based solutions, project teams can access and share real-time data seamlessly, removing silos that otherwise impede collaboration. This end-to-end visibility helps stakeholders coordinate better, enhancing the decision-making process and removing the possibility of gaps. Besides, the standardised data can be employed to formulate training materials and guidelines for developing an informed workforce that can leverage these advanced technologies. Overall, IoT-based SHM completely redefines safety and maintenance processes, paving the way for better opportunities for the construction sector in terms of efficiency as well as sustainability (Figure 7.13).

Figure 7.13. Impact on safety and maintenance practices.

Source: https://www.supplyht.com/ext/resources/Issues/2024/02-February/Columns/SHT-0224-Eye-Feat-Slide1-1170x658.webp?t=1708397676.

7.4.3. Examples of Successful Internet of Things (IoT)-Based Building Monitoring Projects

The application of IoT technologies in building monitoring projects has reflected outstanding progress in structural health monitoring (SHM). One of the best examples is the application of Radio Frequency Identification (RFID) and Long Range (LoRa) technologies combined with sensor networks and Building Information Modelling (BIM). This new method was tested in prefabricated construction (PC) projects, where real-time data was tracked to record the performance of structures during installation phase. RFID technology was employed for tracking and strain sensors for performance data, as the system provided critical data updates that improved project management and safety. The BIM model hosted on the cloud played a crucial role in facilitating stakeholders' ability to visualize and make access to real-time data, thus affirming the effectiveness of this comprehensive methodology in improving project performance and delivering structural integrity within a changing environment, as summarized in the work of Liu et al. (2019). Another successful use of IoT in monitoring buildings is present within data-driven algorithms for damage detection within structures. These techniques rely heavily on sensor information and enable timely identification and assessment of structural defects. Algorithms driven by data are robust since they have the ability to deliver real-time solutions, hence they are appropriate for the majority of SHM applications. These systems can use advanced analytics as well as machine learning procedures to analyze enormous amounts of data fed into them from structural sensors to identify potential damage, localization, and fault classification. Models developed for this specific purpose have the capability to adapt according to the structures particular characteristics, thus yielding more accurate judgments of the state of health. This capacity demonstrates the ongoing evolution of IoT applications in SHM, whereby technologies continue improving the methodology employed in structural safety monitoring, as exemplified in Cherta et al. (2020). In brief, the successful implementation of IoT technologies within building monitoring schemes illustrates a paradigm shift of structural health monitoring. The synergistic use of RFID, LoRa, and sensor networks indicates the path for real-time data to contribute to efficiency and safety improvement in the construction sector. Similarly, data-driven approaches have emerged as a stronghold for rapid damage inspection, illustrating machine learning's power to revolutionize SHM practice. Both examples serve to illustrate the necessity of constant innovation and adaptation within the field. As such technology evolves, it promises not just to enhance the monitoring process itself, but also the longevity and security of the built environment at large, ultimately contributing to the wider discourse of sustainable construction practices and smart infrastructure management.

7.5. CASE STUDY: DAMS AND WATER INFRASTRUCTURE

The application of the Internet of Things (IoT) in structural health monitoring (SHM) offers great opportunities in managing dams and water infrastructure. Dams have very important roles, including flood control, irrigation, and the generation of hydroelectric power. While important to operate properly, their structural health is crucial to maintain functionality as well as safety. IoT-based SHM systems enable the acquisition of real-time information regarding the physical status of dams, enabling proactive maintenance and vulnerability identification in the initial phases. The new approach remodels the traditional evaluation methods, transforming the latter into predictive models instead of reactive ones. These advances align with the notion of everywhere models, under which localized, context-dependent modeling of environmental structures is considered fundamental to determining their particular attributes and managing uncertainties appropriately (Ahmed et al., 2019). Engineers can enhance decision-making capabilities through IoT technologies while significantly reducing aging infrastructure-related risks. The dynamic context of water infrastructure necessitates continuous adaptation to emerging threats, including climate change and population growth. To effectively overcome such challenges, stakeholders in the water industry must conduct an in-depth analysis of current trends and vision for the future. Water Services Association of Australia (WSAA) emphasizes innovation in terms of sensing opportunities for disruption and potential opportunities for improvement in the urban water industry. A forward-thinking plan would touch upon how IoT-based SHM can be the focal point in creating sound water systems aligned to adapt to future requirements.

With the WSAA calling for proactive reform within the industry, there is a clear link towards using IoT technologies in an effort to construct more sustainable and adaptable water systems (Turner et al., 2017). Dealing with innovative technologies allows for end-to-end understanding of the water infrastructures, thus allowing decision-makers to react both proactively and responsively. Furthermore, the emerging potential of IoT technologies can allow for inter-disciplinary integration of knowledge across various fields of dam management. Through a trans-disciplinary methodology, environmental science experts, engineers, and data scientists can collaborate in solving pressing research questions about water infrastructure. This collective strategy can supplement the fundamental elements of effective SHM practices, and in turn help develop improved dam management practices. Advances in remote sensing technologies, data analysis, and cloud computing are central to creating a dynamic system that adapts with new knowledge. Since the concept of models of everywhere suggests, this method must learn and modify on a regular basis with localized data, creating a connection between the physical and virtual space of water infrastructures (Ahmed et al., 2019). In addition to improving infrastructure assessment, this interaction provides a sustainable future addressing several problems in the water sector adequately.

7.5.1. Importance of Monitoring for Dam Safety

The significance of monitoring for dam safety cannot be overstated, considering the presence of thousands of dams worldwide that pose potential dangers to humans and the environment through structural failure. Utilization of effective monitoring systems plays an essential role in ensuring the stability and reliability of the structures. Recent developments in Internet of Things (IoT)-based Structural Health Monitoring (SHM) technologies have transformed traditional monitoring practices, offering real-time data acquisition and analysis. By merging sensors and algorithmic data-driven inputs, key metrics of structural well-being can be tracked in real time and assessed. Such a proactive approach enables early detection of abnormalities and enables work on maintenance to be prompted in due time, thereby reducing the prospect of catastrophic collapse (Cherta et al., 2020). Routine inspection also leads to a better comprehension of a dams life cycle, informing decision-makers of work that must be upgraded and maintaining public confidence in facility safety. Application of wireless sensor networks as a framework for IoT-based SHM systems enhances the efficiency of dam monitoring activities. The networks are made up of resource-constrained sensors that supply important data on stresses, strains, and other environmental parameters affecting the operation of dams. The systems, however, must overcome inherent shortcomings such as energy consumption to maintain their operability for a long period. Technologies like duty cycling demonstrate a feasible way of attaining energy efficiency in pursuit of the need for continuous data acquisition. By having sensors operate continuously without power supplies draining, researchers can develop greener monitoring systems that can provide longer, sustained monitoring of dam safety (Udoh et al., 2019). In that vein, energy consumption, monitoring accuracy, and system resilience tradeoffs remain fundamental challenges in the development of next-generation SHM technologies. Moreover, continued study that seeks to develop more advanced data analysis techniques continues to significantly contribute to the enhanced performance of dam safety surveillance systems. Using data-driven algorithms, scientists can more effectively detect, locate, and define potential damage to infrastructure. This analytical capacity transforms huge amounts of data gathered by sensors into actionable insights, which not only assist in the detection of dominant structural conditions but also assist in vulnerability prediction (Cherta et al., 2020). Integration with advanced machine learning techniques allows for prediction maintenance schedules, capable of significantly mitigating risks associated with dam operations. Consequently, as IoT-based SHM evolves, the potential for better-informed decisions regarding infrastructure management becomes an essential element of public safety and environmental protection in dam operation (Figure 7.14).

Figure 7.14. Importance of monitoring for dam safety.

Source: https://sisgeo.com/wp-content/uploads/2023/08/Ouldjet-Mellegue-Dam-1.jpg.

7.5.2. Internet of Things (IoT) Technologies Used in Dam Monitoring

The application of IoT technologies to dam monitoring is a breakthrough in structural health monitoring (SHM). Traditionally, dam monitoring was done with periodic surveys and manual data collection that used to favor slow responses to potential dangers. With IoT technologies, real-time data collection is now possible with the placement of a network of sensors that can continuously monitor parameters such as water levels, pressure, and structural health. Data-based algorithms do the lion's share of effort in processing this gush of data and providing actionable intelligence to guide decision-making. This real-time data processing is essential for effective risk management, particularly in locations of high frequencies of extreme weather conditions, where timely intervention can prevent catastrophic failure (Cherta et al., 2020). With the concept of models of everywhere gaining popularity in environmental sciences, its implementation with IoT technology in dam monitoring is highly promising. This is the principle of moving away from general models to tailored models that capture the individual characteristics of dams. IoT technologies enable such models to adapt in real-time through the incorporation of live inputs, thereby adapting to changing conditions and uncertainties. This is not just better for predictive power but also allows engineers and policymakers to quantify risks with greater precision. Cloud computing and data analysis trend enables this shift in the direction of adaptive modeling through a complex interaction between tracking in real-time and predictive research (Ahmed et al., 2019). Additionally, employing IoT technologies for dam monitoring is accompanied by challenges. Data security, reliability in sensor networks, and interfacings of heterogenous data types provide challenges that need a robust information management structure. Technological innovation in data-based algorithms has the capability to mitigate some of these challenges by improving the precision and reliability of damage detection and response mechanisms. These advancements in technologies reinforce the necessity for a multidisciplinary approach blending engineering, data science, and environmental policy. Future research agendas must work towards harnessing the potential of IoT overcoming current limitations, ultimately pushing towards an integrated strategy for dam safety that protects communities as well as environments (Cherta et al., 2020).

7.5.3. Lessons Learned from Internet of Things (IoT) Implementations in Water Infrastructure

IoT implementation in water infrastructure has brought into focus the urgent necessity to manage energy in sensor operation. IoT networks rely on wireless sensor networks, which are battery-powered. Energy efficiency is thus critical in designing and deploying systems. One significant challenge lies in balancing security with power consumption. While encryption and authentication enhance confidentiality and integrity of data, they put more energy load on sensors. Conversely, when there is no security provided, the infrastructure is exposed to energy drain attacks, i.e., denial of sleep attacks, that can adversely affect sensor lifespan and operation reliability. Therefore, a dual approach combining security and energy management becomes unavoidable to ensure IoT application is sustainable and effective in water infrastructure management (Udoh et al., 2019). Moreover, the lessons developed through case studies of IoT implementations identify flexibility in the deployment of technology to enhance system performance. Following several water infrastructure issues, the demand for off-site and modular construction practices has emerged as a primary focus. Through this collective approach, the integration of new technologies can facilitate a streamlined supply chain process like that of a manufacturing plant. The ability to incorporate new construction techniques can significantly reduce operations complexity, costs, and enhance service delivery in water management systems. The emphasis of such innovations accounts for how the application of existing technologies within an open-to-change ecosystem can lead to enhanced infrastructure resilience as well as efficiency (Gosling et al., 2017). Effective IoT deployments in water infrastructure also reflect the importance of end-to-end stakeholder engagement throughout the project life. Stakeholder communication that involves government agencies, local communities, and industry professionals ensures that the implemented IoT solutions address actual real-world issues effectively. This co-creative effort results in mutual understanding of expectations and needs, leading to informed decision-making and optimum solutions. Engagement with various stakeholders not only makes the development process more dynamic but also helps in foreseeing expected challenges and hence enhances the adoption and implementation of IoT technology in existing water management practice. These lessons emphasize that user-centric strategies combined with new technologies have the potential to lead to dramatic energy efficiency and security enhancement, eventually transforming water infrastructure management (Gosling et al., 2017; Udoh et al., 2019).

7.6. FUTURE TRENDS IN INTERNET OF THINGS (IOT)-BASED STRUCTURAL HEALTH MONITORING (SHM)

The combination of the IoT and SHM is bound to transform the management and maintenance of infrastructure systems. One leading trend is enhanced real-time monitoring capabilities through advanced sensor technology that collects vast quantities of structural integrity data. The sensors can detect infinitesimal variations in the condition of a structure and thereby support predictive maintenance and intervention. In addition, by incorporating Big Data analytics into IoT systems, it is now possible to dynamically process and interpret this information, which in turn improves decision-making regarding structural safety and resilience. Importantly, the application of these technologies also raises novel data privacy and user trust issues, most notably since real-time systems operate under dynamically reconfiguring conditions based on shifting user requirements and environmental contexts (Azorin-Lopez et al., 2018). Another upcoming development in IoT-based SHM is through the application of Cyber-Physical Systems (CPS), which are a new convergence of physical and virtual environments. With the use of CPS, structures are able to learn and respond autonomously to conditions sensed, allowing for a more dynamic means of monitoring and maintenance. This development signals the break-up from traditional models of static infrastructure management towards a more dynamic and adaptive model. Not only do these systems enhance the efficiency in operations but they also facilitate optimization of resources. With predictive maintenance models, organizations can move from being reactive to more proactive approaches, ultimately extending the lifespan of structures while minimizing expenditures. The implications of such a transition extend way beyond engineering, reaching policy frameworks and regulations that must be adaptive to such advanced technologies (Adams et al., 2019). In looking for the future of SHM using IoT, it is critical to tackle the meaning of interoperability between systems and technologies. The ability for various systems to communicate and share information in an efficient and seamless process is critical in ensuring that the effectiveness of IoT implementations in structural health management is maximized. With standards for connectivity and data exchange being solidified, more concerted cooperation among engineering teams, IT specialists, and even regulatory organizations is becoming necessary. This collaborative process can introduce innovations in design and application, enabling the development of comprehensive SHM systems that not only consolidate separate achievements but also encompass them as a seamless totality. Hence, the IoT-based SHM future is characterized by an integrated system with technology, users, and infrastructure converging towards optimum outcomes in safety and sustainability (Azorin-Lopez et al., 2018).

7.6.1. Emerging Technologies and Innovations in Internet of Things (IoT)

The IoT represents a revolution, particularly for Structural Health Monitoring (SHM). As infrastructure networks age, the need for robust monitoring rises, driving innovations capable of improving durability and service life. New technologies such as self-sensing and self-healing materials have gained significant interest with their ability to change the way construction is done traditionally. Interesting, though, are smart cement-based materials with infused nanostructures that can self-detect damage and repair cracks on their own, as noted in recent research. Aside from the promise of enhanced structural resistance, such advancements also strike a chord with the bigger picture of creating sustainable and efficient infrastructure solutions. The integration of IoT-sensitized sensors with such high-end materials further enhances their productivity by offering real-time information, informing maintenance activity, and minimizing downtime, thereby addressing both safety and operational efficiency in civil infrastructure structures (Leung et al., 2019). Advances in nanotechnology have made it possible to develop nanosensors which are the key for the application of IoT in SHM. These miniature sensors can provide crucial information about the structural integrity by detecting minute changes in their environment and can detect probable failure at the earliest. From the literature review, it is evident that various types of nanosensors, including those based on piezoelectric and conductive natures, have been successfully employed for bridge maintenance. Their calibrated output in terms of stress, strain, and other environmental factors make them invaluable tools in the hands of engineers. Their use, however, is not obstacle-free, particularly in terms of signal processing and data interpretation. As researchers proceed towards their solution, the expectation is that integration of nanotechnology and IoT platforms will redefine structural health monitoring's nature, ultimately leading to safer, more reliable infrastructure systems (Han et al., 2023). Moreover, the combination of IoT technologies with smart materials holds the prospect for a new generation in structural engineering, with applicability extending from monitoring to predictive maintenance and autonomous decision-making. Emerging IoT applications in this domain employ machine learning algorithms to analyze data readings from sensors and, therefore, enhance prediction regarding degradation of infrastructure. When organizations and cities invest in smart city solutions, the combination of intelligent, self-sensing materials and data analytics becomes important in safeguarding infrastructure resilience. With such technologies, stakeholders can proactively manage infrastructure health while maximizing resource use for maintenance and repair. The integration of these technologies not only develops a culture of sustainability but also has immense economic benefits as knowledge-based decision-making can greatly reduce operation costs and extend the life of infrastructure projects, infusing urban developments with resilience by design (Leung et al., 2019).

7.6.2. Potential for Machine Learning and Artificial Intelligence (AI) Integration

Synergism between machine learning and AI in the context of Structural Health Monitoring (SHM) using Internet of Things (IoT) has enormous potential to revolutionize the field. By analyzing vast amounts of structure-mounted sensor data, these technologies are able to make instantaneous decisions regarding structural integrity. Algorithms based on machine learning can be used to detect anomalies, predict failures, and schedule maintenance on an optimal basis, thereby ensuring higher safety and reduced operation costs. The ongoing advancement of computational capacity and the availability of large amounts of data ensure the feasibility of these integrations, as evidenced by their successful applications in various fields of building performance. It still comes with challenges, including the requirement of robust training sets and transferability of models, which hinder the mass implementation of such smart systems in real applications (Hong et al., 2020). Also, the interfacing of machine learning and big data analytics adds to the believability of SHM systems. Through this, it is made possible to employ advanced modeling methods that are able to handle and interpret complex patterns in the structural response. For instance, models can learn from history in order to predict future conditions and potential failure points accurately. Integrating AI-driven decision-making models presents a way of computerized response to structural defects identified, hence maximizing maintenance processes. However, this transition from theoretical application to practical use continues to be hampered by factors like the lack of massive labeled datasets and cries for concrete cost-benefit analysis in order to advance these technologies' widespread adoption (Doctor et al., 2020). Effective transcendence of these constraints is necessary for unleashing the full potential of AI for SHM. In summary, fusion of machine learning, AI, and IoT can bring about revolutionary possibilities in structural health monitoring but must address some fundamental challenges. As the industry continues to evolve, there will be a need for multidisciplinary collaboration using data science, building science, and engineering disciplines to develop credible, transferable models. Higher levels of collaboration among practitioners and researchers in the field are required, laying even more solid ground for the merging of these technologies in SHM implementation. Through overcoming the current hurdles and continuing to build on the understanding of the interaction between advanced computational methods and structural monitoring, stakeholders are able to reveal innovation that enhances safety, effectiveness, and resilience in built structures (Doctor et al., 2020; Hong et al., 2020) (Figure 7.15).

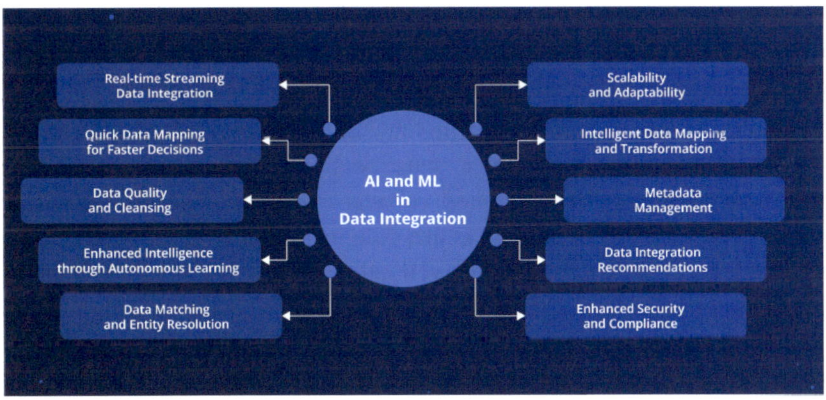

Figure 7.15. Potential for machine learning and artificial intelligence (AI) integration.

Source: https://d3lkc3n5th01x7.cloudfront.net/wp-content/uploads/2023/11/02034836/AI-and-ML-in-data-integration-1.svg.

7.6.3. Predictions for the Future of Structural Health Monitoring (SHM) in Infrastructure

The future of SHM in infrastructure is becoming more and more associated with innovations in Internet of Things (IoT) technologies. As the need for real-time information about the structural health of critical infrastructure grows, data-driven algorithms are becoming essential. These algorithms utilize sensor information to identify damage, enabling speedy decision-making by infrastructure managers. The evolution of these technologies also will make possible record levels of sensitivity in damage detection, localization, and rehabilitation options. Most importantly, the move toward smart infrastructures, through the integration of sophisticated sensor technologies, will make possible a greater capacity for continuous monitoring and to counteract potential failures in their incipience before they become compromising. This synergism can lower the maintenance cost while extending structures' lifespan, making the built environment more sustainable. All these developments are expected to support data-driven methodologies outlined in recent publications and to optimize the overall efficiency of SHM systems for future applications (Cherta et al., 2020). Future predictive maintenance methods will be transformed by machine learning and artificial intelligence being integrated into SHM systems. These technologies are capable of analyzing bulk datasets built up over time to identify patterns and predict future structural behaviors under varying conditions. Predictive analysis not only facilitates easier decision-making but also makes adaptive pathways (APs) required in tackling the complexities brought by climate change and other stressors to infrastructure.

Rising technology provides instruments that enhance the degree of reliability of measurements and predictions, thus enabling more solid infrastructure management methods. With the aid of adaptive pathways from continuous monitoring, infrastructure managers are able to establish interventions needed in advance, so that structures are well adjusted to dynamic environmental demands (Achillopoulou et al., 2023). The future of SHM is thus about making use of these technologies to create systems not only that are responsive but also proactive. As the field of resilience-based engineering develops, SHM will play a central role in ensuring that infrastructure is resilient to increasingly prevalent hazards. Real-time performance monitoring will be fundamental to SHM systems, allowing constant monitoring of structural health in constantly evolving circumstances. Being proactive in this way will allow stakeholders to implement timely mitigation measures and adjust maintenance strategies accordingly. Furthermore, the new paradigms that are combining IoT and the aspects of resilience and sustainability will lead to a more holistic approach to managing critical infrastructure. The integration will be the industry standard, which will ensure that SHM evolves in concert with emerging threats. As the growth advances, the emphasis on integration of technology and engineering concepts will be pivotal in building resilience and adaptability within our built environment (Achillopoulou et al., 2023).

SUMMARY

It is evident that applications of IoT technologies have significantly transformed traditional infrastructure management practices. The case studies evince the various methodologies adopted in various settings, which testify to a common theme: the need for timely data analytics to ensure structural stability and safety. The innovation not only maximizes the decision-making process but also facilitates proactive maintenance practices, thereby reducing long-term costs resulting from infrastructure collapses. As stressed in recent research on data reduction techniques, efficient processing as well as real-time monitoring capability is paramount in ensuring effective use of IoT systems in SHM. Such reliance on such advances speaks to a shift towards more responsive and adaptive means of addressing public safety and infrastructure resilience. Moreover, challenges in the field of IoT-based SHM cannot be overstressed. Technical limitations in terms of computation cost and storage of data restrict the performance of such systems, particularly with low-cost deployments. The research involving parallel execution of streaming History PCA proves the ability to circumvent these limitations through data compression and improved processing speed with no unacceptable data quality losses. It is not just impressive but also required to make IoT solutions deployable and useful across a variety of scales of infrastructure, in order to achieve tenfold compression with minimal degradation in the signal. This is part of a larger trend where technology enables capabilities that enable systems to function better under constricted conditions, enabling IoT to become possible for wider deployment in SHM. Finally, as we look to the future of SHM in the context of IoT, it is essential to be mindful of the social and infrastructural implications of these technologies. The benefit of heightened safety and reduced maintenance costs must be weighed against the ongoing need for ethical consideration of data privacy and security. The open-access nature of many discoveries, such as those published under the Creative Commons Attribution License, encourages collaboration and sharing of ideas, which are vital towards accelerating innovation in this field. Encouraging more research and development activities in parallel with addressing ethical issues will be essential for IoT-based SHM solutions' long-term success. As the case studies included herein demonstrate, making the crossroads of technology and ethics stronger will ensure that such advancements serve the requirements of society's infrastructure.

Incorporating IoT technologies into SHM systems has yielded significant results in various case studies. A remarkable study examined incorporating RFID and LoRa technologies into enhancing the performance of PC projects. The research demonstrated how an RFID and strain sensor-supported sensor network was able to capture real-time data on structural members in the course of installation. The data was seamlessly pushed to a cloud-based BIM model that allowed stakeholders to observe key details remotely. These innovations, besides increasing project performance, also facilitated enhanced decision-making by virtue of reliable data accumulation, thereby testifying to the promise of IoT in revolutionizing construction methods in SHM applications (Liu et al., 2019). Another significant discovery came from the application of Robotic Process Automation (RPA) for SHM, which is relatively new in civil engineering. By cutting down on the use of manual intervention in obtaining data from low-cost accelerometers, RPA reduced man-hours spent on system maintenance. Real-time fault detection and recovery, together with continuous monitoring of structural integrity, were attained through this automation. As a result, the study proved a model that not only maximized working efficiency but also made sure that any potential errors were rectified in real time, hence cutting downtime (Aguilera et al., 2022). These advancements demonstrate that RPA and IoT integration has the potential to further optimize SHM processes with a stronger infrastructure management strategy. All these case studies collectively demonstrate the revolutionary potential of IoT-based technologies adoption in SHM. By effective utilization of innovative solutions such as RFID, LoRa, and RPA, construction and engineering sectors have the potential to achieve maximum efficiency, reliability, and data-driven decision-making. The implications are that future research should look to further advance these technologies and review their deployment in more inclusive SHM systems. Furthermore, the continued growth of IoT in building construction makes it a pivotal area for creating sustainability and cost-effectiveness—both paramount in overcoming contemporary engineering challenges. As more evidence accumulates, the stakes for industrial practice and policy will surely be high, opening the way towards a new age of smart infrastructure management.

REFERENCES

1. Abdelgawad, A., Allen-Zhu, Z., Hardt, M., Li, J., Mitliagkas, I., Reyer, M., Rooshenas, A., et al. (2019). Embedding principal component analysis for data reduction in structural health monitoring on low-cost IoT gateways. https://core.ac.uk/download/578214124.pdf (accessed on 2nd October 2025).
2. Achillopoulou, D. V., Kromanis, R., Makhoul, N., Stamataki, E., et al. (2023). Adaptive pathways using emerging technologies: Applications for critical transportation infrastructure. https://core.ac.uk/download/604656019.pdf (accessed on 2nd October 2025).
3. Adams, R., Amafabia, D., Barker, T., Beka, A., et al. (2019). Proceedings of abstracts engineering and computer science research conference 2019. https://core.ac.uk/download/287582059.pdf (accessed on 2nd October 2025).
4. Agis Cherta, D., Gomez Vargas, R. C., Pedraza, C., Pozo Montero, F., et al. (2020). Damage identification in structural health monitoring: A brief review from its implementation to the use of data-driven applications. https://core.ac.uk/download/288625955.pdf (accessed on 2nd October 2025).
5. Aguilera, E., Atencio, E., Komarizadehasl, S., Lozano Galant, J. A., et al. (2022). Using RPA for performance monitoring of dynamic SHM applications. https://core.ac.uk/download/534172616.pdf (accessed on 2nd October 2025).
6. Ahmed, S., Atzori, L., Hankin, B., Basco-Carrera, L., Bastin, L., Beven, K., Beven, K., et al. (2019). Models of everywhere revisited: A technological perspective. https://core.ac.uk/download/322489447.pdf (accessed on 2nd October 2025).
7. Aish, S., Crawley, D., Eastman, C., Eastman, C., Eastman, C., Engelbart, D., Salmerón-Manzano, E., et al. (2019). Bibliometric maps of BIM and BIM in universities: A comparative analysis. https://core.ac.uk/download/286590556.pdf (accessed on 2nd October 2025).
8. Akyildiz, I. F., Alvarado, C., Ashton, K., Banerjee, A., Cabral Pinto, M. M., Chou, C., Mora, H., et al. (2013). Sensing as a service model for smart cities supported by Internet of Things. http://arxiv.org/abs/1307.8198 (accessed on 2nd October 2025).
9. Alonso Zárate, J., Alonso Zárate, L. G., Kalalas, C., Laya, A., et al. (2016). Goodbye, ALOHA! https://core.ac.uk/download/81571287.pdf (accessed on 2nd October 2025).

10. Andrawos, M., Aßmann, U., Bass, L., Behara, D. G. K., Betts, T., Bianco, P., Bryant, D., et al. (2019). Microservice transition and its granularity problem: A systematic mapping study. https://core.ac.uk/download/426749569.pdf (accessed on 2nd October 2025).

11. Azorin-Lopez, J., Bradley, D., Hehenberger, P., Milne, D., et al. (2018). Mechatronics & the cloud. https://core.ac.uk/download/228178466.pdf (accessed on 2nd October 2025).

12. Baaziz, A., Kleiner, A., Feigelson, E. D., Waller, M. A., Boyd, S., Vandermerwe, S., & Zhou, Z. (2015). Big data and the Internet of Things. http://arxiv.org/abs/1503.07092 (accessed on 2nd October 2025).

13. Barbarov-Rostán, G., Cejudo Ramos, P., Márquez, M., Eduardo, S., et al. (2018). Smart vest for respiratory rate monitoring of COPD patients based on non-contact capacitive sensing. https://core.ac.uk/download/161255152.pdf (accessed on 2nd October 2025).

14. Britchenko, I., Cherniavska, T., & Machashchik, P. (2018). New trends in development of services in the modern economy. https://core.ac.uk/download/323058294.pdf (accessed on 2nd October 2025).

15. Castedo, L., Fernández-Caramés, T. M., & Fraga-Lamas, P. (2017). Towards the Internet of smart trains: A review on industrial IoT-connected railways. https://core.ac.uk/download/95054341.pdf (accessed on 2nd October 2025).

16. Doctor, F., Iqbal, R., Mahmud, S., More, B., et al. (2020). Big data analytics: Computational intelligence techniques and application areas. https://core.ac.uk/download/228156656.pdf (accessed on 2nd October 2025).

17. Ewart, P. (2019). A comparison of processing techniques for producing prototype injection molding inserts. https://core.ac.uk/download/322841895.pdf (accessed on 2nd October 2025).

18. Gosling, J., Koskela, L., Kumar, M., Tezel, A., et al. (2017). From construction to production: Enablers, barriers and opportunities for the highways supply chain. https://core.ac.uk/download/145234527.pdf (accessed on 2nd October 2025).

19. Han, D., Hosamo, H., Nie, R., Ying, L., et al. (2023). A comprehensive review and analysis of nano sensors for structural health monitoring in bridge maintenance: Innovations, challenges, and future perspectives. https://core.ac.uk/download/590878491.pdf (accessed on 2nd October 2025).

20. Hong, T., Luo, X., Wang, Z., Zhang, Y., et al. (2020). State-of-the-art on research and applications of machine learning in the building life cycle.

https://core.ac.uk/download/288432582.pdf (accessed on 2nd October 2025).

21. Leung, C., Mishra, D., Yu, Y., et al. (2019). Self-sensing and self-healing 'smart' cement-based materials–A review of the state of the art. https://core.ac.uk/download/237905115.pdf (accessed on 2nd October 2025).
22. Leung, K. K., Leung, V. C. M., Liu, C. H., Moreno, W., et al. (2016). Context-awareness for mobile sensing: A survey and future directions. https://core.ac.uk/download/30610426.pdf (accessed on 2nd October 2025).
23. Liu, Z., Mbachu, J., & Zhao, L. (2019). Development of intelligent prefabs using IoT technology to improve the performance of prefabricated construction projects. https://core.ac.uk/download/231841646.pdf (accessed on 2nd October 2025).
24. Shi, X. (2017). Recent advances in sustainable winter road operations – A book proposal. https://core.ac.uk/download/286999895.pdf (accessed on 2nd October 2025).
25. Stahl, E., & Dupuis, N. (2017). Trends in smart city development. https://core.ac.uk/download/83111937.pdf (accessed on 2nd October 2025).
26. Turner, A. J., & White, S. (2017). Urban water futures: Trends and potential disruptions. https://opus.lib.uts.edu.au/bitstream/10453/77441/1/WSAA-ISF-URBAN%20WATER%20FUTURES-2017-01-31.pdf (accessed on 2nd October 2025).
27. Udoh, E., & Udoh, E. (2019). An energy aware and secure MAC protocol for tackling denial of sleep attacks in wireless sensor networks. https://core.ac.uk/download/305119264.pdf (accessed on 2nd October 2025).

CHAPTER 8

SECURITY AND RELIABILITY IN SHM SYSTEMS

Contents

INTRODUCTION

Increased reliance on structural health monitoring (SHM) systems in various industries underscores the supreme significance of security and reliability. SHM systems have a significant role to play in ensuring the structural well-being of infrastructure, from bridges and skyscrapers to offshore wind turbines, with an aging infrastructure and more uncertain environmental conditions. SHM technologies' deployment must not only focus on data buildup but also guarantee that the acquired information is safeguarded from malicious attacks and remains stable in operation under diverse, sometimes hostile operating environments. In light of this, for instance, wireless sensor networks, a major component of SHM, rely principally on retaining functionality in adverse conditions, and thus highlight the importance of robust security protocols (Kerkhoff et al., 2009). As the costs of safe operation mount, so do the issues with implementing SHM systems that are compliant with security and reliability standards. An example is the development of renewable energy sources, particularly onshore wind farm development. Considering that the UK is set to realize ambitious renewable energy goals, ensuring the safety of such a facility becomes paramount since any form of damage to its structure would have grave implications not only on energy production but also to public safety. The necessity for specialized SHM solutions that are not only cost-effective but also dependable in the surveillance of structural health is thus apparent (Currie et al., 2012). Taking these aspects into account enhances the potential performance and lifespan of energy systems and thereby solidifies their presence within next-generation infrastructure. The integration of new technologies into SHM systems, such as the design of adaptive and intelligent monitoring solutions, also provides the potential to boost security and reliability. As these systems progress, they will be required to integrate with adaptive algorithms that can respond to changing environmental conditions and potential cybersecurity threats. A judicious focus on the physical robustness of SHM devices and the integrity of their software is a built-in approach to monitoring. Having the ability for anticipation and mitigation of failures in advance of their occurrence is critical to maintaining operating security and public trust in infrastructure systems. Improving the physical and cyber threat resilience of SHM systems not only guarantees their longevity but also that of the assets they are designed to protect, hence its development is key for use in the future across various fields.

Definition of Structural Health Monitoring (SHM)

Structural Health Monitoring (SHM) is a systematic process focused on civil infrastructure health evaluation and management. This multidisciplinary discipline entails continuous monitoring of structures through the incorporation of sensors and analysis systems. By monitoring real-time parameters such as strain, vibration, and displacement, SHM provides vital data regarding the state and stability of various infrastructures such as bridges, buildings, and dams. The applicability of SHM becomes evident considering the growing need for reliable infrastructure maintenance, particularly in light of deteriorating infrastructure and rising environmental concerns. The reliability and accuracy of SHM systems are important since these play crucial roles in ensuring that such buildings can withstand unforeseen stress factors and progressive wear and tear over time. The underlying principles of SHM are based on advanced technologies for increasing structural data acquisition and processing. They are enabled through the deployment of wireless sensor networks, fiber optic sensors, and advanced data acquisition systems with the ability to monitor at high resolution over extended periods. The application of machine learning and data analytics also aids the interpretability of data collected, supporting predictive information regarding structural performance as well as failure. Notably, these technologies not only improve the efficiency of maintenance practice but heavily impact SHM system security and reliability. These advances warrant close attention, since the efficacy of SHM has a direct impact on safety and longevity of key infrastructure, and therefore on public confidence and safety. From the security and reliability point of view, also of concern are the definition and deployment of SHM taking into account lurking vulnerabilities in the monitoring systems themselves. This involves not only protecting the physical buildings from the environment, interference, and damage but also implementing robust data integrity and cybersecurity measures. Through the use of digital infrastructures to transfer and process the data, compromise could make SHM results unreliable, with the potential to cause harmful consequences. It is relevant to analyze the intersection of SHM and cybersecurity in light of the ongoing discussion regarding the safeguarding of such vital monitoring techniques from external interference. It is the responsibility of organizations to be proactive, necessitating extensive security protocols to protect the collected data as well as SHM working models to facilitate the overall success of these vital systems (Bush et al., 2011; Dong et al., 2010) (Figure 8.1).

Figure 8.1. Structural health monitoring (SHM) represents a systematic approach.

Source: https://sisgeo.com/wp-content/uploads/elementor/thumbs/SHUTTE1-2-qdbxyvv6t8crw02t6vm6aj0fkilv1vz738ntrxu6bs.jpg.

Importance of Security and Reliability in Structural Health Monitoring (SHM) Systems

Security and reliability are of the highest priority when it comes to SHM systems, particularly when the systems are deployed in high-consequence applications. The deployment of SHM in environments such as aging infrastructure or offshore wind farms highlights the need for robust security features. As infrastructure ages, the incidence of natural and human-made hazards increases, and systems are required that not only ensure structural integrity of the assets but also safeguard sensitive data collected under surveillance. High reliability in those systems is strictly linked with public security, as latent structural failure holds immense dangers. Dependence on trustworthy SHM systems rests on the integration of sound security measures to circumvent unauthorized access, thereby system functionality and data integrity. Harsh environments subject SHM systems to additional vulnerability. For instance, wireless sensor networks, which are the norm in remote monitoring, need to be robust against potential interference or tampering. These vulnerabilities can lead to security intrusions that impact not only data collection but potentially the safety of public in general. The literature stresses the need to ensure that there are robust protocols in place to ensure consistency of operations at less than optimum conditions, and particularly in applications for application in wind power systems and similar critical infrastructure. As seen, security and safety are the pillars of success in renewable energy ventures in the UK, where structural strength is one of the keystone factors toward ensuring secure supply of energy. The creation of monitoring solutions that specialize in handling such specific needs is imperative for sustaining excellence in operation and fostering the confidence of the public in the use of SHM (Currie et al., 2012). Compilation of an overall profile of the significant parameters of SHM systems would demand an interdisciplinary approach based on risk assessment and enhancement of reliability with emerging technologies. Application of advanced monitoring technologies such as condition-based monitoring systems would be able to facilitate early detection of likely failures, hence allowing timely intervention or maintenance. These anticipatory measures not only improve the robustness of SHM systems but go a long way in promoting the lifespan of critical structures. Furthermore, high robustness should guarantee uninterrupted availability over the systems' entire service life, a requirement emphasized by real-world practice where environmental stresses can cause quite a lot of stress. It is from this security and reliability viewpoint that the investors are able to reinforce investment in SHM technologies such that these systems remain operational and trustworthy (Kerkhoff et al., 2009).

8.1. OVERVIEW OF STRUCTURAL HEALTH MONITORING (SHM) SYSTEMS

SHM systems form a significant aspect of today's engineering practice, particularly in ensuring the security and reliability of infrastructure. These systems apply various technologies to track the condition of structures such as bridges, buildings, and wind turbines throughout their life. With sensors to detect parameters such as strain, displacement, and vibration, SHM facilitates easy detection of potential defects before they turn into major issues as well as enhancing the maintenance process. This avoids premature interventions, thereby extending the life of structures as well as enhancing safety. Increased complexity of structures necessitates the integration of SHM systems into design and operation methodologies to preclude possible failures. The application of SHM is particularly essential in high-reliability and security industries, such as renewable energy. For onshore wind turbines, however, the need for dependable SHM systems is driven by the randomness of wind loads and operation loads. As highlighted in the latest studies, structural integrity is key to making wind power feasible as a renewable source. Low-cost customized monitoring solutions for testing turbine foundation structural condition are especially needed to give early warning of structural failure. The incorporation of such tailored systems improves the efficiency of operations and guards against expensive downtime, thereby solidifying the security and reliability of the energy supply chain as exemplified in Currie et al. (2012). Moreover, the advancement of SHM technologies continues to progress with the establishment of new methodologies that improve their effectiveness. Applications of artificial intelligence and machine learning in data analysis have changed the process of analyzing structural data and making decisions. The change guarantees predictive maintenance, whereby structural issues are predicted based on patterns from previous data. Since structures are still being made more complex, there is a requirement for robust SHM frameworks to deal with the resultant risk. Better data collection and web monitoring capability enable decision-makers to decide in such a way that safety is ensured while costs are minimal. Lastly, ongoing development of SHM systems is a sound foundation for achieving greater security and dependability in various industries based on their key roles in future infrastructure management based on Bush et al. (2011).

8.1.1. Components of Structural Health Monitoring (SHM) Systems

Among the systems of SHM, there are several primary factors which collaborate in harmony to offer security and reliability of the infrastructure. Sensors are at the center of SHM systems because sensors are responsible for providing critical data regarding the functionality of structures. Sensors, including strain gauges, accelerometers, and fiber optic sensors, are mounted across a structure to continuously take data regarding stress, vibration, and temperature. By processing this data, engineers are able to analyze the structural integrity and any issues before they turn into significant failures. The integration of these sensors in various components of infrastructure helps in achieving an overall monitoring configuration, essential for the assurance of operating integrity and longevity. Data analysis and management packages are also another essential component in SHM systems. Once raw data is collected via sensors, it is analyzed intensively using algorithms and software platforms specifically designed to decipher complex datasets.

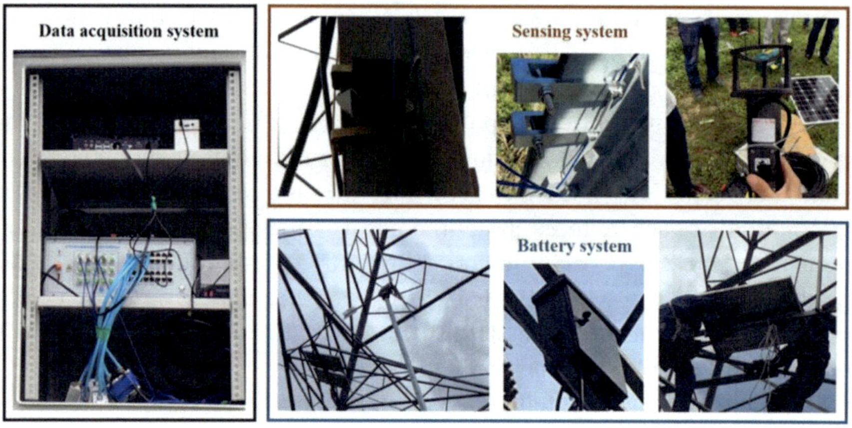

Figure 8.2. Components of structural health monitoring (SHM) systems.

Source: https://www.researchgate.net/publication/348123151/figure/fig5/AS:1133015455543299@1647143343919/Components-of-the-SHM-system.jpg.

Advanced techniques such as machine learning and predictive analytics are increasingly being employed to identify patterns and predict structural performance under varying loads. Not only does this analytical function render the outcome more dependable, but it also makes possible the making of quick maintenance and repair decisions. A correct interpretation of sensor readings is very important in applications like wind turbine condition monitoring, where randomness of the loads requires accurate predictive models to react ahead of time to structural vulnerabilities (Currie et al., 2012). The integration of communication systems also makes SHM systems stronger, enabling vital information on the structure's health to be sent in real time to engineers and maintenance crews. Today's SHM systems prefer to employ Internet of Things (IoT) platforms for wireless information exchange, hence remote monitoring and prompt response to any abnormalities felt. Such connectivity brings about openness and accessibility of structural health information and promotes proactive maintenance and repairs. Along with the smart materials that can sense and react to the surroundings, these communication systems form a strong network that is vital for intelligent infrastructures. These technologies not only increase overall security within infrastructure but are required due to changing urban conditions to meet present environmental and engineering requirements (Angelis et al., 2025) (Figure 8.2).

8.1.2. Types of Structural Health Monitoring (SHM) Technologies

The SHM technologies category has a diversity of approaches employed to guarantee infrastructure integrity. Passive and active monitoring systems are the two broad types of SHM technologies. Passive systems, more times than not, employ sensors to record changes in structural behavior without actively giving external stimuli. For instance, acceleration gauges and accelerometers quantify vibrations and deformations primarily from known structural responses. Active SHM systems, however, employ external excitations, for instance, controlled vibration, in quantifying in-depth details about a structure's condition. This distinction in methodology highlights the niche applications of SHM technologies, where passive systems excels best in documented conditions, while active systems excel in quantifying minor damage during inspection processes. Data-driven algorithms are a notable milestone in SHM technologies, a sign of the transformation of structural assessment being performed. By making use of data that is collected from an array of sensors, such algorithms facilitate real-time analysis and decision-making capabilities. Described here, such techniques can be decentralized, relying on computing power to handle large data sets collected from continuous monitoring processes. This approach not only enhances the accuracy of damage detection but also enhances operational effectiveness in maintenance and management processes. Deservingly, data-driven methods can be grouped into some sub-categories, including machine learning and statistical techniques, each with varying advantages in structural damage detection, localization, and classification, thereby complementing the overall safety of infrastructure networks (Cherta et al., 2020). Furthermore, the use of smart materials and technologies within SHM systems is transforming the practice. The development of smart sensors that can sense and react to changes in the environment is a concept that is revolutionary. Smart systems utilize embedded algorithms that allow for data analysis to be conducted locally, reducing the necessity for central processing and enhancing real-time response. Moreover, wireless sensor technology allows the improvement of the dependability and security of SHM systems, with less risk of losing data when there is compromised connectivity. As such, the evolving improvements in these technologies ensure that utilization of advanced materials and monitoring processes represents a highly firm ground for ensuring structural integrity and thus addressing severe issues in the field of infrastructure management and maintenance (Dong et al., 2010).

8.1.3. Applications of Structural Health Monitoring (SHM) in Various Industries

Applications of SHM range across various industries, highlighting the significance of SHM in maintaining safety and reliability. In civil engineering, SHM systems find application for structural condition monitoring in bridges, dams, and buildings. They employ data-driven algorithms to analyze real-time sensor data, which allows them to identify impending structural failure at an early stage. As noted by Cherta et al. (2020), the efficiency of data-driven algorithms is important because they aid in quick decision-making procedures through consistently acquired data. By integrating the monitoring systems, engineers can effectively track the health of structures, undertaking maintenance in a timely manner before catastrophic collapse. This proactive approach of infrastructure management not only enhances safety but also extends the lifespan of critical assets, providing evidence of SHMs significant value to public safety and economic performance. The aerospace industry also uses SHM technologies to a large extent, namely for aircraft maintenance and safety. Advanced SHM systems are used to monitor airframe and component condition continuously, providing critical information directly to maintenance schedules and safety inspections. This decreases the probability of in-flight failures and optimizes operational efficiency by keeping aircraft in condition-based maintenance as opposed to indiscriminate time-schedule-based maintenance. The use of SHM in aviation is also warranted on the grounds of potential cost savings in terms of unnecessary maintenance and loss of aircraft time, an argument expressed in the cross-industry requirement for capability stated in Pierson-Balik et al. (2005). The emphasis on communication skills and teamwork also raises the productivity of SHM in this sector, as planning for maintenance and analysis of data through collaborative processes can lead to more sophisticated decisions regarding aircraft well-being. In the energy sector, SHM can play a big part in monitoring main infrastructures such as wind turbines and power plants. These facilities must operate under tight safety regulations since there are risks associated with energy generation system failure. SHM systems are employed to monitor continuously the structural condition of components, detecting damage indicators like fatigue or degradation that can lead to system failure. The application of data-driven algorithms not only helps in damage detection but also in prognosis and predictive maintenance strategies, as mentioned in Cherta et al. (2020). Through maintaining high reliability in energy infrastructures, SHM supports not only security but also offers a stable energy supply, which is in turn interconnected with economic stability. Thus, the application of SHM technologies in various sectors demonstrates the critical contribution of SHM in increasing security and reliability in critical systems (Figure 8.3).

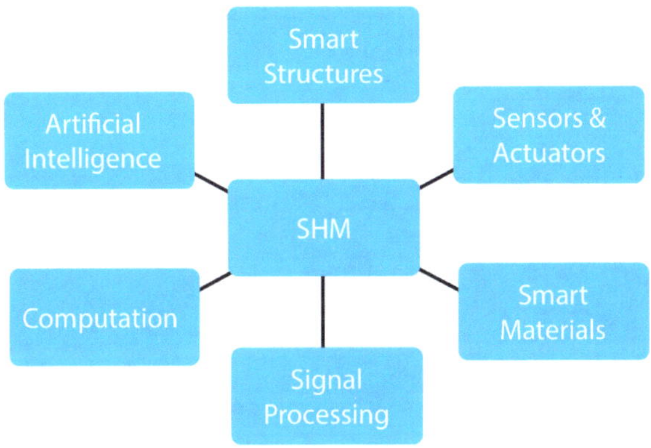

Figure 8.3. The applications of structural health monitoring (SHM).

Source: https://www.mdpi.com/buildings/buildings-13-01903/article_deploy/html/images/buildings-13-01903-g002.png.

8.2. SECURITY CHALLENGES IN STRUCTURAL HEALTH MONITORING (SHM) SYSTEMS

Integration of SHM systems in infrastructure has presented the potential for data-driven decision-making. These systems are, however, faced with various security threats that must be thoroughly mitigated to ensure their reliability and effectiveness become a certainty. They include cybersecurity-related threats such as unauthorized access and modification of data, insider threats, and they are prominent threats to SHM systems that lead to tampering with data. For instance, if an attacker were to manipulate sensor outputs or alter monitoring logs, the impact would be calamitous, destroying confidence in the systems output and ultimately undermining the safety of users and assets. It is thus important that high-security controls are put in place that encompass encryption, online surveillance, and intrusion detection systems that will safeguard against potential breaks that can exploit system vulnerabilities. Additionally, the dependence of damage identification in SHM systems on data-driven algorithms highlights a key paradox between performance and security. Such algorithms are critically dependent on the integrity and accuracy of arriving data, which can be compromised if left unsecured. As noted in recent literature, while data-driven methods are beneficial for real-time analysis as well as decision-making, they require high-performance processors with adequate computational capabilities. This reliance on computational capability can offer strong points of vulnerability to cyber-attacks if proper security systems are not in place.

Figure 8.4. Security challenges in structural health monitoring (SHM) systems.

Source: https://www.resensys.com/Blog/wp-content/uploads/2025/02/Digital-Twin-Technology.png.

Therefore, a balance between computing competence, data accuracy, and robust security systems has to be achieved to enhance reliability and minimize risks associated with utilization of SHM systems within critical infrastructure. Additionally, the installation of SHM systems comes with the installation of many sensor devices that, in addition to collecting critical data, pose new security concerns. Since there are varied sensors employed, they are susceptible to all forms of threats from physical to cyber-attacks, which necessitate a comprehensive security policy for these threats. Cherta et al. (2020) highlights the importance of robust monitoring and assessment mechanisms based on data-driven algorithmic frameworks for effective damage identification. However, as greater reliance is placed on these systems, so too must their underlying technologies be protected against misuse. A multi-layered security approach that combines hardware defenses, secure communication protocols, and software protections is needed in fortifying SHM systems against future threats, thus ensuring the reliability and safety of our civil infrastructure (Figure 8.4).

8.2.1. Cybersecurity Threats to Structural Health Monitoring (SHM) Data

Cybersecurity threats must focus more with the integration of SHM systems in critical infrastructures. These systems, which capture and relay sensitive data regarding the status and security of buildings, are vulnerable to cyber-attacks. The agents of threats can exploit the interconnectivity of SHM devices, as they tend to rely on wireless mechanisms of transmission that can be intercepted or compromised with ease. These breaches can have calamitous consequences, ranging from compromising not only the integrity of the information but even lives and infrastructure. As the European Safety, Reliability and Data Association (ESReDA) emphasizes, the resilience and preparedness of infrastructures depend heavily on the effectiveness of cybersecurity in safeguarding important data pertaining to SHM systems (Mohamed et al., 2017). Thus, understanding the various forms of cybersecurity threats becomes essential in ensuring the reliability of such systems. Besides, the advent of the Industrial Internet of Things (IIoT) has revolutionized the landscape of SHM in the aspect of real-time data capture and processing via interconnected devices. Nevertheless, it has the natural tendency to improve cyber vulnerabilities. As the railways and other industries adopt high-end communication technologies, they are also vulnerable to sophisticated cyber-attacks that may reduce data integrity and system performance. Research indicates that predictive maintenance practices and data management can minimize cyber-attack risks, and the overall security response of SHM systems (Castedo et al., 2017). Thus, a clear comprehension of potential cybersecurity threats and putting in place robust security measures is required to ensure reliability and effectiveness of SHM systems. Furthermore, the impact of cyber-attacks on cybersecurity extends beyond in-built data integrity problems; it undermines public confidence in the integrity of infrastructures as well. The consequences of successful cyber-attacks on SHM systems can lead to colossal economic losses, regulatory repercussions, and stakeholder trust loss. For example, railroad networks employing outdated communication protocols may be more exposed to attacks, which can disrupt operations or compromise the safety controls. By engaging with current cybersecurity paradigms and continuous assessment of evolving threats, industries can make their systems more resilient to cyber threats. Overall, reacting to cybersecurity threats is not only quintessential to protecting SHM data but also to supporting long-term sustainability of critical infrastructure systems and maintaining public safety and trust in these essential services (Figure 8.5).

Figure 8.5. Cybersecurity threats to structural health monitoring (SHM) data.

Source: https://www.miniorange.com/blog/assets/2025/cyber-security-threats.webp.

8.2.2. Physical Vulnerabilities of Structural Health Monitoring (SHM) Sensors

Physical vulnerabilities involving SHM sensors play a vital contribution towards the overall reliability and security of SHM systems. The sensors are likely to be fixed in areas with exposure to harsh conditions and may be negatively impacted by environmental factors like temperature fluctuations, humidity, and vibration. These physical vulnerabilities not only risk sensor components' longevity but also compromise data integrity, potentially creating faulty estimates of structural health. There must be a fine balance between the design of these sensors and the environmental state they will be in; inappropriate protection or faulty installation can magnify threats. Hence, it is required to support innovations that enhance the robustness of SHM sensors against these vulnerabilities so that they can operate smoothly, even under harsh conditions. Also, the integration of SHM sensors into larger infrastructural systems poses other challenges as concerns susceptibility to physical tampering or damage. A ready availability of monitoring equipment exposes them to malicious actions, including vandalism or sabotage, which poses significant challenges to the maintenance of systemic integrity. This is particularly the case in sectors such as transport, where SHM sensors form part of safety and reliability provision in bridges and railways. To mitigate against these threats, enhanced security provision should be performed alongside maintenance practice to safeguard these resources. As indicated in many of the studies, the application of robust architectures that deal with physical attacks and prospective cyber vulnerabilities is extremely critical for the effective functioning of SHM systems to render them trustworthy in the context of critical infrastructure applications. Finally, as SHM technologies evolve further, an integrated approach must be developed to deal with their physical vulnerabilities. Technological advancements in wireless sensor networks (WSNs), for example, offer such potential benefits of scalability and real-time reporting but raise the issue of signal integrity and sensor resilience in harsh environments. The railway industry, in particular, stands to gain from its improved SHM methods that leverage the Industrial Internet of Things (IIoT) to maximize monitoring and predictive maintenance while mindful of physical vulnerabilities that accompany such integration. Through active research and development, stakeholders can build improved sensor resilience and total system reliability, as proven by recent reviews of intelligent infrastructures in high-risk environments (Castedo et al., 2017; MOHAMED et al., 2017). Such proactive measures will eventually safeguard the infrastructures that support modern society, dampening threats emanating from physical and environmental threats.

8.2.3. Risks Associated with Data Transmission

Data communication in Structural Health Monitoring (SHM) systems is vulnerable to numerous threats that can compromise reliability and security to a great extent. These threats originate from various sources, such as data interception during communication and unauthorized access. As SHM systems usually involve remote sensor data gathering, opportunities for cyber-attack are a significant concern. Malicious users can exploit weakness in communication channels, which leads to data tampering or unauthorized system operation modification. Effective and prompt information is necessary in order to ensure the upkeep and security of buildings, and any interference that arises from data transmission risks has serious real-life consequences that not only undermine the structural integrity but also public safety. Furthermore, integrity and quality of passed data are essential to maintain effective decision-making in SHM implementations. Data corruption may be achieved through degraded network performance, sensors' inherent faults, or cyber-attacks, thereby complicating damage identification procedures. If untrusted data is combined with analytical models, the achieved outputs can lead to wrong conclusions about a structures state of health, thereby degrading the effectiveness of maintenance decisions. This is supplemented by reliance on data-based algorithms, which are likely to need high-performance processors capable of processing data in real-time, as indicated in Cherta et al. (2020). Hence, data transmission reliability should be a point of concern for researchers and professionals engaged in SHM.

Finally, countering the vulnerabilities of data transmission entails establishing robust security protocols and mechanisms. Encryption, secure communication, and authentication capabilities are basic requirements for the protection of transmitted information and its integrity. As emphasized by Bush et al. (2011), the use of these capabilities is highly efficient in minimizing the possibility of unauthorized access and enhancing the overall security position of SHM systems. Further, routine checks and updates on the security infrastructure can allow for the adaptation to emerging threats and vulnerabilities and ensure unbroken protection of critical infrastructure. The inclusion of such practices not only ensures protection of integrity of data but also gives confidence to the effectiveness of SHM systems, ultimately culminating in safer and more sustainable architectural designs.

8.3. RELIABILITY ISSUES IN STRUCTURAL HEALTH MONITORING (SHM) SYSTEMS

The safety and integrity of infrastructures, particularly in high-risk environments like wind farms, are heavily dependent on the dependability of SHM systems. SHM systems apply various sensors to continuously monitor structural conditions and raise an alarm for any probable failures. However, the variability of load regimes makes it difficult to predict structural integrity in the long term. Studies indicate that affordable specialized monitoring solutions play a key role in slowing down reliability problems, especially with the growing use of renewable energy sources. The effectiveness of such systems not only affects the operational efficiency but also the sustainability of renewable energy plants over the long term and therefore the necessity of developing effective monitoring protocols capable of predictively diagnosing imminent failures before they become calamitous hazards (Currie et al., 2012). One of the most important considerations of SHM systems is the damage identification methodology that is used, which can have a major impact on reliability results. Two of the most common strategies that are used in this case are data-driven algorithms and model-based analyses. Data-driven approaches rely significantly on gathering and processing sensors' data to provide real-time information about the structure's performance. While these algorithms, in general, are likely to be favored for their ability to provide actionable intelligence in real time, they put heavy computational resource demands at the cost of challenging real-time decision-making capacities' reliability (Cherta et al., 2020). While model-based methods, being less likely to be resource-intensive, are subject to being limited by the accuracy of the models to be implemented, there is a need for the application of appropriate methodology depending on certain operational scenarios. Other than methodological concerns, selection and deployment of sensors significantly impact the reliability of SHM systems. Richness of sensors in terms of availability may lead to fluctuations in the quality and nature of data obtained, which is crucial for accurate detection and measurement of damage. Sensors are to be chosen wisely based on their sensitivity, lifetime, and robustness in environmental conditions to function uniformly over a long time. Apart from that, the use of advanced sensor technologies, such as wireless networks and intelligent monitoring systems, can make data acquisition more effective and improve system reliability. In totality, solving these sensor issues is vital for improving security and reliability of SHM systems to enable them to operate effectively under fluctuating conditions and protect infrastructure investments (Cherta et al., 2020) (Figure 8.6).

Figure 8.6. The reliability of structural health monitoring (SHM) systems.

Source: https://sisgeo.com/wp-content/uploads/2023/08/San-Siro-stadium.jpg.

8.3.1. Sensor Accuracy and Calibration

Sensor accuracy in structural health monitoring (SHM) systems is essential in providing reliable data collection and interpretation. Sensor accuracy has a direct influence on the effectiveness of damage detection and assessment techniques, which are important in maintaining the structural stability of buildings and infrastructures. High accuracy of sensor measurements ensures that engineers and operators receive consistent data to make decisions upon real-time data. According to research, damage identification process can be optimized through data-driven algorithms that are highly reliant on accurate sensor data. These data not only contribute to damage localization as well as to the estimation of severity but also facilitate the development of smart structures that can conduct real-time self-assessment. Thus, ensuring sensor accuracy through stringent calibration processes is indispensable to meet the needs of intelligence for modern SHM systems, thereby enhancing the overall security and reliability of monitored structures (Cherta et al., 2020). Calibration is an essential process of achieving accuracy and dependability in sensor measurements, which needs systematic practices that are compatible with the specific operational context of SHM systems. Effective calibration compensates for potential variations caused by environmental factors, production tolerances, and operational wear and tear, thus enabling sensors to provide consistent results over time. Sensor networks' complexities, typically being part of larger cyber-physical systems (CPS), also require periodic calibration inspections.

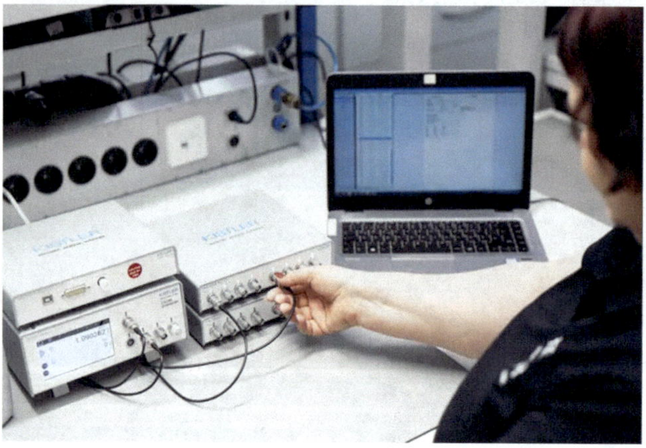

Figure 8.7. Sensor accuracy and calibration.

Source: https://kistler.cdn.celum.cloud/SAPCommerce_CMSGalleryStandard_720x480/kistler_AndreasHenn-006-DSA05967-Bearbeitet-m_16414.webp.

Wireless sensor networks (WSNs), enabling CPS to be used in SHM, require highly stringent verification techniques to ascertain their credibility. By application of sophisticated calibration techniques, operators can improve the reliability of continuous monitoring systems, hence reaffirming their ability to transmit precise data efficiently. It is important in enabling timely action where structural deviations are achieved, ensuring enhanced safety and performance integrity (Duquennoy et al., 2013). The interplay of sensor calibration and data processing technology also plays a significant role in the improvement of SHM systems efficacy. With enhanced algorithms in handling data from sensor output, individual sensor calibration becomes increasingly vital towards achieving appropriate damage classification and prognosis. Sensors do not only find application in real-time monitoring by such algorithms but also in predictive maintenance, which can significantly lower operating costs and enhance infrastructure lifespan. This symbiosis represents a significant step toward intelligent infrastructure management, where accuracy of the data obtained from calibrated sensors stimulates the decision-making process for structural safety measures. Therefore, improving calibration protocols is not merely an operational need; it is a strategic need that enables the applicability and efficacy of SHM systems in contemporary engineering settings (Cherta et al., 2020; Duquennoy et al., 2013) (Figure 8.7).

8.3.2. Environmental Factors Affecting Reliability

Environmental factors that can impact sensor performance and structural integrity have a great effect on the reliability of Structural Health Monitoring (SHM) systems. Environmental conditions like temperature variation, humidity, and chemical corrosion directly impact the physical components of SHM systems and the information they generate. As an example, excessive temperatures can lead to sensor distortion, which distorts the signals recorded and results in incorrect judgments for structural conditions. High humidity levels can accelerate corrosion in sensors and structural materials and induce further errors in data interpretation. The integrity of SHM systems relies on accurate data collection for informed maintenance decisions and thus environmental changes come with inherent difficulties to be closely monitored during the design and implementation of such systems to ensure their reliability in structural health monitoring in the long run. Apart from atmospheric conditions, external forces such as seismic and vehicular heavy traffic loads are accompanied by real threats to the integrity of SHM systems. Such factors can drastically change the structural dynamics that sensors are designed to track. For instance, in seismically active areas, the capacity of SHM systems to provide rapid estimates of damage through data-driven algorithms becomes pivotal. The algorithms need to deal with real-time data so they can rapidly provide stakeholders with information regarding a structures safety after an event has occurred. Incorporation of advanced computing techniques is of utmost concern in this regard, as they enhance the capability of SHM to not just detect but classify and diagnose damage competently. An understanding of environmental stressor-SHM system capability relationship ensures that such systems are resilient, especially in disaster-prone areas, as pointed out in research evaluating transit systems and infrastructure performance against environmental stressors (Bartin et al., 2017). Finally, materials and design used in the fabrication of SHM systems are also prime drivers of reliability under environmental adversity. The provision of adequate protection against environmental degradation through weathering should be considered to enhance the operating life and accuracy of such systems. The application of materials identified to offer resistance against harsh weather, for example, is capable of reversing the effects of moisture and corrosion and hence maintaining sensor integrity. Continuous sensor technology and algorithmic developments are required as they support improved data acquisition and processing capability needed for effective monitoring. Therefore, compensation of the many environmental factors affecting SHM systems is vital to providing uniform security and functional effectiveness in infrastructure monitoring, which ultimately feeds into improved maintenance and disaster preparedness decision frameworks in structural health monitoring activities (Cherta et al., 2020).

8.3.3. Data Integrity and Processing Challenges

Data processing and integrity problems are such crucial impediments in the domain of Structural Health Monitoring (SHM) systems, particularly as the usage of technology increases. Proper data collection and transmission is necessary to ensure the reliability of the SHM systems because even slight inaccuracies could lead to cataclysmic failure of monitoring infrastructure. This requirement is especially important with renewable energy infrastructure like onshore wind turbines, where information about the structural integrity calculated by sensors is paramount to the functionality and safety of the infrastructures. Under the stochastic and dynamic load conditions that are encountered by wind turbines, data inconsistencies not only prevent monitoring but also undermine maintenance strategies, exposing systems to potential vulnerabilities that can cause structural failure and resultant financial loss (Currie et al., 2012). Processing issues are not just brought about by the volume of data generated by SHM systems but also by the need to process the data in real-time to ensure the best decision-making. Application of complex algorithms and machine learning can enhance data processing capability but comes with possible problems in the form of system complexity and data corruption or data breach. Because SHM systems are likely to be deployed in remote or unmanned environments, chances of outside interference or failure of data transmission mechanisms are new susceptibilities. High data integrity, thus, has become synonymous with safeguarding SHM systems from security breaches as good quality data are at the heart of decision making that safeguards infrastructure (Boujemaa et al., 2009). Moreover, the creation of SHM technology must be based on a sound paradigm for managing data integrity and processing challenges. The paradigm will not only encompass advanced data validation techniques but also redundancy mechanisms to guarantee that failure is avoided. Condition-based monitoring can facilitate an early warning system for structural failure, particularly in high-risk operations such as wind turbine operations, where intervention has to occur early. Since focus has been laid in case studies, the deployment of new monitoring systems can strengthen faith in the durability and lifetime of renewable energy systems to a large extent. Thus, strategical intervention with data integrity and processing problems can be at the core of reinforcing the future of SHM systems, especially in the emerging sector of renewable energy (Currie et al., 2012) (Figure 8.8).

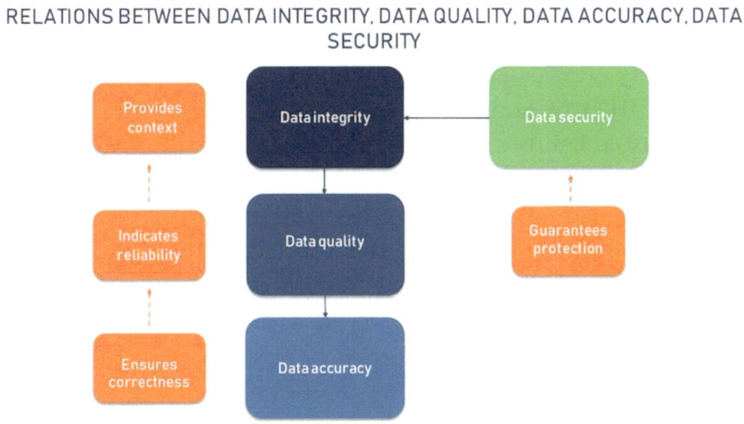

Figure 8.8. Data integrity and processing challenges.

Source: https://www.altexsoft.com/static/blog-post/2023/11/b4438bf6-50c3-47a4-a448-acbe91078b93.jpg.

8.4. IMPORTANCE OF DATA SECURITY IN STRUCTURAL HEALTH MONITORING (SHM)

The importance of data security in SHM systems cannot be overstated, particularly as reliance on such technologies continues to grow within sectors of critical infrastructure. With the mounting deployment of sensors that are embedded in structures, the collected data can reveal important information about structural integrity as well as probable modes of failure. But this data is prone to tampering and unauthorized access, which could compromise the performance and safety of the structures being monitored. Security breach in data would mislead stakeholders regarding the status of an asset and consequently lead to catastrophic failures. By robust data protection controls, organizations can better safeguard against the risk of cyber threats, thereby enhancing the reliability of SHM systems and supporting public confidence in the safety of critical infrastructure. Moreover, the impact of compromised data extends beyond immediate safety concerns into economic and legal consequences. For critical industries like renewable energy, where asset performance and stability are key concerns, data integrity is necessary for ensuring reliability in operations. For instance, as emphasized by recent research, structural integrity in the case of wind turbines has a direct relationship with the data gathered by SHM systems and impacts their overall performance. Inadequate protection of this information can lead to false assessment and tampered maintenance decisions, ultimately affecting the turbines efficiency and lifespan. Integration of safe data management practices thereby not only minimizes risks, but long-term sustainability of renewable energy sources as well, as in the case presented in Currie et al. (2012). Overall, SHM system data security is paramount in order to provide safety, reliability, and public trust in strategic infrastructure. The threat landscape surrounding such technologies is evolving continuously, and thus necessitates an aggressive approach towards cybersecurity measures that provide data integrity and bar unauthorized access. With the need for effective SHM increased in fields like transportation and energy systems, organizations must assign top priority to investing in advanced security protocols. These should be designed not only to safeguard against current cybersecurity risks but also to innovate against future vulnerabilities. Prioritizing data security will ultimately contribute towards further improving SHM systems on an ongoing basis, assisting in making the transition to more trustworthy infrastructure solutions, as suggested in existing research (Bush et al., 2011) (Figure 8.9).

Figure 8.9. The importance of data security in structural health monitoring (SHM) systems.

Source: https://d2ds8yldqp7gxv.cloudfront.net/Blog+Explanatory+Images/What+is+Data+Security+1.webp.

8.4.1. Protecting Sensitive Infrastructure Data

Integrity and confidentiality of sensitive infrastructure information is a high-level concern within the context of Structural Health Monitoring (SHM) systems. As these systems increasingly depend on digital technology to gather real-time structural information, the potential for data breaches and unauthorized access increases exponentially. Traditional intrusion- and exfiltration-based security models are now no longer adequate in the current dynamic environment. With the emergence of concepts like de-parameterization, organisations need to redefine how they protect infrastructure data. This new reality necessitates a dynamic system that involves technological innovation and organizational discipline to properly manage security, thereby providing levels of protection that take into account dynamic threats and vulnerabilities (Cleeff et al., 2008). To address the challenge of protecting sensitive infrastructure data, SHM systems need to emphasize transparency as well as user awareness. With data analysis techniques increasingly being part of data collection processes, consumers are regularly in the dark regarding how their own individual data is protected and utilized. This could lead to acute privacy concerns, particularly with the increasing use of interconnectivity. To determine how such complexities can be handled, there needs to be an active response. This entails the employment of privacy-protecting approaches that not only secure data integrity but also ensure stakeholders are actively involved when it comes to data utilization habits. This can enhance users' and stakeholders' confidence, thus encouraging a safer environment within the context of SHM (Aghasian et al., 2018). Generally, the protection of sensitive infrastructure information is not merely a technological challenge but also a regulatory and ethical requirement. Therefore, companies dealing with SHM systems must have their security operations align with lawful policies and moral principles that govern data protection. The conversation between technical controls and compliance forms the ground for reliable SHM systems capable of resisting malicious external attacks. The principle of re-parameterization also offers a ground for companies to continuously reassess their security mechanisms so that they can adapt to emerging threats in due course. Thus, creating a comprehensive security plan that addresses both technical and regulatory dimensions of data protection makes SHM systems more robust and creates stakeholder confidence (Aghasian et al., 2018; Cleeff et al., 2008).

8.4.2. Legal and Regulatory Implications

Legal and regulatory considerations of Structural Health Monitoring (SHM) systems are significant in establishing frameworks to provide security and reliability to infrastructure. These systems operate under a maze of legislation and standards, usually having to satisfy national and international regulation. For example, following standards such as the American Society of Civil

Engineers (ASCE) standards ensures that SHM systems not only monitor the structural health effectively but also comply with safety standards important for public health. The problem lies in maintaining such regulations in terms of the current technological advancements. Regulatory bodies must have constant consultation with technical experts to address changing threats posed by SHM technologies, particularly as they become increasingly entangled with networked infrastructure systems. This synergy of legal mandates and operational deployments is essential in the effective deployment of SHM systems to ensure structural integrity and public trust. Moreover, the integration of SHM systems is with severe legal consequences for liability and accountability. As such systems increasingly rely on advanced technologies such as Internet of Things (IoT) devices and artificial intelligence, responsibility demarcation in the occurrence of system failure or mis monitoring becomes more complex.

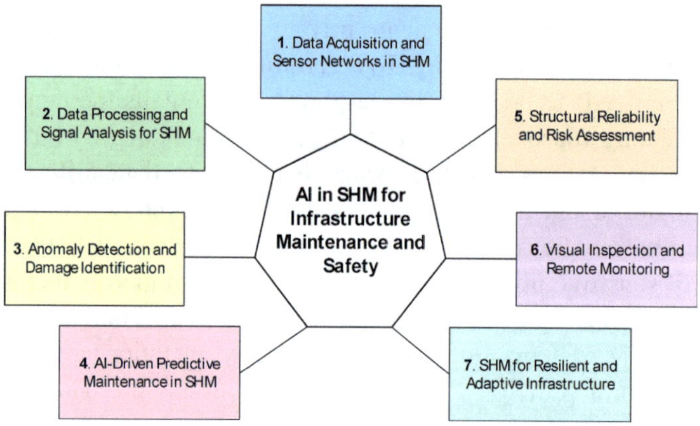

Figure 8.10. The legal and regulatory implications of structural health monitoring (SHM) systems.

Source: https://www.mdpi.com/infrastructures/infrastructures-09-00225/article_deploy/html/images/infrastructures-09-00225-g001-550.jpg.

Legal frameworks must evolve to handle potential conflicts over data integrity, ownership, and use rights. For instance, stakeholders such as governments, private institutions, and the general public have to negotiate contracts that guarantee data protection while providing openness. In addition, there must be policies instituted to harmonize data sharing behavior among organizations to enable collaboration and decision-making grounded in SHM expertise. These are needed legal underpinnings to establish the climate where SHM systems are able to operate successfully and limit the capability of legal challenges that will sap their effectiveness. In addition, aligning SHM systems with regulatory demands not only increases the level of reliability but also promotes accountability culture and ethical behavior among practitioners. Consistent application of best practices in handling data, reporting, and response procedures should be complemented by regulatory demands that compel stakeholders to take security as their highest priority in operations. As attested to by the incremental evolution of governance systems across countries, such as the evolving state governance model in Ukraine, the integration of citizen participation within oversight roles has the potential to significantly enhance SHM programs. Improved citizen participation creates the channel for provision of feedback and enables better understanding of the problems facing SHM systems. Therefore, firming up the legal and regulatory underpinnings of SHM is warranted to guarantee that such systems not only meet the needs of the present but also adapt to meet future technological and infrastructure management developments (Britchenko et al., 2017a, 2017b) (Figure 8.10).

8.4.3. Consequences of Data Breaches in Structural Health Monitoring (SHM)

The effects of data breaches in SHM systems are not limited to financial loss but also impact organizational confidence and public security. Organizations are severely damaged by their reputation as a result of a data breach, leading to irreversible loss of stakeholder trust. As an illustration, in SHM, a high reliance on data availability and integrity pertaining to infrastructure condition, a breach can undermine confidence within the public's capability to provide accurate and timely information using monitoring systems. The study shows that hospital patients' trust is a good predictor of their reported safety, and organizational factors are important in providing assurance of security perceptions (Lo et al., 2018). The same is applicable in the environment of SHM; stakeholders must be assured of data integrity and security to engage fully with monitoring solutions. Data leaks can also prompt regulatory scrutiny and exorbitant fines, particularly in highly regulated industries that require high compliance, such as civil engineering and public infrastructure management. The legal repercussions can compound even more because companies can be held liable for damages caused by leaked data. This is compounded during an era of de-parameterization, where the traditional security paradigms are reassessed to accommodate the evolving technology landscape. Organisations are increasingly being called upon to employ robust security mechanisms on multiple platforms, which proves challenging and labor-intensive (Cleeff et al., 2008). Increased demands for vigilance necessitate organizational culture shift towards prioritizing data protection and adherence, adding to operational realities for SHM professionals. Finally, the operation impact of data breaches on SHM systems forces organizations to reconsider their security practices with great caution. An incidence of a breach will typically force them to reconsider accessible monitoring technologies as well as security practices and provoke short-term disruptions as well as long-term changes. These changes may include the use of complex encryption techniques as well as strict access control mechanisms to enhance data integrity. Besides, as an organization is forced to navigate through the consequences of a breach, they are bound to suffer from internal resource shortages and low morale among the staff. Hence, a comprehensive policy solution is imperative in an effort to reduce risks of data breaches, establish a safe monitoring environment practice, and finally offer the reliability demanded of SHM systems.

8.5. STRATEGIES FOR ENHANCING SECURITY IN STRUCTURAL HEALTH MONITORING (SHM)

The evolving situation of SHM demands strong steps to secure these systems. Intelligent infrastructures are one of the most effective ways to achieve this with the aid of advanced technologies. The emerging technologies revolve around the combination of electronics and smart materials in a bid to enhance the monitoring system's performance, subsequently enabling engineers to obtain more comprehensive understanding of structure health. As research progresses globally, these step-by-step technologies will lead to infrastructures that have complete SHM, allowing real-time data acquisition and analysis that facilitates better decision-making and assessment of structural integrity (Angelis et al., 2025). By embedding security measures within these technological systems, stakeholders can eliminate vulnerabilities as well as related risks from data compromise or system failure. Another primary approach is multidisciplinary work by engineers, environmental scientists, and policymakers. The interactive relationship between environmental factors and structural safety requires full comprehension of relevant data streams. It is necessary to combine environmental, economic, and socio-cultural data to develop effective SHM systems that take into account suggested hazards such as climate change or natural disasters. These are multifaceted challenges that are addressed by consideration, resulting in higher security of SHM, particularly in transition areas or where significant infrastructural challenges are present (Daniela Luminita Constantin, 2025). This collaborative environment not only promotes knowledge sharing but also fosters creative solutions that improve system reliability against complex threats. Moreover, constant capacity building and development of professionals in the field of SHM is a must for security development. Due to changing technologies, professionals working within the field of SHM must be competent enough to effectively make use of new tools and technology. Periodic workshops, training sessions, and knowledge-share initiatives guarantee that personnel are well-positioned to meet future challenges and threats. Also, developing a proactive security culture within organizations highlights the importance of close monitoring of upholding the integrity of SHM systems. By investment in human capital, organizations are able to establish an adaptive environment that not only enhances the performance of SHM but also enhances its security against potential disruptions caused by evolving risks (Angelis et al., 2025).

8.5.1. Implementation of Encryption Techniques

Deployment of encryption techniques in Structural Health Monitoring (SHM) systems is critical for ensuring confidential information is not accessed or tampered with by unauthorized individuals. As there is increased reliance on networked equipment and cloud offerings, threats posed by cyber-attacks increase as well. By the incorporation of robust encryption techniques, SHM systems are able to ensure data confidentiality and integrity captured from sensors and communicated via networks. Such an approach lessens vulnerabilities for adversaries to attack, hence making the overall system more dependable. Further, as the organizations get de-parameterized, the traditional security models do not meet the needs, so they must resort to the use of advanced encryption methods to create a secure operating environment. The transition points to the need for adaptive and scalable security solutions in line with changing IT threat dynamics as defined by Cleeff et al. (2008). In the SHM systems scenario, encryption is not only a security feature but also a major enabler of regulatory compliance with various regulatory frameworks. Business entities must find themselves in a position where data protection laws require strong protection of personal and sensitive data. Application of encryption methods ensures compliance with regulations and, at the same time, develops stakeholder and user confidence. As construction management and infrastructure operations increasingly utilize digital solutions, the complexities of data utilization become evident, particularly with regard to the sensitivity of data aggregation procedures for users. Encryption thus becomes an essential consideration in addressing issues of privacy in SHM buildings in order for organizations to remain ahead of risk in data misuse while tapping the potential in monitoring and analysis, as in observations regarding online data systems by Aghasian et al. (2018). Besides, encryption techniques can enhance the resistance of SHM systems to cyberattacks, which are becoming progressively more sophisticated. Employing multilayer encryption standards can fortify data integrity collected and processed to guarantee that even where there is security infringement, the impact on system performance can be significantly minimized. This approach is necessary because organizations experience phases of speedy de-parameterization when connectivity and sharing of data among various stakeholders increase. By applying encryption at various stages—ranging from data-at-rest to data-in-transit—SHM systems are capable of countering the threats while accessing real-time functionality and response. It provides this protection in line with the industry best practice and facilitates the movement towards predictive security measures that keep pace with evolving digital contexts so that the SHM systems are constantly reliable and dependable under evolving operating conditions (Figure 8.11).

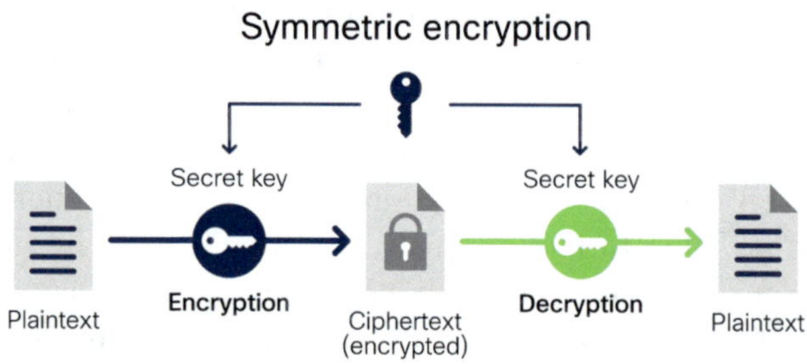

Figure 8.11. The symmetric encryption.

Source: https://www.cisco.com/content/dam/cisco-cdc/site/images/legacy/assets/swa/img/1100/symmetric-encryption-1256x706.png.

8.5.2. Regular Security Audits and Assessments

One of the cornerstones of guaranteeing security and reliability in Structural Health Monitoring (SHM) systems is the undertaking of regular security reviews and audits. The audits serve a dual role: the identification of loopholes in the system infrastructure and ensuring adherence to specified security guidelines. As technology advances, so do threats, and as such, it is imperative that SHM systems undergo systematic reviews that uncover any vulnerabilities. This proactive approach helps in minimizing risks that are connected to unauthorized access and data compromise, instilling a culture of continuous improvement in the organization. Periodic audits not only enhance protection of sensitive information but also provide various stakeholders, including management and customers, with a better perception of the integrity of systems and hence a better trust in the reliability of gathered data. The effectiveness of security audits is also supported by employing community engagement methods. By using the knowledge from social safety net programs, local data can enrich the appraisal process through the identification of specific risks in line with distinctive surroundings, much like how beneficiary targeting can be affected by community groups (Conning et al., 2025). Local stakeholders involved allow the assessments to possess more layers, with contextual knowledge that could otherwise go unnoticed. Second, by including local SHM system users' feedback, organizations can tailor their security controls to address practical problems in the field. A collaborative approach not only improves the findings of the audits but also fosters responsibility and a sense of ownership from the community, which improves overall system dependability. Recalibrating security arrangements in SHM to a more institutionalized framework, as with the governments' reforms in public financial management, can also improve system security (Crnomarkovic et al., 2017). Emphasizing the importance of security procedures based on international standards leads to better decision-making and resource allocation in SHM systems. By comparing security measures to global best practices repeatedly, organizations can achieve a higher level of operational efficiency while reducing the perils of fiscal impropriety accompanying data compromise. The use of solid and clear security tests stands as a precursor to improved fiscal outcomes, while ensuring that organisations are able to make effective use of public and private investments into SHM technology, ultimately contributing towards greater reliability and trust in the systems.

8.5.3. Development of Secure Communication Protocols

Development of secure communication protocols is key to the resolution of the special challenges faced by Structural Health Monitoring (SHM) systems, precisely due to the widespread use of wireless sensor networks. These networks, on which data acquisition and transmission in SHM are based, typically involve energy restricted sensors. The application of security mechanisms, such as authentication and data encryption, enhances the data's integrity and confidentiality while being transmitted. However, as discussed in the models introduced by Udoh et al. (2019), these applications inadvertently introduce the energy burden on these resource-constrained devices. Therefore, there must be a delicate balance; while data must be protected, employing energy-efficient protocols will minimize the risks of energy drain attacks, de facto extending the lifespan of SHM systems. With the dynamic landscape of technology where connectivity ever-evolving and escalating, it is imperative to reevaluate traditional security measures in SHM systems. De-parameterization is a theme of recent security discourse that calls for the de-construction of rigid security boundaries where trusted assets can be strictly isolated from untrusted ones. As the argument presented in Cleeff et al. (2008) says, this trend towards de-parameterization heralds an intensification of the requirement for adaptive security mechanisms attuned to various organizational and technological contexts. While conventional security systems rely on the static security perimeter, SHM systems must incorporate adaptive communication protocols that are able to respond to changing threats and adapt to diverse network topologies. This flexibility not only enhances security but also works to enhance the dependability of data transmission within heterogeneous operating environments. Additionally, integrated solutions that integrate security and energy efficiency play a pivotal role in SHM system futures. With research continuing to develop new Medium Access Control (MAC) protocols and other communication systems, emphasis is laid on the development of systems that are attack resilient, such as denial of sleep. Energy conservation can be improved without compromising data integrity by optimizing duty cycle and simulating various MAC protocols, as explained in Udoh et al. (2019). This two-fold focus on security and energy efficiency maintains SHM systems effective and functional over time. At the end of the day, promoting secure communication in SHM through novel protocol design not only guards data but also maintains the integrity of structural health monitoring operations, hence safeguarding public safety and infrastructure reliability (Figure 8.12).

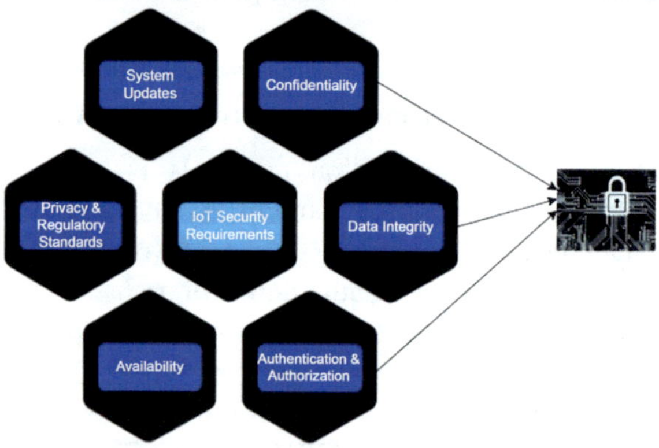

Figure 8.12. The development of secure communication protocols.

Source: https://media.springernature.com/lw685/springer-static/image/chp%3A10.1007%2F978-3-031-82826-3_6/MediaObjects/637413_1_En_6_Fig5_HTML.png.

8.6. STRATEGIES FOR IMPROVING RELIABILITY IN STRUCTURAL HEALTH MONITORING (SHM)

Increased reliability in SHM systems is crucial for ensuring the safety and longevity of infrastructure such as bridges that are central to the transportation system. Routine monitoring practices, such as the utilization of sensors to capture real-time measurements of structural performance, are a viable method. For instance, through a Bayesian Belief Network (BBN) approach, it is achievable to shift away from conventional visual inspections to an advanced, data-driven monitoring procedure. The method allows the health state of each bridge component to be dynamically assessed, thereby facilitating early detection of deterioration before failures occur at critical levels. Early identification of issues not only enhances reliability but also lowers secondary economic expenses by minimizing maintenance planning and enhancing structure durability (Andrews et al., 2017). Another important way of increasing reliability in SHM is leveraging advanced materials and technologies that enhance the monitoring capabilities of infrastructure systems. The new technologies, including smart materials and embedded electronics, support the direct incorporation of sensing capability into the structural components. The new approach makes it easier for engineers to obtain and monitor data in real time and ultimately leads to active maintenance. The combination of piezoelectric and fiber optic materials, for example, can offer the information about structural change that would otherwise go undetected. This shift towards smart infrastructures is an evolution in SHM, which not only makes spaces more responsive to physical alterations but also one that can streamline operating performance as well as maintenance strategies (Angelis et al., 2025). Beyond that, encouraging interdisciplinary collaboration among engineers, data scientists, and environmentalists is instrumental in bolstering the legitimacy of SHM systems. This partnership can lead to the development of coupled models that encompass a host of environmental variables affecting the lifespan of infrastructure. Through the synergy of multi-disciplinary expertise, adaptive monitoring techniques that take into account not just the structural responses, but also extrinsic factors such as weather conditions, traffic loading, and geologic movements are made possible. This synergistic approach not only enriches the data set available for analysis but also enhances predictability in structural health. As infrastructure systems become more subject to environmental and operational loads, these collaborative efforts will be essential to making SHM systems resilient, reliable, and effective in protection of public safety and investment (Andrews et al., 2017; Angelis et al., 2025).

8.6.1. Regular Maintenance and Sensor Updates

In Structural Health Monitoring (SHM) systems, maintenance is critical to both security and reliability with periodic maintenance and regular updating of sensors. As SHM systems advance in the application of technology, so too does the complexity of running them. Standard maintenance processes confirm the functionality and viability of the sensors that form the foundation for such systems. Without proper maintenance, sensors can yield erroneous readings, leading to faulty structural integrity analysis. Besides, as is seen in the ongoing advancements in technology and automation in aviation, preventive risk reduction measures are becoming more of a need. This requires increased focus on surveillance, assessment, and future-oriented interventions to manage risks effectively while keeping SHM systems in their optimal condition under evolving operational conditions (Ancel et al., 2025). Sensor replacement is crucial to keep SHM systems viable and functional. As technology proceeds at a rapid pace, legacy sensors not only have the potential to degrade data quality but also expose systems to security breaches.

Figure 8.13. The importance of regular sensor maintenance keeping.

Source: https://nexgensensors.com/wp-content/uploads/2024/12/The-Importance-of-Regular-Sensor-Maintenance-Keeping-Your-Sensors-in-Top-Shape.jpg.

Upgrading sensors means carrying out firmware updates, adding new legacy modules, or enhancing the analytic frameworks utilized in data interpretation. These are critical in adapting to emerging challenges posed by advanced industrial environments, as upcoming developments in Industrial Wireless Sensor Networks (IWSNs) underscore. These networks, while bringing numerous advantages, also involve issues that can hinder performance, particularly in hostile or dynamic settings if not serviced and updated on a regular basis (Aslam et al., 2018). A systematic approach to sensor updates is thus paramount in enhancing the security protocol for SHM systems. Finally, routine maintenance and sensor updates bring about a robust foundation for SHM systems that supports long-term reliability and security. By incorporating routine checks and updates into operating procedures, organizations can sustain high performance and resilience in monitoring. This integrated approach not only addresses potential weaknesses but also enhances overall accuracy in data gathering, leading to improved decision-making regarding structural safety. As the complexity of SHM systems continually increases, a compliance with frequent maintenance and regular updating will become essential in defending against potential future attack and optimal functioning. Thus, establishing clear processes and procedures of regular maintenance and sensor updating becomes an essential factor in enhancing security and reliability in SHM systems (Ancel et al., 2025; Aslam et al., 2018) (Figure 8.13).

8.6.2. Use of Redundant Systems for Data Verification

In the context of Structural Health Monitoring (SHM) systems, redundant systems for data verification purposes serve as a basis towards realizing security as well as reliability. The redundant configurations are essential to maintaining preservation of vital data pertaining to structural integrity against potential system failures or faults. Redundancy can take many forms such as twin sensor arrays or redundant processing units that allow cross-validation of data. In using these systems, anomalies can be detected with greater precision, thus enhancing overall reliability of SHM systems. This is particularly critical in situations where environmental and operational interference compromises data integrity. In that manner, redundant systems not only act as a safeguard against the potential loss of data but also enhance stakeholders' trust in the information provided by SHM systems. Additionally, the issue of handling multiple error situations in control systems, especially in harsh environments, emphasizes the need for robust methods within data validation. For instance, systems operating under hostile conditions, as provided in the case of electromagnetic disturbances, exhibit the impact of redundancy on reliability. In face of threats of upsets, such as software-dependent error modes without damage to components, redundant systems can go beyond cognitive insight into such adversity through empirical validation. The approach empowers engineers to design tailored approaches that enhance the resilience of SHM systems against unforeseen perturbations. Under such an environment, the intersection of redundant data validation systems and real-world use, e.g., in aerospace or transportation industries, will be the most important factor to make sure that functionality and safety are never compromised by performing operations under varying conditions, as the study conducted by Rockwell International Corporation and NASA illustrates. Finally, the future of SHM systems will rely on ongoing refinement of redundant technologies and cross-pollination advances in analytics. As more is required for operational efficiency and integrity of real-time information, redundant systems will become a necessary aspect not only in construction and infrastructure upkeep but also in aerospace technologies, where reliability standards must be more or less strictly met. Insights from research on multichannel control systems can be applied to designing these redundant systems, offering a matrix that will ensure a high level of reliability in adverse environments. Usage of these methods in conjunction with SHM systems aligns with the overall safety and operational excellence objective in diverse applications. In this regard, investment in double-checking data benefits the performance of SHM technologies in addition to contributing to the formulation of safety procedures in a significant number of high-risk applications to a great extent, ensuring that such critical systems are utilized effectively in all scenarios, as reflected in recent literature on aerospace reliability.

8.6.3. Advanced Algorithms for Data Analysis

Advanced algorithms for data analysis play a key role in making Structural Health Monitoring (SHM) systems more reliable and secure. With infrastructure resilience being the top priority in a world today, advanced algorithms enable data obtained from an array of sensors implanted into structures to be analyzed and interpreted in real-time. Using data-driven methods, SHM systems can accurately monitor changes in structural health over time, allowing for the early detection of anomalies that will render structures unsafe. The algorithms have the potential to do much more than merely collect data; they analyze complex datasets to determine the health of structures and predict potential failures. Thus, the integration of advanced algorithms in SHM underscores the trend of increasingly intelligent, data-intensive monitoring systems whose preoccupation is the safety of users and the lifespan of structure. The efficiency of SHM systems greatly relies upon data analysis accuracy, which is typically achieved through intricate algorithms. Algorithms exist in data-driven and model-based categories. The former tends to employ algorithms to filter sensor data and derive helpful inferences, while the latter is identified with the use of pre-defined models to assess structural performance. More developments have leaned towards data-driven algorithms due to their ability to process big datasets and deliver real-time insights to support decision-making. However, Cherta et al. (2020) observes that these computation-demanding frameworks need high-performance computational infrastructure, emphasizing the necessity to select appropriate algorithms that meet analytical richness against run-time efficiency. These advancements not only enhance SHM system performance but also support them in keeping pace with evolving challenges in infrastructure monitoring. For the purpose of better explaining the importance of advanced algorithms, their applicability also extends to other SHM systems frameworks, especially via linked Internet monitoring systems. These systems, as outlined in Rius et al. (2013), leverage advanced data management capabilities to promote the monitoring of infrastructure, especially in cases like the maintenance of a bridge. By combining data from an independent monitoring system with an Internet-based monitoring interface, operators receive full information regarding structural performance. The application of these algorithms promotes enhanced decision-making by systematic condition assessment, permitting proactive infrastructure management. Finally, effective use of advanced algorithms within SHM systems offers a more secure and dependable environment, which is a huge improvement in the processes applied for structural maintenance and monitoring.

SUMMARY

All in all, safety and dependability of SHM systems are of paramount importance in ensuring infrastructure integrity and public safety. With increased reliance on technology in structural assessment, it is increasingly imperative to establish robust protocols to protect information from intrusive attacks and ensure real-time monitoring systems' correctness. A deep understanding of the vulnerabilities and potential threats for SHM will support the development of legitimate countermeasures. This not only raises the efficiency of monitoring but also enhances stakeholders' confidence in the structures they oversee as being reliable, thereby security becomes an imperative aspect of SHM applications. Additionally, the algorithm selection used to analyze information is key to SHM system reliability. The literature indicates that algorithms based on data are advisable for their ability to process and analyze sensor data in real time, hence ensuring decision-making within a timely fashion due to detected structural irregularities. Dependency on such sophisticated algorithms, however, needs high-end processors due to their computational complexity. As underscored by recent reviews, while data-driven methods are pivotal to efficient damage detection, balance between algorithmic sophistication and strength of the parameters applied in damage assessment must also be attained. This is a multicriterion issue that must be dealt with through sustained research and development. Ultimately, the future of SHM systems depends on the marriage of novel technologies and methods that maximize security and reliability. The shift towards smart structures, as discovered in latest studies, is indicative of a growing trend towards the implementation of smart sensors and proactive maintenance strategies. Through an in-depth exploration of data-driven and model-based approaches, researchers are unlocking doors towards innovations that optimize damage detection and identification. As seen through the extensive review of SHM applications, convergence of these methods has the potential to transform structural performance monitoring. Therefore, embracing interdisciplinary collaboration and constant innovation in SHM technology is instrumental to improving infrastructure resilience against future challenges. Under Security and Reliability in SHM Systems, resilient damage identification procedures become a crucial aspect of the overall efficiency of these systems. The debate regarding these methodologies primarily describes two approaches: data-driven algorithms and model-based analysis. Data-driven methods employ data collected from sensors and describe real-time feedback on the structural health of what is being analyzed. They find their best choice in their immediate and speedy operation, fitting high-speed decision-making environments typical for engineering applications. The model-based method depends on the precision of the inspections prediction models, which utilize massive amounts of qualitative and quantitative data pertaining to what is being analyzed structurally. Awareness of these techniques is the bedrock that enables creation of enhanced SHM systems that prioritize security and reliability simultaneously. There is much discourse directed at the union of advanced sensor technologies and computational techniques in enhancing the reliability of SHM systems. Data acquisition via multimodal sensors cannot be stressed too much, as they are the drivers of effective damage detection, localization, classification, and prognosis. Existing literature demonstrates a growing shift towards the utilization of high-performance processors that have the potential to manage complex data-driven analysis. This innovation enables SHM systems to interpret and analyze masses of data generated by real-time monitoring, which ultimately translates into a more preventive approach towards infrastructure maintenance. Literature indicates that it is vital to meet the computational costs of data-driven methods in order to optimize the performance of SHM systems for them to meet the high demands of reliability in modern-day engineering applications (Cherta et al., 2020). Additionally, the establishment of smart structures is a vital transforming trend in SHM systems that has an effect on their security and reliability. As these systems mature and advance, the creation of integrated frameworks for damage detection, along with the latest in sensor and data analysis technology, becomes of highest priority. Merging these elements allows not only a higher sense of structural health condition awareness, but also preventive action against future failure. As the intricacy of modern-day infrastructures continues to increase, the development of analysis programs capable of tailoring their result based on the individual structural characteristics ensures that the endeavor of monitoring becomes both timely and precise. The above discussion serves to highlight the manner in which the ongoing developments in sensor technology as well as data improvement strategies are likely to revolutionize the terrain of SHM systems and enhance their security and reliability levels (Dong et al., 2010).

REFERENCES

1. Aghasian, E., Garg, S., & Montgomery, J. (2018). User's privacy in recommendation systems applying online social network data, a survey and taxonomy. http://arxiv.org/abs/1806.07629 (accessed on 2nd October 2025).
2. Agis Cherta, D., Gomez Vargas, R. C., Pedraza, C., Pozo Montero, F., et al. (2020). Damage identification in structural health monitoring: A brief review from its implementation to the use of data-driven applications. https://core.ac.uk/download/288625955.pdf (accessed on 2nd October 2025).
3. Ancel, E., Corbetta, M., Darafsheh, K., Dill, E., et al. (2025). Architecture and information requirements to assess and predict flight safety risks during highly autonomous urban flight operations. https://core.ac.uk/download/pdf/288485506.pdf (accessed on 2nd October 2025).
4. Andrews, J., Remenyte-Prescott, R., & Vagnoli, M. (2017). Towards a real-time structural health monitoring of railway bridges. https://core.ac.uk/download/96621252.pdf (accessed on 2nd October 2025).
5. Angelis, C. T., Baclesanu, D., Amariei, D., Gillich, G. R., & Loutas, T. (2025). Intelligent infrastructures systems for sustainable urban environment. https://core.ac.uk/download/pdf/7155299.pdf (accessed on 2nd October 2025).
6. Aslam, N., Cao, Y., Hussain, S., Khan, M., et al. (2018). A critical analysis of research potential, challenges and future directives in industrial wireless sensor networks. https://core.ac.uk/download/96885354.pdf (accessed on 2nd October 2025).
7. Bartin, B., Deka, D., El-Awar, A., Lou, Y., et al. (2017). Performance measures to assess resiliency and efficiency of transit systems. https://core.ac.uk/download/129532928.pdf (accessed on 2nd October 2025).
8. Belcastro, C. M. (2025). Laboratory test methodology for evaluating the effects of electromagnetic disturbances on fault-tolerant control systems. https://core.ac.uk/download/pdf/42825418.pdf (accessed on 2nd October 2025).
9. Boujemaa, N., Compañó, R., Dosch, C., Geurts, P., et al. (2009). *CHORUS Deliverable 3.4: Vision Document*. https://core.ac.uk/download/11434795.pdf (accessed on 2nd October 2025).
10. Britchenko, I., & Danshina, Y. (2017). Adaptation of domestic state governance to international governance models. https://core.ac.uk/download/322492655.pdf (accessed on 2nd October 2025).

11. Britchenko, I., & Danshina, Y. (2017). Net structure of subject-to-subject relations in the management of the system of administrative services provision. https://core.ac.uk/download/322492656.pdf (accessed on 2nd October 2025).
12. Brown, J., Chen, I., Earhart, R., Ehrlich, A., et al. (2025). Advanced Manned Launch System (AMLS) study. https://core.ac.uk/download/pdf/42811271.pdf (accessed on 2nd October 2025).
13. Bush, S., Henning, T., McCarten, P., Omenzetter, A., et al. (2011). Risk based bridge data collection and asset management and the role of structural health monitoring. https://core.ac.uk/download/29465888.pdf (accessed on 2nd October 2025).
14. Casas Rius, J. R., Miyamoto, A., & Motoshita, M. (2013). Development of an integrated remote monitoring technique and its application to para-stressing bridge system. https://core.ac.uk/download/41777708.pdf (accessed on 2nd October 2025).
15. Castedo, L., Fernández-Caramés, T. M., & Fraga-Lamas, P. (2017). Towards the Internet of smart trains: A review on industrial IoT-connected railways. https://core.ac.uk/download/95054341.pdf (accessed on 2nd October 2025).
16. Conning, J., & Kevane, M. (2025). Community based targeting for social safety nets. https://core.ac.uk/download/pdf/6223757.pdf (accessed on 2nd October 2025).
17. Constantin, D. L. (2025). Environmental data and indicators and human security related issues in Romania: An institutional approach. https://core.ac.uk/download/pdf/7036350.pdf (accessed on 2nd October 2025).
18. Crnomarkovic, A., Fuchs, S., Ganguli, R., et al. (2017). *Serbia–Public Sector Accounting Review: Report on the Enhancement of Public Sector Financial Reporting.* https://core.ac.uk/download/189169506.pdf (accessed on 2nd October 2025).
19. Currie, M., Quail, F., & Saafi, M. (2012). Development of a robust structural health monitoring system for wind turbine foundations. https://core.ac.uk/download/9558383.pdf (accessed on 2nd October 2025).
20. Dong, Y., Liu, H., & Song, R. (2010). Bridges structural health monitoring and deterioration detection synthesis of knowledge and technology. https://core.ac.uk/download/286607584.pdf (accessed on 2nd October 2025).
21. Duquennoy, S., Höglund, J., Misra, P., et al. (2013). Supporting cyber-physical systems with wireless sensor networks: An outlook of software and services. https://core.ac.uk/download/17037990.pdf (accessed on 2nd October 2025).

22. Eid Mohamed, Kopustinskas, V., Simola, K., & Zutautaite, I. (2017). *Critical Infrastructures: Enhancing Preparedness & Resilience for the Security of Citizens and Services Supply Continuity: Proceedings of the 52nd ESReDA Seminar Hosted by the Lithuanian Energy Institute & Vytautas Magnus University*. https://core.ac.uk/download/132627183.pdf (accessed on 2nd October 2025).
23. Kerkhoff, H. G. (2009). Dependable reconfigurable multi-sensor poles for security. https://core.ac.uk/download/pdf/11478693.pdf (accessed on 2nd October 2025).
24. Lo, M. C., Peikari, H., Ramya, T., et al. (2018). Patients' perception of the information security management in health centers: The role of organizational and human factors. https://core.ac.uk/download/228156431.pdf (accessed on 2nd October 2025).
25. Pierson-Balik, D., Kass, H., Cleary, J., Dixon, K. A., Harrington, L., Berger, P., & Gyarfas, S. (2005). Building skills and alliances to meet demand in New Jersey's labor market. https://core.ac.uk/download/71341756.pdf (accessed on 2nd October 2025).
26. Udoh, E., & Udoh, E. (2019). An energy aware and secure MAC protocol for tackling denial of sleep attacks in wireless sensor networks. https://core.ac.uk/download/305119264.pdf (accessed on 2nd October 2025).
27. van Cleeff, A., & Wieringa, R. (2008). De-parameterization as a cycle: Tearing down and rebuilding security perimeters. https://core.ac.uk/download/pdf/11468848.pdf (accessed on 2nd October 2025).

CHAPTER 9

DESIGN AND IMPLEMENTATION OF AN IOT-SHM SYSTEM

Contents

INTRODUCTION

10 years have passed since the Internet of Things (IoT) was deployed in Structural Health Monitoring (SHM) systems, and ever since then, it has been a game-changer to monitor and maintain infrastructure. The systems employ advanced sensor technologies to obtain real-time data regarding structural condition, enabling engineers to make appropriate decisions on repair and maintenance. Implementation of IoT-SHM systems not only adds to safety and

life spans but also optimizes resource utilization and costs savings on traditional inspection. With the demand for effective monitoring solutions continuing to grow, the necessity for innovation and development of new IoT-SHM systems is more apparent in the context of aging infrastructure and global warming (cumulative stress to structural integrity). The development of such systems involves a deep comprehension of the structural intricacies of buildings and the dynamic capabilities of advanced data-driven algorithms. A critical step in the process is deciding and employing proper methods of data collection, usually through the use of sensors that measure key information regarding the status of the infrastructure. The strength of data-driven techniques lies in their ability to handle vast amounts of data, which enables real-time assessment and prompt interventions. Moreover, these algorithms enhance damage detection techniques, which enable the right assessments of structural condition. Although effective, these algorithms have high computational needs and pose challenges to implementing them, especially in the constrained environment typical of IoT deployment. Therefore, research on memory-efficient algorithms, such as those that can improve the principal component analysis (PCA) method, is essential to advancing the efficiency of IoT-SHM systems (Abdelgawad et al., 2019). As research on IoT-SHM continues to evolve, adapting to new ways of data processing and analytics becomes essential to ensure the efficiency of such systems. Real-time damage detection, localization, and prognosis increasingly rely on smart models that could potentially use sensor information effectively. Surprisingly enough, while data-driven and model-based approaches are available, the former's advantage of faster quicker solutions has established them as preferred in contemporary applications over traditional methods. High-performance processing implies the necessity for further research into scalable solutions that will try to address computational cost without sacrificing accuracy. This IoT-SHM development environment is characterized by a commitment to the achievement of more improved structural integrity monitoring but toward fostering an environment that is friendly to the evolution of smart structures, ultimately opening up pathways for safer, stronger infrastructures (Cherta et al., 2020).

Definition of IoT (Internet of Things)

IoT is a new paradigm in technology that links an enormous network of sensors, systems, and devices to exchange data and talk to one another through the internet. IoT provides an ability to embed sensors and actuators in regular devices, hence enabling the devices to record and forward information, thus enhancing their performance and providing valuable information. This IoT network of connected devices forms the core of most applications, such as industrial automation, medicine, and home automation, where it automates operations and enhances efficiency. For SHM systems, the IoT plays a significant role by enabling real-time monitoring from sensors mounted in structures to provide constant monitoring and assessment of the stability of structures. As such, being aware of the definition and importance of IoT becomes part of the understanding of its impact on modern engineering practices.

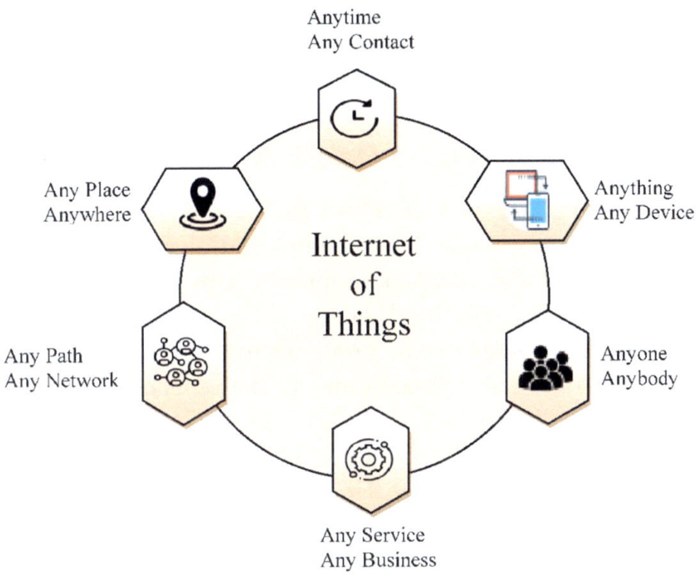

Figure 9.1. Concept of internet of things IoT.

Source: https://www.researchgate.net/publication/347844179/figure/fig5/AS:1095936654753804@1638303068946/Concept-of-internet-of-things-IoT.png.

The use of IoT technologies in SHM systems enables advanced monitoring strategies that can highly improve the maintenance and safety of the structure. Voice-controlled sensors and data analytics algorithms derive meaning from the data received, enabling predictive maintenance and reducing risks associated with structural failure. Implementation of a memory-optimized streaming History PCA algorithm enhances the data processing capability in such IoT-based systems with which big datasets accumulated using structural sensors can be analyzed efficiently in real time (Abdelgawad et al., 2019). Further, the versatility of IoT systems allows them to be developed based on particular requirements of diverse structures and making the system more customized, thereby providing a customized monitoring solution that contributes value to performance as well as reliability. The application of IoT in construction projects is also demonstrated by its integration with other emerging technologies such as Radio Frequency Identification (RFID) and cloud computing. When these technologies are integrated with IoT, automated monitoring and real-time analysis of prefabricated units are made possible, significantly simplifying the process of project operation and management (Liu et al., 2019). For instance, the ability to sense in-situ components and monitor their structural health during installation with IoT facilitates better resource planning and project efficiency. Overall, IoT definition and significance enhance the design and implementation of SHM systems as well as redefine monitoring and maintaining infrastructure in an increasingly connected world (Figure 9.1).

Overview of SHM (Structural Health Monitoring)

The significance of SHM is that it can assure the integrity and safety of infrastructure assets from when they are built until their lifespan is complete. With age and repeated exposures to various stresses from environmental and usage aspects, opportunities for hidden damage increase. SHM employs a range of sensors and technologies to have structures monitored continuously, where real-time data informs maintenance strategy and enhances safety. With the integration of advanced data analysis techniques, for instance, those employed in IoT platforms, SHM systems not only detect structural anomalies but are also able to predict eventual failures. This pre-emptive action enables decision-makers to effect prompt interventions and ensure adherence to safety protocols, save lives, and avoid unnecessary resource utilization. Emerging technological innovations have dramatically enhanced the performance of SHM systems. The application of IoT technologies, including sensor networks in conjunction with cloud computing, offers real-time data capture and analysis. For instance, innovative means of utilizing Radio Frequency Identification (RFID) and Long Range (LoRa) technologies have been found to be effective in tracking components of pre-fabricated construction (PC). Liu et al. (2019) points out that not only is the technology applied in component location tracking, but it also verifies the structural performance of components upon their installation. These advances make SHM systems more consistent and effective, enabling real-time monitoring that plays a crucial role in the timely and effective making of maintenance decisions within the construction industry. To further enhance the analytical capability of SHM systems, advanced data reduction techniques like PCA are being deployed on the terms of embedded IoT platforms. While computationally expensive standard PCA algorithms, with the advent of improvements like memory-efficient parallelized streaming History PCA, present viable low-cost solutions for IoT gateways. This technique, as indicated in Abdelgawad et al. (2019), demonstrates a staggering 10x compression ratio and extreme memory savings but with high fidelity in data representation.

The ability to carry out PCA on low-end devices not only expands the range of SHM technologies into other applications but also facilitates effective and efficient data management. Therefore, these techniques are crucial in the overall reliability of structural health assessments in IoT-based systems.

Importance of Integrating Internet of Things (IoT) with Structural Health Monitoring (SHM)

Merging IoT with SHM systems revolutionizes the process of assessing and guaranteeing the integrity of infrastructure and buildings. With the capability of IoT, such as real-time data gathering and distant monitoring, SHM can become a more cost-efficient and proactive process. Utilizing various sensors allows continual evaluation of structural status, backing up timely decision-making for fixing and maintaining them. Additionally, IoT enables data-driven algorithm automation so that there is greater understanding of structural behavior, which can be utilized to detect anomalies likely to denote future failure. Data-driven solutions, as defined in recent publications, enable real-time analysis that increases the responsiveness of SHM systems to changing environmental conditions and structural behaviors substantially, which is crucial to ensuring safety and longevity of infrastructure (Cherta et al., 2020). Further, the merging of IoT technologies with analytics technologies encourages advanced damage detection and classification in SHM systems. Sensor data can be daunting in volumes, but IoT systems ease the handling of the data through enabling real-time acquisition and processing of information. The combination of Machine Learning (ML) algorithms enhances the ability to detect pattern trends that indicate structural degradation, facilitating better predictive maintenance plans. Consequently, the intersection of data-driven and model-based approaches can lead to more accurate estimations of the health of structures. The systematic review of damage identification techniques illuminates how such systems leverage IoT to optimize decision-making efficiency and enhance civil infrastructure resilience (Ciabattini et al., 2024). This intersection of technology not only enhances the science of structural dynamics but also enhances the field of civil engineering as a whole. The effects of IoT integration extend beyond data collection; they encompass end-to-end infrastructure health management. With the establishment of an end-to-end IoT-SHM ecosystem, the stakeholders are able to achieve enhanced accountability and transparency of asset management. This has far-reaching implications for public safety and economic considerations, as timely interventions based on credible data can prevent disastrous failures. For instance, identifying structural anomalies early through constant monitoring can reduce maintenance cost and lengthen asset life. As the need for sustainability grows, adopting IoT for SHM is in line with contemporary imperatives for efficiency and innovation in infrastructure management. The suggested IoT-based SHM framework not only emphasizes the need for high-level data analysis and model fusion but also anticipates future development in technology that will transform and enhance structural health monitoring practice (Ciabattini et al., 2024).

9.1. BACKGROUND OF STRUCTURAL HEALTH MONITORING (SHM)

Structural Health Monitoring has emerged as the key to ensuring the safety and longevity of civil engineering infrastructure. This involves perpetual structural integrity monitoring with varied technologies and processes that allow collection and analysis of real-time information. Via sensors, cameras, and advanced computational techniques, engineers are in a position to monitor and rectify potential structural failures prior to these turning into catastrophic issues. The importance of SHM is also emphasized by the increasing frequency of natural disasters and the aging of infrastructure, which cause the need to create new solutions that ensure public safety and reduce maintenance costs. In the process of further development of the construction industry, SHM systems become key components of smart cities and sustainable development and are the start of a new epoch in infrastructure management. One of the most advanced techniques for SHM is the integration of Internet of Things (IoT) technology, which has revolutionized monitoring and data acquisition systems. IoT-SHM systems utilize networked sensors that gather and transmit structural data in real time, allowing remote monitoring by engineers and managers. This not only enhances the efficiency of maintenance programs but also enables predictive analysis to forecast potential structural failures by using historical patterns in data. Combining IoT with higher-grade algorithms, such as Principal Component Analysis (PCA), surpasses data-processing challenges in resource-constrained situations. Recent developments in PCA implementations have highlighted impressive advancements in memory useability and computational cost, thus enhancing the feasibility of using these systems on low-cost IoT gateways (Abdelgawad et al., 2019). Deployment of novel technologies like Radio Frequency Identification (RFID) and Long Range (LoRa) network functionality has brought with it innovative opportunities for SHM applications, particularly prefabrication construction (PC). Sophisticated systems coupling RFID with sensor networks enable location tracking and structural performance monitoring at the stage of PC component installation. In addition to enabling real-time data transmission on site, these systems enhance project management via cloud-based Building Information Modeling (BIM) to provide stakeholders with immediate access to information. The application of such technologies is not just representative of enhanced efficiency in operations but also reduced waste, therefore a reflection of sustainable construction practices (Liu et al., 2019). The ever-evolving SHM, facilitated by these technologies, is a new era of monitoring and maintenance of infrastructure in the modern world (Figure 9.2).

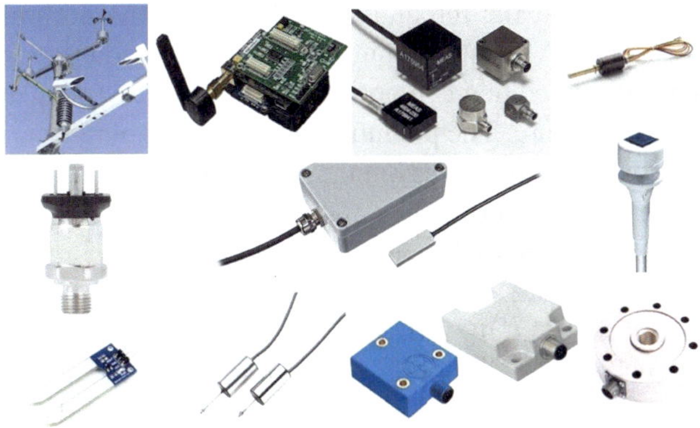

Figure 9.2. Sensors for vital parameters.

Source: https://arihantelectricals.com/wp-content/uploads/2024/08/Sensors-for-vital-parameters.jpg.

9.1.1. Historical Development of Structural Health Monitoring (SHM) Technologies

Evolution of SHM technologies over history has been immense from simple visual inspections to sophisticated advanced systems based on existing technology. Structural assessment in the early days relied heavily on manual checks and subjective assessment, which in most cases were not adequate to detect minute deficiencies. Over the years, advancements in sensor technologies, wireless communication, and computational methods have turned SHM into a more intense, automated sector. In particular, integration of these technologies facilitates real-time monitoring, encouraging rapid response to developing structural issues. While the world demand for infrastructure continues to rise, the evolution of SHM technologies has become more and more vital in a bid to deliver safety and longevity, yet fostering a climate where proactive maintenance can prevent failures of monumental scale. During the last decades, the union of SHM and information technology has ushered in an age of data-driven decision and real-time evaluation. The development of Internet of Things (IoT)-based solutions has transformed structural health monitoring into a new form, with the ability to mount a network of sensors that transmit data in real time to centralized systems to be evaluated. This transition from the traditional methods is a paradigm change, where everywhere models and data analytics are employed to converge and integrate to build an integrated picture of structural conditions (Ahmed et al., 2019). With increased advances on this front of research, it is evident that the integration of IoT elements not only enhances the dependability of SHM systems but also enables predictive maintenance techniques to be formulated that can optimize resource utilization and optimize the lifespan of infrastructure. The potential of SHM technologies is not just that they can perceive existing structures but that they can impact future designs and constructions. The utilization of smart materials with internalized self-sensing functions is a pioneering breakthrough. These innovations signify a paradigm shift toward a resilience and adaptability-based design regime in infrastructure development (Britchenko et al., 2018). Second, as the engineering field more and more integrates a trans-disciplinary thinking, the field of overlap among disciplines such as civil engineering, computer science, and environmental science can propel novel strategies to SHM deployment. This interdisciplinary setting not only enhances the strength of structures but also enables sustainable principles that are today more essential in our postindustrial world.

9.1.2. Key Components of Traditional Structural Health Monitoring (SHM) Systems

A fundamental component of traditional SHM systems is the inclusion of data acquisition sensors. Accelerometers, strain gauges, and transducers of displacement as sensors are needed to apply in order to acquire meaningful metrics regarding a structure's response to various physical stressors. Utilizing both wired and wireless technology, traditional SHM systems can transfer acquired data for analysis. The accurate positioning of the sensors is most important as they serve to be the eyes and ears of the monitoring system, providing real-time signals of the structural condition. The integration of different sensors ensures a comprehensive assessment, whereby deviation from normal function implies likely damage or failure. Therefore, traditional SHM systems offer a grounding template for structural behavior learning, serving as a stepping stone to the advanced features in IoT-based systems. Traditional SHM systems, excluding the data acquisition sensors, involve robust data processing and analyzing techniques. Processed data is essential in determining the health and safety of the structure, which helps the stakeholders make decisions on maintenance. These systems typically employ a combination of model-based approaches and data-driven algorithms in such analysis. The model-based approach relies on established engineering knowledge, whereas data-driven algorithms are focused on detecting anomalies and patterns in the sensor data. Surprisingly, the latter has gained enormous popularity with its ability to handle enormous amounts of data in real time, ultimately facilitating proactive maintenance schedules. With the evolution of SHM systems, the integration of process-based and data-driven approaches brings into focus a shift toward more sophisticated methodologies improving the detection of structural failures, thereby elevating the overall safety of infrastructure within the built environment (Cherta et al., 2020).

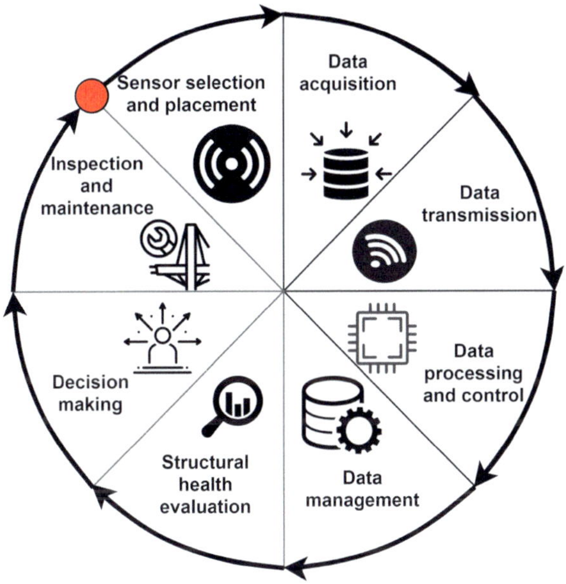

Figure 9.3. Key components of traditional structural health monitoring (SHM) systems.

Source: https://journals.sagepub.com/cms/10.1177/14759217211036880/asset/f756ffcc-7747-46a9-8ca4-eb1f6e3c82f8/assets/images/large/10.1177_14759217211036880-fig1.jpg.

Lastly, communication systems are critical in traditional SHM systems to enable the transmission of data from sensors to an analysis center. Earlier, communication systems were wired; however, with the development of wireless technologies such as Radio Frequency Identification (RFID) and wireless sensor networks, data availability is better and real-time monitoring is greatly advanced. Communication over such channels facilitates instant reporting of data and reaction, which is important in applications of critical infrastructure. The incorporation of cloud computing platforms further upgrades the functionality of conventional SHM systems through the facilitation of remote access to data and collaborative analysis of project stakeholders. Through the use of a cloud-based strategy, stakeholders have access to vital information from anywhere, enhancing the quality of decision-making processes (Liu et al., 2019). Last but not least, evolution of communication strategies within conventional SHM is paving the way for plug-and-play integration of IoT capabilities, and a new era of structural monitoring is dawned (Figure 9.3).

9.1.3. Limitations of Conventional Structural Health Monitoring (SHM) Approaches

Classic SHM techniques are often characterized by their reliance on fixed arrays of sensors that may limit flexibility and responsiveness to changing structural conditions. Classic systems tend to be founded upon using fixed sensors mounted across a structure and monitoring data at pre-specified intervals. This inflexibility can limit the ability to monitor dynamic changes resulting from environmental conditions or structural degradation with time. In addition, conventional SHM systems have to deal with integrity issues related to data since failure of sensors or loss of communication can lead to lost data points, thereby altering analysis results. The failure to adapt in reacting to unexpected issues signifies a critical deficiency, signifying the need for improved adaptive, real-time monitoring technologies through the utilization of state-of-the-art technologies such as IoT integrations, which provide improved monitoring capabilities and improved data gathering methods (Mottola et al., 2011). Moreover, conventional techniques of SHM tend to struggle with integrating intricate data-processing methods needed for accurate structural assessments. Such traditional systems tend to heavily depend on engineers' judgment in interpreting the gathered data, which causes responses to potential issues to be slow. The data fragmentation due to different sensor technologies also tends to rule the analysis, causing decision-making inefficiencies. The limitations in capacity can also make it more difficult to identify minute changes in structural health. On the other hand, IoT-based architectures have the potential to leverage cloud computing, real-time analytics, and machine learning methods for data processing optimization. This innovation promises to make the monitoring procedure efficient, thus making timely interventions a possibility in case structural anomalies do arise (Ewart et al., 2019). In an attempt to bridge these challenges, researchers in the field of SHM are investing more length into the possibility of hybrid frameworks that combine traditional and emerging methodologies. The integration of IoT-based systems enhances not only the sensing and processing capacities for data but also promotes interoperability among different types of sensor platforms. Such cross-disciplinary integration provides enhanced awareness of structural conditions by harmoniously marrying numerous streams of information. Nevertheless, the installation of such systems has its challenges, primarily in safeguarding data security and confidentiality. With increased device interconnectivity, matters of unauthorized access to vital structural data must be solved as a matter of priority. As such, ongoing research is essential in developing effective methodologies that guarantee data integrity in the most effective way possible while fully harnessing the potential of IoT-SHM systems in the long run, paving the way for even more robust infrastructure management solutions (Mottola et al., 2011).

9.2. INTERNET OF THINGS (IOT) TECHNOLOGIES AND THEIR RELEVANCE

The arrival of the IoT technologies is a revolution in various industries, particularly structural health monitoring (SHM). IoT allows for the development of integrated systems that provide real-time data acquisition, transmission, and analysis, all of which are critical in assessing the integrity of structures. In the case of SHM, IoT devices like sensors and smart meters are especially important for continuously monitoring infrastructure health, be it bridges, buildings, or roads. As discussed in recent literature, a seamless combination with sophisticated data management methods is imperative for maintaining data integrity and protecting vital measurements from tampering and falsifications. Taking advantage of existing architectures, such as four-layered ones on the MODRON platform, possibilities for effective and secure SHM applications open much wider, stressing the feasibility of IoT technologies in ensuring safety and effectiveness of operation in infrastructure management (Felice et al., 2022). Furthermore, the impact of IoT technology extends beyond the simple monitoring; they open up innovative solutions that ensure sustainability and enhance lifespan of operations within the transportation infrastructure sector. A systematic review of Industry 4.0 technologies reveals that there is significant interest in bringing IoT together with the thinking of sustainability, particularly environmental sustainability in transport schemes. The discussion categorizes IoT applications as being pivotal to the reduction of sustainability problems, reinforcing the essence of technologies that facilitate energy measurement, sustainable planning, and asset management. Notably, various studies establish that IoT sensors have the potential to augment the decision-making process, lengthen predictive maintenance schedules, and reduce environmental footprints, hence moving towards the greater vision of infrastructural sustainability in the context of SHM systems (Abbasnejad et al., 2025). Furthermore, the combination of IoT solutions and blockchain technology creates a new door for enhancing the transparency and traceability of SHM data. Blockchain combined with IoT-based SHM frameworks offers robust authentication mechanisms for data integrity, particularly because SHM systems would usually involve various stakeholders and sources of data. By publishing key SHM data on blockchain networks, participants can develop an immovable record of measurements and enhance accountability and trust in infrastructure management procedures. The innovations proposed in the employment of W3C Web of Things standards still uphold the security and interoperability of such systems, demonstrating how IoT technologies are not only facilitators of real-time monitoring but also components of more sophisticated and reliable data management and communication frameworks required for the success of SHM initiatives (Felice et al., 2022) (Figure 9.4).

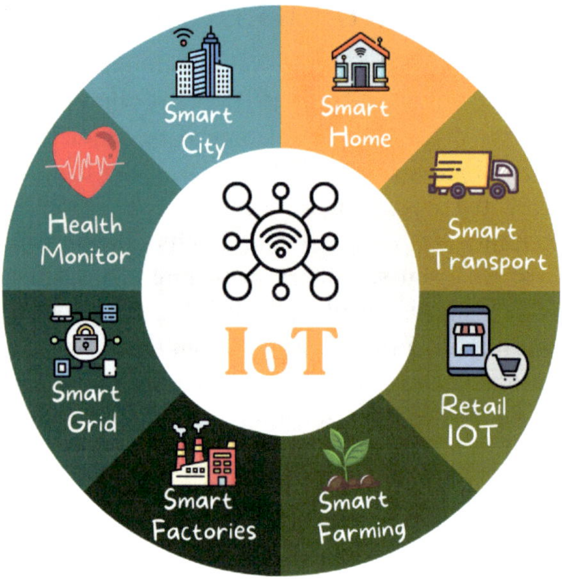

Figure 9.4. The advent of the Internet of Things (IoT) technologies marks a transformative era in various industries.

Source: https://www.mdpi.com/sensors/sensors-22-09271/article_deploy/html/images/sensors-22-09271-g001.png.

9.2.1. Overview of Internet of Things (IoT) Architecture

The IoT architecture forms a critical foundation for the effective design and implementation of smart monitoring systems, particularly in structural health monitoring (SHM). In its basic structure, IoT structure consists of a number of layers that work in concert to collect, transport, and process data from various sources. The layers typically consist of perception, network, and application layers, each supporting unique functions. The perception layer involves devices and sensors used in collecting real-time data on structural and environmental parameters. Concurrently, the network layer facilitates efficient data transmission via various communication protocols, connecting the physical world to processing systems. Lastly, the application layer interprets this data into comprehensible results, offering actionable insights invaluable for ongoing monitoring and decision-making processes in SHM applications, staying current with the ongoing need for greater infrastructural safety and efficiency. In addition, another notable characteristic of IoT architecture is how it combines cutting-edge technologies such as RFID and cloud computing with the purpose of further integrating data management within building projects. For instance, RFID- and LoRa-based projects enable tracking of pre-fabricated components on-site with precision while simultaneously analyzing their structural response. As noted by Liu et al. (2019), the new approach utilizes sensor networks in collecting the vital information, which is pushed to a cloud-based Building Information Model (BIM). The strategy not only ensures the availability of complete information but also offers real-time visualization and remote access for the project stakeholders. Therefore, IoT application in this case enhances the overall efficiency of the construction process, emphasizing the architecture of worth contribution to developing smart, responsive environments in SHM systems. In addition, the sensor data fusion component in IoT architecture cannot be overemphasized, considering that it significantly enhances the data reliability and accuracy. The sensor data fusion process is bringing together data from different sensor types to construct an integrated perception of the target environment. Perera et al.)2015) define this ability as the basis for smart city applications, which equally apply to structural health monitoring by pooling data into aggregate form. Through processing and merging diverse datasets, effective sensor fusion creates robust decision-making procedures, allowing engineers and stakeholders to identify potential structural integrity issues prior to their occurrence. Overall effectiveness of IoT architecture coupled with advanced data fusion algorithms allows professionals to create more resilient infrastructure, hence ensuring security and sustainability in cities.

9.2.2. Types of Sensors Used in Internet of Things (IoT) Applications

Among the uses of the IoT, sensors exist in various types and play important functions in capturing real-time information and facilitating improved decision making. Environmental sensors lead the list of types and are tasked with monitoring factors such as temperature, humidity, and air quality. Environmental sensors find their optimal usage in smart city applications, where they contribute to creating sustainable cities by collecting data that inform urban planning and public health initiatives. The integration of environmental sensors with cloud-based applications provides for smooth data transfer and processing, facilitating a more proactive approach to environmental issues. In an IoT-SHM system, the sensors ensure that stakeholders are constantly aware of the environmental conditions that would affect the structural health, therefore allowing for proactive intervention and maintenance actions. Another significant class of sensors is structural health monitoring (SHM) sensors that have the crucial role of quantifying infrastructure integrity.

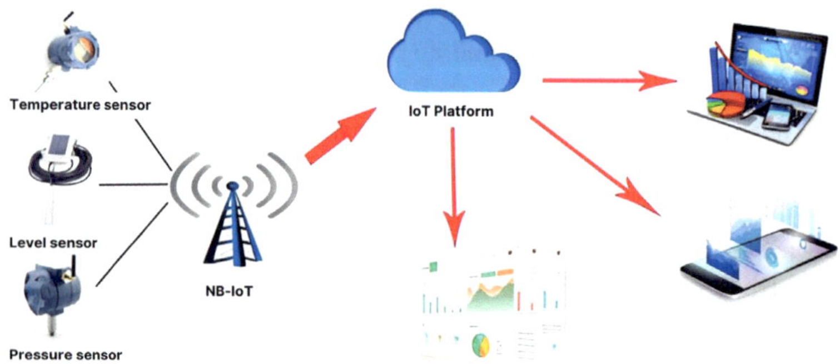

Figure 9.5. Types of sensors used in Internet of Things (IoT) applications.

Source: https://www.renkeer.com/wp-content/uploads/2021/06/IoT-and-sensors.jpg.

Displacement sensors, accelerometers, and strain gauges are used extensively to determine the physical state of structures in order to understand potential weaknesses or structural failure. Strain sensors, for instance, can detect minor material deformations, prompting regular checks before they develop into serious safety concerns. Integration of the above sensors into IoT platforms allows continuous data creation, which is further processed to provide an overall condition of a structure. Data transmitted via secure communication technologies such as LoRa allows stakeholders in question to receive vital information in real-time, which facilitates on-time decision-making in building and safety interventions (Liu et al., 2019). The integration of diverse sensor data significantly enhances the performance of IoT applications. With the implementation of advanced algorithms in cloud computing, data from various sources may be merged to generate rich insights that a sensor cannot detect on its own. Sensor fusion may be applicable in smart transportation systems where a combination of sensor inputs—traffic cameras, radar, and collision sensors—is merged to offer optimal traffic flow and enhance safety measures. The sensor data fusion framework in IoT is expanding, addressing challenges such as data accuracy and integration of heterogeneity (Perera et al., 2015). Last but not least, the composite applications of sensors and their ability to fuse data are a core element of constructing efficient and robust IoT-SHM systems, which have the potential to revolutionize infrastructure monitoring and management (Figure 9.5).

9.2.3. Communication Protocols in Internet of Things (IoT) Systems

The application of communication protocols in IoT systems is essential for facilitating the exchange of data between devices, particularly in applications such as Structural Health Monitoring (SHM). Because IoT systems consist of numerous interlinked sensors, actuators, and other devices, the selection of adequate communication protocols impacts not only data integrity and latency but also the overall energy consumption of the network. Protocols like MQTT (Message Queuing Telemetry Transport) and CoAP (Constrained Application Protocol) are also widely adopted options since they are light in weight and can function efficiently even in constrained networks. The inclusion of these protocols in an IoT-SHM system enables the seamless transferring of sensor measurements while enabling the differing operational loads typical of the environment under which such systems function. Therefore, understanding how communication occurs in IoT systems is essential for enhancing the performance and reliability of SHM applications. Energy efficiency is equally a critical factor in designing communication protocols for IoT systems, particularly given that most devices have batteries and must be powered by limited sources of energy. Research has indicated that a significant percentage of the energy usage occurs at the level of communication and particularly in the radio transceivers of IoT devices. All these advancements have concentrated mainly on achieving low-power circuitry designs as well as power-efficient physical layers in order to minimize the energy consumed per information bit in communication. However, such innovations can be undermined by ineffective network management and resource allocation strategies like performance degradation in multiple directions across throughput and responsiveness. Addressing such challenges calls for the deployment of an energy-centric optimization framework under which processing styles are coordinated with communication activities to guarantee that the general efficiency of the IoT-SHM system is optimized (Biason et al., 2017). In practical use, the application of context-aware communications protocols is crucial in respecting evolving operational conditions and performance requirements. Context awareness allows IoT devices in an SHM system to utilize self-adaptive techniques to dynamically shift their communication policies based on the prevailing environmental conditions, such as data importance and network status. This flexibility not only allows energy saving but also permits timely and accurate transfer of critical structural health data. The integration of adaptive signal processing with energy-saving optimized channel access protocols has been shown to be effective in providing an energy consumption balance against the requirements of SHM systems' performance. As evidenced by ongoing research, such actions enable IoT systems to prolong their operating life while maintaining fidelity in monitoring tasks, which is critical for effective infrastructure management and safety monitoring (Biason et al., 2017) (Figure 9.6).

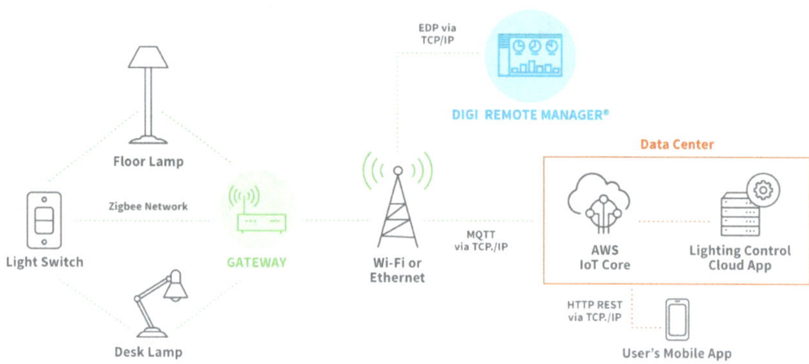

Figure 9.6. The deployment of communication protocols in Internet of Things (IoT) systems.

Source: https://www.digi.com/getattachment/Blog/post/How-Do-IoT-Devices-Communicate/digi-wireless-communications-lamp.jpg?lang=en-US.

9.3. DESIGN CONSIDERATIONS FOR INTERNET OF THINGS (IOT)-STRUCTURAL HEALTH MONITORING (SHM) SYSTEMS

The considerations for IoT-based SHM system design are a key interface between technology and engineering concepts. Among the most significant considerations in designing such systems is that of choosing the sensors, which must fulfill specific specifications, including accuracy, reliability, and cost-effectiveness. For instance, the use of capacitive sensing technology has been able to show enormous potential in remote monitoring of health parameters, as evidenced in devices used for patients with Chronic Obstructive Pulmonary Disease (COPD) (Barbarov-Rostán et al., 2018). The sensors not only need to acquire data but also need to transmit the data well, so the whole monitoring system is responsive. In addition, the design should also consider the physical environment in which such sensors operate, as factors such as temperature, humidity, and vibration may have a far-reaching impact on data integrity. As such, the application of appropriate sensors in conjunction with environmental resilience is paramount to successful IoT-SHM systems. Additionally, the inclusion of communication protocols also plays a pivotal role in IoT-SHM system design. Sensor-derived data needs to be sent to a center platform for making decisions and analyzing them. This necessitates an appreciation of many communication technologies like Wi-Fi, Zigbee, and LoRa with their own strengths and weaknesses. A middleware layer is typically proposed for communication between sensors and processing elements, as in traditional distributed computing systems (Mottola et al., 2011). The inclusion of a middleware layer can facilitate programming tasks while ensuring interoperability between different system components. Furthermore, the middleware has the capability to enhance reliability and security in data transfer, thereby improving the system robustness of the IoT-SHM. Such careful attention to communication planning and middleware integration must therefore take priority in developing an effective structural health monitoring framework. Finally, data analytics is the cornerstone in developing and deploying IoT-SHM systems. Effective data processing and analysis are necessary to reach meaningful conclusions from the vast amounts of data that monitoring systems produce. By applying machine learning methods, predictive capability of these systems can be significantly enhanced, allowing the system to anticipate probable structural failures before they occur. As depicted in the research on smart vest prototypes for respiratory rate monitoring, rigorous data validation and error checking must be employed to ensure measurement precision (Barbarov-Rostán et al., 2018). Moreover, utilizing advanced analytics platforms with in-built real-time processing will facilitate timely assessment and intervention. In general, IoT-SHM system design needs to focus not just on hardware and communication but also be specifically designed for sophisticated data handling approaches to facilitate effective contributions toward the safety and maintenance of infrastructure.

9.3.1. System Architecture and Framework

Good system architecture is essential to IoT-based Structural Health Monitoring (SHM) system design and deployment. This architecture facilitates seamless interaction among the various components to provide seamless communication and data exchange. An IoT-SHM system, in its basic form, consists of sensor nodes that record real-time information from the structural members, such as strain, temperature, and vibration. Sensors feed the information to a centralized server, where the incoming data is analyzed via cloud computing technologies. The architecture allows for data storage, analysis, and visualization, and this allows engineers and stakeholders to monitor the structural health accordingly. Use of this kind of architecture not only allows for timely detection of any faults but also enhances decision-making regarding repair or further testing when it is necessary.

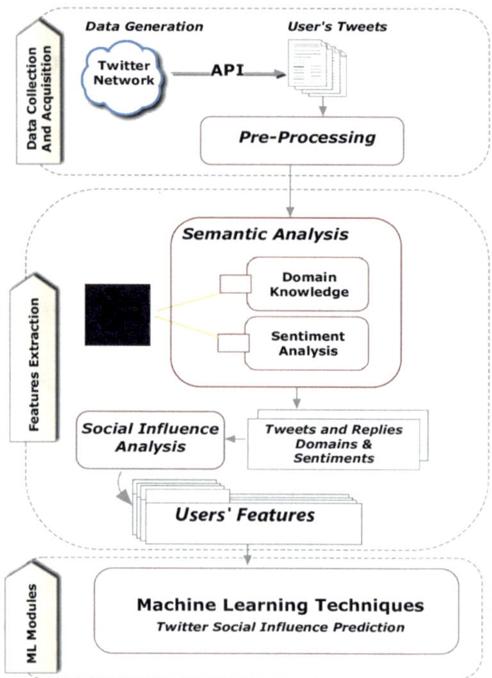

Figure 9.7. System architecture and framework.

Source: https://www.researchgate.net/profile/Bilal-Abu-Salih/publication/338762802/figure/fig1/AS:850348294033408@1579750240029/System-Architecture-Development-Framework.png

Integration of sophisticated technology into the system architecture significantly enhances the efficiency of an IoT-SHM system. For instance, employing Radio Frequency Identification (RFID) and Long Range (LoRa) technologies is of tremendous benefit when it comes to accurately tracking prefabricated building components and remotely monitoring their performance.

The system proposed in previous research proves how RFID and strain sensors can collect critical information during the construction phase and deliver it through a cloud-based Building Information Modelling (BIM) model. This model functions as a dynamic gateway for project team members to access real-time information through any mobile device and thereby develop a nimble and responsive project environment. These innovative methods not only improve project performance but also optimize the use of resources and reduce waste, as per modern construction practices (Liu et al., 2019). Moreover, the sensor fusion capability of the IoT-SHM framework optimizes analytical precision and facilitates enhanced decision-making. Through the application of information from diverse sources, the system can efficiently integrate information to provide an overall description of structural conditions. The frameworks analysis of current sensor data fusion methods highlights some of the most significant parameters that affect performance in actual application. Detailed examination of these methods reveals some of the existing challenges, particularly handling inconsistency in data and real-time processing. Resolving these challenges will not just increase the accuracy and reliability of the IoT-SHM system but also inform more widespread smart city applications that increasingly heavily rely on robust sensor fusion in order to operate at their optimal levels (Perera et al., 2015). As technological innovation in this area continues to develop, the future for transformative impacts on infrastructure monitoring and management becomes increasingly evident (Figure 9.7).

9.3.2. Data Acquisition and Processing Methods

Data acquisition and processing methods are an important factor in the design and creation of an Internet of Things-based Structural Health Monitoring (IoT-SHM) system. Effective data acquisition begins with the application of multiple sensor technologies to monitor structural parameters such as strain, temperature, and vibration. These sensors collect significant amounts of information in real-time, which are essential to ascertain the structural integrity and functionality of infrastructure. The integration with advanced communication protocols, e.g., Long Range (LoRa), ensures maximum efficiency in data transmission to enable real-time monitoring remotely. This real-time access to essential data facilitates timely decisions regarding structural repair and maintenance, thereby achieving safety and durability in construction activities. The processing techniques for the enormous amounts of data from IoT sensors need to be handled with care in order to achieve optimal performance and resource efficiency. Methods like PCA have the capability of greatly reducing the number of dimensions in the data, making analysis possible while maintaining critical structural data. Nevertheless, the computational complexity of PCAs presents a challenge, particularly if running on low-capability embedded devices. To combat this issue, recent advances have introduced memory-bounded variants of PCA, such as the streaming History PCA algorithm. The new solution achieves substantial gains, like a 10x compression factor and a 59x reduction in memory consumption at no significant loss of signal quality, thus being deployable in low-cost IoT gateways (Abdelgawad et al., 2019). Besides, incorporating technologies like Radio Frequency Identification (RFID) contributes more towards the collection of data through enabling precise tracking of prefabricated components on site. The incorporation of RFID, sensor networks, and cloud computing offers a robust platform for automatic collection and analysis of real-time data on structural members. These systems not only monitor the performance of these components during installation but also provide related data regarding the Building Information Modelling (BIM) and make it accessible and up-to-date for stakeholders within the projects. The integration enables construction teams to be able to manage resources effectively while recording ideal safety and performance levels during the construction (Liu et al., 2019). Lastly, the integration of such data gathering and analysis methods into an IoT-SHM framework assists in a meaningful way in enhancing operational efficiencies and optimizing project performance.

9.3.3. User Interface and Visualization Tools

User interface and visualization tools play crucial roles in the design and implementation of an Internet of Things-based Structural Health Monitoring (IoT-SHM) system since they enhance the interaction of the user and comprehension of data. A user interface optimized provides the intuitive navigation that allows users to communicate effectively with the system and make the decision-making process easier in real-time. These tools not only show the status of structural components but also include various visualization techniques, such as 3D models and heat maps, to create a proper representation of likely issues. By implementing the latest user experience design principles, the developers are able to bring important information into instant visibility so that users are able to perceive data and make decision-making choices about structural health conditions quickly (Duquennoy et al., 2013). Also, tools for visualization are a key element of data interpretation within an IoT-SHM platform. The extremely large amount of information generated by wireless sensor networks (WSNs) is daunting in character; thus, innovative visualization methods are needed to summarize this information into useful insights. Technology such as dynamic dashboards, with refreshment at real-time, can be instrumental in the detection of trends and anomalies within structural response. This allows stakeholders to react preemptively rather than respond to likely failures. Moreover, incorporation of data mining in visualization software can enhance analysis functionality of the IoT-SHM system, thus leading to increased accuracy in determining the integrity of the structure over time (Sharma et al., 2018). Incorporation of user interface and visualization software into the IoT-SHM system is not merely a technological advancement; it represents a paradigm shift in translation and utilization of data in structural monitoring. As the users utilize such systems, they benefit from advanced features that contextualize information, such as alert notifications of established thresholds and predictive analysis. Contextualization proves vital in structure integrity-dependent sectors as it dictates maintenance time schedules and necessary repairs, eventually extending infrastructure lifespan. Through continued development of these tools through adding user experiences and evolving technology, developers can make more functional IoT-SHM systems with an emphasis on usability and enhance the safety and reliability of a building (Duquennoy et al., 2013).

9.4. IMPLEMENTATION STRATEGIES

There should be a well-defined implementation strategy for the successful integration of an IoT-based Structural Health Monitoring (SHM) system. The key to this strategy lies in the application of cutting-edge technologies such as Radio Frequency Identification (RFID), Long Range (LoRa) communication, and sensor networks. With these elements included, the IoT-SHM system will be capable of effectively collecting real-time data from various structural members. For instance, the utilization of RFID technology allows tracking of materials to be precise at construction sites to save costs and avoid delays. It also allows the long-range communication aspect of LoRa technology, whereby sensor data are transferred to a central server without having to establish extensive infrastructure. This new approach not only optimizes working efficiency but also provides project teams with real-time evidence of the structural behavior of prefabricated components, thereby facilitating enhanced decision-making during construction (Liu et al., 2019). Utilization of data-driven algorithms for damage analysis and detection in the SHM system is another critical element of implementation plans. These algorithms utilize sensor data to provide real-time analysis of a structure's condition, finding potential damages prior to when they develop into serious problems. Data-based methods rely on the use of high-performance processors capable of processing massive amounts of computational loads. The ability to analyze real-time information allows stakeholders to make swift decisions, eventually leading to increased efficiency in maintenance operations. Damage detection data-driven not only allows for efficient monitoring processes but also enhances the quality of the generated information, which in turn helps to localize and classify structural flaws accurately (Cherta et al., 2020). Responsiveness in real time is crucial for ensuring structural integrity and safety in the future. Finally, integration of cloud computing into the IoT-SHM infrastructure enhances the usability and availability of the data collected. With the incorporation of cloud technology in Building Information Modeling (BIM), project teams can remotely access critical information, making it easier to collaborate among engineers, architects, and construction workers. This move toward digitalization enables coherent communication and provides all involved with updates on project development and potential structural problems. Additionally, data storage within a cloud-based BIM model ensures history data can be referenced for future analysis and referential use. Therefore, this strategic combination of cloud computing, IoT, and BIM not only enhances project performance but also sets industry standards higher by making the construction industry more flexible and robust. With these prudent implementation tactics, the design of an IoT-SHM system can be made effective and sustainable successfully (Liu et al., 2019) (Figure 9.8).

Figure 9.8. A well-defined implementation strategy.

Source: https://www.teamly.com/blog/wp-content/uploads/2021/12/Creative-strategy-development-and-implementation.png.

9.4.1. Selection of Appropriate Hardware and Software

The selection of suitable hardware and software is fundamental in the design and implementation of an Internet of Things (IoT)-based Structural Health Monitoring (SHM) system. It has a significant impact on the system's ability to effectively acquire, process, and transmit data. The hardware, e.g., sensors, data acquisition modules, and communication modules, should be able to manage the real structural conditions and environmental conditions present within the monitoring area. For instance, sensors should be chosen based on how sensitive they are, the range in which they can pick up signals, and their durability in order to support extreme environments in measuring correctly. Utilizing hardware that meets the projects' requirement ensures seamless integration and functionality, particularly relevant when working with data originating from various sources as highlighted by the research concerning sensor data fusion evaluation frameworks (Perera et al., 2015). Software selection is also critical as it manages data processing, analysis, and visualization. It should be the kind of software that can support real-time data acquisition and have a user-friendly interface for analysis and monitoring. Additionally, the architecture of the software must be scalable to support compatibility with future growth or modifications because IoT systems are dynamic in nature. Cloud computing capability is required to process and manage large volumes of data from the sensors, as outlined in a number of studies that explore IoT integration in smart city solutions (Adams et al., 2019).

Figure 9.9. The selection of appropriate hardware and software.

Source: https://witekio.com/wp-content/uploads/2023/07/Hardware-and-software-support-800x800.jpg.

Furthermore, having robust security protocols incorporated into the software design can protect sensitive data from fraud and illegal usage, which is a concern for IoT platforms. Lastly, hardware and software selection interaction is the foundation upon which an effective IoT-SHM system is based. An in-depth understanding of both elements is needed to attain optimized performance and reliability, enabling the system to become dynamic in adapting to the individual monitoring needs of various structures. Continuous innovation in hardware technologies, in the form of improved sensor functionalities as well as miniaturization, coupled with innovative software solutions, will enhance the general performance of IoT-SHM systems. With advancements in the area, ongoing investigation of the sensor data fusion and integration processes will continue to provide new information regarding the selection of optimal hardware and software configurations for future systems to guarantee that structural health is properly and effectively monitored (Perera et al., 2015) (Figure 9.9).

9.4.2. Integration of Sensors and Data Transmission

The use of sensors in an IoT SHM system is fundamental in ensuring continuous data acquisition and analysis. Installation of varied sensors such as strain sensors and temperature sensors within structural members makes monitoring continuous. Sensor data provides critical information on the structural condition and performance over time, required for preventive maintenance and safety assessment. These sensors are networked using advanced data transmission systems, e.g., LoRa, which facilitate long-range communication without significant power consumption. Blending these systems allows for rapid identification of anomalies and facilitates the proactive maintenance of infrastructure, in effect revolutionizing culture towards maintenance in the construction industry. Data communication is crucial in enhancing the capability of IoT-SHM systems to relay data gathered using sensors to a central system for analysis. With the volumes of data generated by different sensors running the risk of becoming overwhelming, improved communication technologies must be employed to process and manage the data in the right manner. The use of LoRa technology, as identified in the study, allows reliable data transmission across wide distances with low power usage, thereby making it especially suitable for real-time applications. Its integration not only allows for instant access to performance details of a structure but also enables quick response to any problems by project teams, thus ensuring increased success in projects. Furthermore, redundancy and loss of data are reduced, thus the monitoring system is more reliable. More importantly, emerging technologies such as Radio Frequency Identification (RFID) coupled with data transmission technologies enhance the reliability of the IoT-SHM framework as a whole. The use of RFID tags integrated with sensor networks enables the precise positioning and tracing of structural components throughout their life cycle. This facilitates end-to-end monitoring, with possibilities of views at different stages of the project, ranging from installation and overall performance monitoring. Cloud-based integration with Building Information Modelling (BIM) systems elevates this integration further by making it possible to visualize data in real time. As can be seen in the research, the approaches proposed here have far-reaching effects on improving project performance by providing real-time views that improve decision-making and project efficiency in general. Thus, the integration of sensors and data transmission technology in IoT-SHM systems is a paradigm shift in the monitoring of construction and infrastructure, bridging the gap between traditional techniques and high-technology digital solutions (Figure 9.10).

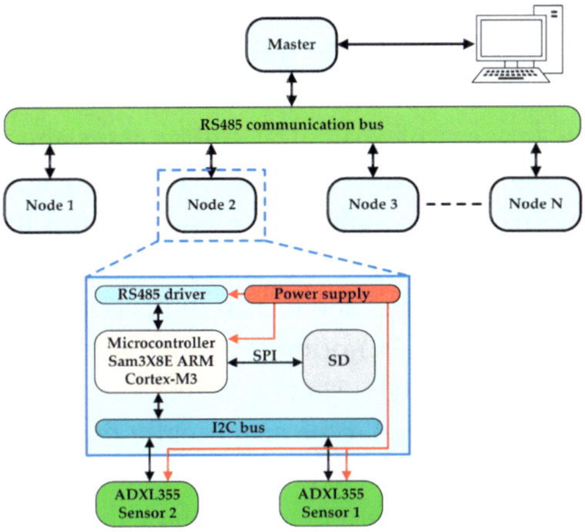

Figure 9.10. The integration of sensors within an Internet of Things (IoT) structural health monitoring (SHM) system.

Source: https://www.mdpi.com/sensors/sensors-20-04908/article_deploy/html/images/sensors-20-04908-g002.png

9.4.3. Testing and Validation of the Internet of Things (IoT)-Structural Health Monitoring (SHM) System

Testing and validation of the IoT SHM system are crucial to ensuring that it remains effective and reliable. An end-to-end methodology for evaluation involves both real-world deployment and simulation methods, which dynamically probe the system's ability to monitor structural health accurately. During the process of validation, algorithms developed for data-driven damage detection need to be rigorously tested against datasets collected from sensors for a variety of conditions. These data, resulting from measurement in real time, present an idea of the way the system can identify, locate, and classify potential structural damage. As indicated by the literature, the performance of algorithms based on data depends on the quality and accuracy of the data being processed, so thorough testing is essential to determine confidence in the SHM systems capacity (Cherta et al., 2020). Also, the application of advanced technologies such as Radio Frequency Identification (RFID) and Long Range (LoRa) networks enhances the resilience of the IoT-SHM system. These technologies enable seamless collection and communication of data from the built environment to facilitate continuous performance monitoring. Another recent study integrated an RFID technology and strain sensor system that demonstrated significant improvement in project performance through accurate on-site positioning of prefabricated components and real-time monitoring during installation (Liu et al., 2019). Cloud-based Building Information Modeling (BIM) enables convenient storage and retrieval of essential project data and further improves the decision-making process of structural safety assessment. This method of validation not only assures the reliability of IoT-SHM systems but also shows how they are capable of revolutionizing the construction industry. Success of the IoT-SHM system is ultimately measured by its performance in operational conditions when it detects damage and has an impact on maintenance decisions beforehand. Validation tests must not only include the performance criteria of the systems but also opinions from users regarding ease of use and accessibility. Real-world implementations need to be specifically designed to mimic various environmental conditions and mixed structural characteristics. This extensive testing procedure is utilized to fine-tune the algorithms and system elements so that fine-tuning may be employed to enhance response times in addition to reducing error rates during damage detection. Continuous testing assists with guaranteeing the IoT-SHM system remains adaptable to innovative technologies and approaches, embracing innovation and robustness to the structural engineering field. The collaborative wisdom of testing and validation phases brings about an open door to a greater reliable, efficient, and user-friendly IoT-SHM system.

9.5. CASE STUDIES OF INTERNET OF THINGS (IOT)-STRUCTURAL HEALTH MONITORING (SHM) APPLICATIONS

The use of IoT technology in SHM systems has transformed how engineers assess and maintain infrastructure at a rapid rate. One interesting case study involves the use of a cutting-edge system utilizing Radio Frequency Identification (RFID) and Long Range (LoRa) technologies in prefabricated construction (PC) projects. As highlighted by recent research, the method not only promotes cost and energy efficiency but also addresses effectively issues of traditional PC project monitoring. Through the use of sensor networks, data gathering is achieved in real-time without any interference, allowing stakeholders to monitor PC components on construction sites and assess their structural condition when installed. The result of the system showcases the potential of cloud-based Building Information Modelling (BIM) to provide project information dynamically and significantly enhance decision-making for remote project teams and facilitate ease of work on-site (Liu et al., 2019). A notable example shows how different advances in data processing algorithms can potentially increase the performance of IoT-SHM systems. In particular, the innovative application of streaming Principal Component Analysis (PCA) assists in compensating for the computational limitation of using low-cost embedded platforms in these systems. The development of memory-efficient parallelized realization yields enormous computation time reductions and memory usage at the cost of no loss in data integrity.

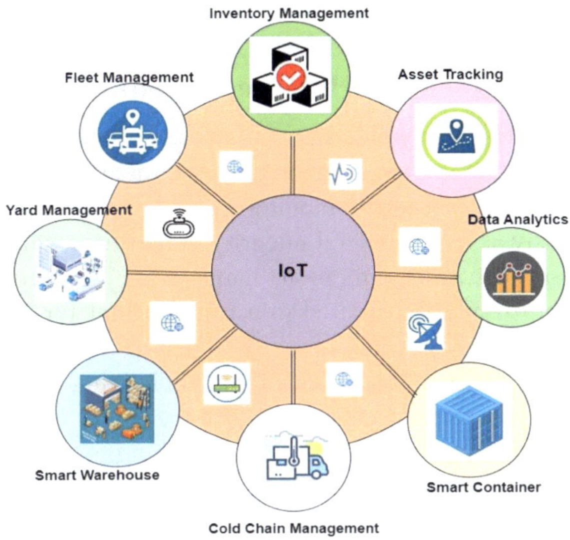

Figure 9.11. Case studies of Internet of Things (IoT)-structural health monitoring (SHM) applications.

Source: https://www.researchgate.net/publication/376258537/figure/fig3/AS:1143 1281209898102@1701896751829/The-use-cases-of-IoT-in-logistics.png.

From the research findings, this algorithm presents a tenfold compression ratio and nearly a sixtyfold memory demand reduction at the expense of minimal loss in signal performance. This enhancement is critical towards allowing SHM systems for IoT applications to operate even when conditions are hardware-limited, thereby enhancing the reliability and effectiveness of structural monitoring projects (Abdelgawad et al., 2019). The use of IoT for SHM is not limited to individual projects but also extends across broader systems of infrastructure. For instance, large case studies have shown the implementation of IoT-SHM for key industries like bridges and buildings, where public safety and maintenance scheduling require real-time monitoring of structure conditions.

These case studies illustrate the capacity of IoT devices to enable real-time data collection, which enables predictive maintenance and timely intervention that can avert risks of structural collapse. Moreover, the use of emerging technologies like machine learning and big data analytics is attracting research in a bid to enable the interpretative capacity of the data collected. This union of IoT and SHM not only encourages sustainable construction but also supports a mature perception of asset management and durability in civil engineering activities (Figure 9.11).

9.5.1. Examples of Successful Internet of Things (IoT)-Structural Health Monitoring (SHM) Implementations

Its deployment in SHM has yielded several successful case studies that show its efficacy in enhancing safety and maintenance strategies. One such example is the deployment of IoT sensors in monitoring bridges and subsequently being able to continuously assess structural integrity in real time. With the inclusion of wireless sensor networks, vibration, temperature fluctuation, and strain data can be collected and transmitted to a central point for processing. This proactive approach not only reduces the risk of disastrous structural failure but also maximizes maintenance schedules, ultimately helping to increase lifespan in infrastructure. Efficient projects have shown that SHM enabled by IoT can significantly improve the processes of decision-making for civil engineering by providing informative data supporting improved infrastructure resilience and safety standards. Another significant use of IoT-SHM is monitoring high-rise buildings, with sensors placed strategically throughout the building to gauge environmental conditions and occupant safety. For instance, through the use of environmental and structural sensors in combination, engineers can analyze wind loads and seismic activity. Data acquired can be utilized to make real-time modifications in building systems to minimize potential damage or hazard. Further, integration with smart data analytics makes predictive maintenance possible, guiding interventions before issues escalate. As underscored by recent research, the worth of adaptive communication protocols, particularly with the emergence of fog computing, can facilitate effective processes of handling data to enhance the overall performance of IoT applications in SHM (Alonso Zárate et al., 2016). This is a growing trend in urban development, with technology playing a direct implication in building resilience. Interestingly, IoT technology has also been used in the maritime sector for SHM application, particularly in monitoring the health of ships and offshore structures. IoT sensors mounted on these structures produce a flow of data continuously, enabling mechanical condition assessment and alerting operators to impending failure. Through the use of a cloud system to process data, organizations are able to assess trends over time and respond appropriately to any anomalies that are detected. In light of the research recommendation, the integration of fog-to-cloud computing improves real-time processing capacity to ensure navigation systems remain responsive and efficient (Carpio et al., 2019). The successful use of IoT-SHM in marine environments validates the ability and versatility of such systems to enhance safety and operational efficiency in different applications in engineering and infrastructure management.

9.5.2. Analysis of Performance Metrics in Case Studies

In the context of Internet of Things (IoT) based Structural Health Monitoring (SHM) systems, performance metric analysis plays a critical role in determining system effectiveness. Researchers can properly analyze how monitoring methods perform in real-world scenarios with the aid of case studies. For example, research into a non-contact portable respiratory monitoring device demonstrated high accuracy in respiratory rate estimation with the mean error being only 0.01 breaths per minute. High precision is essential in remote patient monitoring, particularly for patients suffering from chronic illnesses like Chronic Obstructive Pulmonary Disease (COPD). These results not only validate the technical feasibility of the device but also provide information on its potential applications within an IoT-SHM system and hence indicate the importance of precise performance metrics in case studies (Barbarov-Rostán et al., 2018). Additionally, performance metrics derived from case studies provide vital information about the energy efficiency and security functionalities inherent in IoT-SHM systems. Wireless sensor network analysis is in agreement that energy conservation mechanisms must be optimized along with security protocols. For instance, the deployment of strong security protocols such as encryption and authentication will unwittingly consume more power, which is difficult in sensor deployment lasting several years. In regard to this, there are case studies that focus on efficient energy utilization protocols such as duty cycling, which significantly extends battery-powered sensor life. Testing these protocols in terms of energy load and network performance provides valuable insight into security needs versus operational feasibility within IoT-SHM systems (Udoh et al., 2019). Finally, integration of the performance metrics collected from these case studies into future innovations in IoT-SHM systems is necessary. Iterative refinement of sensor networks based on empirical data allows for accuracy improvement as well as reliability. Experiences learned from previous applications can drive system design breakthroughs to a point where emerging IoT-SHM solutions not only execute flawlessly in structural health monitoring but also continue to be sustainable in practice. This ongoing improvement with performance criteria can continue to yield adaptive systems that are capable of adapting to the future demands in safety as well as regulatory compliance, therefore making the analysis of case studies even more vital for this field. The integration of findings across the different studies forms a common body of knowledge that drives the design and construction of robust IoT-SHM systems to continue being responsive to pragmatic concerns (Barbarov-Rostán et al., 2018; Udoh et al., 2019).

9.5.3. Lessons Learned from Real-World Applications

Implementation of IoT-based Structural Health Monitoring (SHM) systems has provided with key learnings that are vital to their design and operation. One of the key learnings is the significance of robust sensor data fusion techniques for effective monitoring. Experience in real-world applications has proven that multiple sensor modalities, such as accelerometers and strain gauges, must cooperate so as to provide comprehensive information about structural integrity. This is in accordance with Perera et al. (2015), where sensor fusion capability is noted to be vital for smart city application. By using a sophisticated sensor data fusion assessment framework, engineers can enhance the accuracy and relevance of the data and subsequently more accurate structural health evaluations. Thus, lessons learned from actual deployments in IoT-SHM identify the importance of an integrated approach to data gathering that focuses on cross-functional technological inputs and their synergistic value. A second critical lesson gained is energy management in IoT-SHM systems, which directly affects effectiveness and lifespan of deployed sensors. As noted in various projects, a significant share of energy consumption by IoT devices is from their radio transceivers. The results drawn from real-world scenarios bring into focus the need to develop energy-efficient solutions and protocols that counter this problem. The research offered in Biason et al. (2017) brings forward the fact that without planned development in resource management and process mechanisms, even optimally designed low-power components will not be able to handle the situation. Therefore, applying adaptive strategies, such as compressed signal processing and dynamic routing protocols, becomes essential. Such optimizations not only optimize energy consumption but also optimize the overall responsiveness and resilience of IoT-SHM systems, illustrating that energy issues have to be addressed from the initial stage of system design. Furthermore, actual deployments of IoT-SHM systems illustrate that flexibility and adaptability are key to long-term successful operation. Real deployments have demonstrated that environmental and operational conditions can differ heavily, and hence systems have to be able to adjust their parameters correspondingly. Lessons learned mandate that inserting context-aware mechanisms into the system design allows for increased performance under different conditions. As experimental frameworks suggested in Biason et al. (2017) hypothesize self-adaptation abilities of IoT networks through learning architecture and reconfigurable algorithms as irreplaceable, adaptability does not just improve the system's performance to work at its optimal level under different conditions but also enhances network lifetime and reliability. Thus, the encouragement of flexibility in IoT-SHM systems is one significant lesson of real practice and suggests a new paradigm of more adaptive and responsive structural monitoring solutions (Figure 9.12).

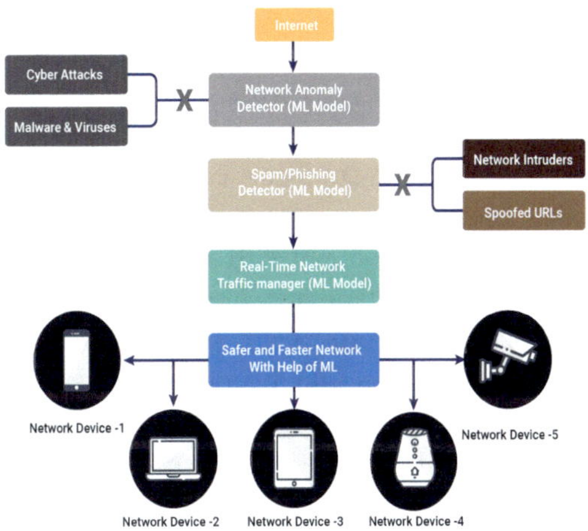

Figure 9.12. Real world applications of machine learning.

Source: https://osmphp.s3.amazonaws.com/619bfad820a24-Screen+Shot+2021-11-22+at+1.15.48+PM.png

9.6. CHALLENGES AND FUTURE DIRECTIONS

The construction and operation of an Internet of Things (IoT) Structural Health Monitoring (SHM) system are faced with numerous challenges to be overcome to ensure effectiveness and reliability. As urban populations continue to grow exponentially, the pressure on infrastructure increases, and sophisticated monitoring systems are needed to avert failures. Unification of several IoT devices to form a cohesive system that talks in harmony with each other and also has the capacity to process and understand enormous amounts of data in an efficient manner is one of the central challenges. Sensor heterogeneity, varied communication protocols, and lack of standards complicate the system architecture and add to inefficiencies. The critical need for intelligent city initiatives further accentuates the necessity of frameworks enabling data capture as much as collaboration, while also improving decision-making processes through real-time feedback on structural health (Akyildiz et al., 2013). IoT-SHM systems' future development routes must provide higher scalability and integration capabilities of monitoring technologies. As has been evident from the recent evolution, the utilization of cloud-based platforms has significantly enhanced resource management and the delivery of analytical capacities, which has streamlined the management of urban infrastructure issues. However, the future requires a robust platform that can adapt to handling advanced sensor technologies and elevating the overall user experience. Particularly, extending the IoT paradigm to include sensor networks for non-invasive monitoring, such as has been achieved in the Internet of Medical Things (IoMT), is a potential way forward. Such convergences could provide end-to-end real-time assessments without saturating users or infrastructure (Barbarov-Rostán et al., 2018). Achieving such milestones will require inter-disciplinary work and R&D spending to establish standards and protocols. Moreover, giving privacy and security concerns top priority in shaping the future of IoT-SHM systems is essential. As data collection becomes increasingly pervasive in urban areas, sensitive information must be safeguarded to secure the trust of the public. The sophistication of cyberattacks poses tremendous risks that will compromise not just personal privacy, but also the integrity of vital systems. It is important to design security mechanisms that not only safeguard information but also enable regulatory schemes compliance. Further, engaging stakeholders like city planners, engineers, and the public will ensure that transparency and accountability are prioritized as technology advances. By addressing these ethical concerns, future IoT-SHM systems can make valuable contributions to urban resilience while building public confidence in smart city initiatives (Akyildiz et al., 2013).

9.6.1. Technical Challenges in Internet of Things (IoT)-Structural Health Monitoring (SHM) Systems

Integration of IoT systems in SHM entails a number of technical problems that may hinder their performance. One primary problem is integrating various hardware and software components, keeping in mind the diversification of sensing technologies and communication protocols. In the majority of applications, the components need to operate in concert to collect and transmit data securely. The reliance on previously built operating systems in order to develop applications for wireless sensor networks (WSNs) is usually to blame for making it complicated, which could be alleviated by the middleware layer. Middleware can provide a standardized communication platform and data abstraction, thereby making the development of interoperable applications easy. However, necessity and viability of introducing such a layer in WSNs have been contested, as it could require additional resources and skills that might not be within the capacity of every developer, which could make IoT-SHM system design and implementation more complex (Mottola et al., 2011). Secondly, the challenge of checking data correctness and consistency for IoT-SHM systems should not be underestimated. Sensor readings are susceptible to various forms of noise and interference, which will lead to erroneous measurements compromising system reliability. Solving these issues is through sophisticated data processing techniques to remove noise and enhance signal quality. Furthermore, given the constraint of low processing capacity in the majority of IoT devices, computational simplicity relative to meaningful data analysis takes center stage. Where sensors are deployed over large infrastructures, data aggregation from heterogeneous sources adds to the complexity. The need for fault-tolerant algorithms with the ability to execute in real-time while guaranteeing efficient utilization of resources poses a significant research challenge in IoT-SHM systems' development (Ewart et al., 2019). Security is also a technical hindrance to IoT-SHM system uptake. The networked and integrated nature of IoT devices makes them vulnerable to cyber-attacks, thus raising issues related to data integrity and privacy of the user. With the possibility of data obtained through structural monitoring being sensitive in nature, it becomes important that stringent security measures are implemented to prevent unauthorized access and leakage of data. Work on end-to-end security frameworks that target hardware and software components is necessary to mitigate these risks. This becomes increasingly challenging given the diversity of devices and platforms used in IoT-SHM systems that may not all have equal security measures in place. Therefore, addressing the multi-faceted security challenges in IoT-SHM systems requires inter-disciplinary investigations and cooperation among researchers, engineers, and cybersecurity experts to create resilient and secure monitoring systems.

9.6.2. Regulatory and Ethical Considerations

The integration of IoT technologies in Structural Health Monitoring (SHM) systems requires a close examination of regulatory and ethical considerations. As more advanced streams of data, like real-time tracking and predictive analytics, join the fray, the potential for data misuse or leakage is intensely greater. Data privacy is of great significance with the possible gathering of personal data during monitoring operations. Regulatory systems need to advance so that the collection of data is aligned with legislation like the General Data Protection Regulation (GDPR), focusing on user consent and data privacy.

Aligning IoT-SHM processes with existing frameworks can help stakeholders build trust among users and parties engaged in the upkeep of infrastructure while offering a secure setup for data handling. Securing ethical use of IoT-SHM systems further complicates the operational environment. As indicated by recent research, ethical concerns are not strictly based on problems of privacy but also on data exploitation transparency and possibilities for system bias. For instance, predictive algorithms can inadvertently favor certain outcomes based on the input data they are fed, thus perpetuating disparities in resource allocation or maintenance tracking. Use of frameworks for continuous permitting and screening of compliance becomes imperative to mitigate these threats. There is a requirement for the stakeholders to be active in identifying and addressing ethical constraints, as suggested in the case of digital twins and their application in construction (Alsaffar et al., 2025).

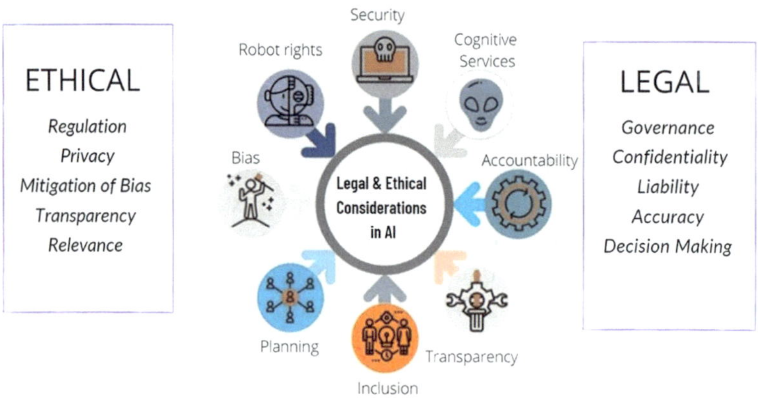

Figure 9.13. Regulatory and ethical considerations.

Source: https://www.frontiersin.org/files/Articles/862322/fsurg-09-862322-HTML/image_m/fsurg-09-862322-g001.jpg.

Collaborations between engineers, data scientists, and ethicists are the best way to address such complex considerations. Moreover, socio-economic consequences of IoT-SHM systems also require serious thought in the context of regulatory and ethical principles. The interaction between innovation and the danger of socio-technological disruptions must be adequately balanced. Advanced technology systems, for instance, are capable of enhancing lifecycle sustainability in construction but can produce entry price obstacles that exclude small players from engagement, thereby deepening the competitiveness gap in the sector. These challenges ring true with studies on Ambient Assisted Living technologies, where maintaining affordability and usability is paramount (Boem et al., 2025). Future work needs to research how to develop inclusivity within the regulatory framework while advancing the technological capability of IoT-SHM systems. This emphasis will ensure that the advantages of these systems are equally shared, ushering in development that benefits society as a whole (Figure 9.13).

9.6.3. Future Trends in Internet of Things (IoT) and Structural Health Monitoring (SHM) integration

With IoT technology being merged with SHM systems, future technology has been poised to change the way we think about infrastructure safety and maintenance. Data-driven algorithms play a crucial role for their capability in processing real-time sensor data in an efficient manner, which is critical in making prompt decisions regarding structural integrity. These algorithms provide rapid damage detection, localization, classification, and prognosis, which ultimately translates into enhanced system responsiveness and reliability. Researchers have more and more come to favor this data-driven approach over other more traditional modeling approaches, not least due to its ability to cope with various structural conditions and the complexity of practical applications. Thus, the use of high-performance processors to cope with computational demands becomes a valuable investment for future IoT-SHM systems aimed at public safety and infrastructure resilience (Cherta et al., 2020). In addition to data-driven approaches, future IoT-SHM systems will also leverage artificial intelligence (AI) and machine learning (ML) technologies to achieve improved predictive analytics. Employing sophisticated data management frameworks, these systems can incorporate huge amounts of sensor data and overcome processing and dataset development problems. The integration of AI makes it easier to merge information from various sources, hence enhancing the damage detection process by highlighting anomalies. This paradigm shift not only contributes towards enhancing efficiency but also towards enhancing damage classification and assessment precision, ultimately resulting in effective proactive maintenance practices that can mitigate risk prior to extensive structural failure (Ciabattini et al., 2024). With continued development of data-centered, AI-driven techniques, the potential of intelligent, autonomous SHM systems becomes substantially bigger. In the future, the interaction between IoT and SHM will likely enable the development of smart infrastructure with self-diagnostic capabilities and automated reporting. These technologies indicate that civil engineering infrastructure can report its health condition in real-time to building managers or maintenance personnel, supporting immediate response action when structural anomalies are detected. Integration of such systems can contribute significantly to the extension of asset life while ensuring safety and regulatory compliance. Future IoT-SHM systems can further incorporate remote access and easy-to-use dashboards that present data in actionable format. As these technologies converge, the industry players in construction and maintenance have to remain committed to their collaboration and creativity to maximally leverage the potentials of merged IoT and SHM solutions (Ciabattini et al., 2024).

SUMMARY

The invention and deployment of an SHM system that is IoT-based is a significantly significant convergence of technology and engineering. The fruition of this work underlines the critical necessity to merge state-of-the-art data fusion techniques for effective monitoring. By enabling real-time data gathering and analysis across different sources, an IoT-SHM system is better able to assess the structural integrity of infrastructure. Such frameworks not only enhance communication and operational efficiency, but they also enable predictive maintenance practices with a massive potential to reduce downtime as well as maintenance expense. As urban areas continue evolving to become smart cities, the application of IoT in SHM becomes increasingly prominent, necessitating ongoing research and investment in leading-edge monitoring technologies. With the issues associated with carrying out advanced algorithms within low-capacity embedded systems, the research necessitates more efficient computational processes. The concurrent execution of History PCA with streaming offers a clear improvement, with outstanding compression and memory consumption without a substantial performance compromise. By transcending the computational limitations of low-cost IoT gateways, the approach demonstrates the capability of enhanced algorithms to maintain high fidelity in structural data analysis. Through this manner, the research not only contributes to technical knowledge but also gives more emphasis to the necessity of enhancing existing methodologies for utilization in real-world applications in smart cities so that SHM systems can be implemented and made practical and accessible. Furthermore, the study of sensor data fusion structures holds the key to future IoT-SHM systems. As intimated, this merging of multiple sources of information has the potential to allow smarter analytics and provide more precise insights into structural health. Addressing open research challenges in this field is critical in further advancing the accuracy and reliability of SHM outcomes. Through focus on new approaches and continuous evaluation, researchers and practitioners can strive toward further evolving the discipline so that IoT-SHM systems remain attuned to the growing requirements of today's infrastructures. The relevance of employing IoT technologies for structural health monitoring cannot be exaggerated because it is the foundation for more secure, more sustainable, and more urbanized living environments.

Merging IoT technology with SHM systems has produced tremendous advancement in construction as well as in infrastructure management. Above all, the study results demonstrate that the application of sensor networks has the potential to enhance the accuracy and reliability of data collection on structural performance considerably. Specifically, the employment of RFID and LoRa technologies has facilitated an in-time monitoring capability, which is crucial for preventive maintenance measures. The study confirms that the installation of such advanced technology not only streamlines the process of recognizing prefabricated components on-site but also facilitates continued monitoring of their condition throughout the course of installation. This indicates a shift towards a more streamlined process that can ultimately lead to risk elimination and improved project success. Tied down in the studies, application of a cloud-based Building Information Model system was an important instrument for managing information derived from sensor networks. Application of the cloud platform allows stakeholders in the project to access and analyze data with no inconvenience from remote locations, which is fundamental to the present project management procedures. The BIM integrates various data acquired points, which improve visualization of present projects and enable decision-making. This incorporation of data into a usable format not just encourages collaborative working across all stakeholders but also signifies the potential of IoT-based systems to enhance the efficiencies of construction. Augmented by BIM, real-time data analytics facilitate decision-making based on accurate and current information rather than deciding based on historical numbers. Lastly, research on sensor data fusion identifies the need to enhance research and development in the same. Comparison of various sensor data fusion platforms has highlighted 10 crucial parameters that should guide these research processes. Notably, the research conclusion is that efficient sensor fusion can effectively conquer the distinctive challenges that define heterogeneous data sources while optimizing IoT application potential for implementation in smart city contexts. As cities continue to evolve and become more technologically advanced, addressing these open research issues in sensor data fusion is crucial. The below partnerships and new strategies emerging from these findings may provide avenues for more adaptive and resilient urban spaces, solidifying further IoT's role in SHM system design and deployment, as pinpointed in the initial study Liu et al. (2019) and extended to smart city scenarios according to Perera et al. (2015).

REFERENCES

1. Abbasnejad, B., Ahankoob, A., Kaewunruen, S., et al. (2025). Industry 4.0 technologies for sustainable transportation projects: Applications, trends, and future research directions in construction. https://core.ac.uk/download/653323966.pdf (accessed on 2nd October 2025).
2. Abdelgawad, A., Allen-Zhu, Z., Hardt, M., Li, J., Mitliagkas, I., Reyer, M., Rooshenas, A., et al. (2019). Embedding principal component analysis for data reduction in structural health monitoring on low-cost IoT gateways. https://core.ac.uk/download/578214124.pdf (accessed on 2nd October 2025).
3. Adams, R., Amafabia, D., Barker, T., et al. (2019). *Proceedings of Abstracts Engineering and Computer Science Research Conference 2019*. https://core.ac.uk/download/287582059.pdf (accessed on 2nd October 2025).
4. Agis Cherta, D., Gomez Vargas, R. C., Pedraza, C., et al. (2020). Damage identification in structural health monitoring: A brief review from its implementation to the use of data-driven applications. https://core.ac.uk/download/288625955.pdf (accessed on 2nd October 2025).
5. Ahmed, K., Atzori, L., Hankin, B., Basco-Carrera, L., Bastin, L., Beven, K., Beven, K., et al. (2019). Models of everywhere revisited: A technological perspective. https://core.ac.uk/download/322489447.pdf (accessed on 2nd October 2025).
6. Akyildiz, I. F., Alvarado, U. R., Ashton, K., Banerjee, S., Cabral Pinto, M. M., Chou, C. T., Mora, H., et al. (2013). Sensing as a service model for smart cities supported by Internet of Things. http://arxiv.org/abs/1307.8198 (accessed on 2nd October 2025).
7. Alonso Zárate, J., Alonso Zárate, L. G., Kalalas, C., et al. (2016). Goodbye, ALOHA! https://core.ac.uk/download/81571287.pdf (accessed on 2nd October 2025).
8. Alsaffar, A., Beach, T., & Rezgui, Y. (2025). Enhancing asset management: Integrating digital twins for continuous permitting and compliance – A systematic literature review. https://core.ac.uk/download/630478718.pdf (accessed on 2nd October 2025).
9. Biason, A., Kelly, M., Pielli, C., et al. (2017). EC-CENTRIC: An energy- and context-centric perspective on IoT systems and protocol design. https://core.ac.uk/download/84922330.pdf (accessed on 2nd October 2025).
10. Boem, F., Mandalari, A. M., Ritchie, M. A., et al. (2025). An interdisciplinary

overview on ambient assisted living systems for health monitoring at home: Trade-offs and challenges. https://core.ac.uk/download/639101762.pdf (accessed on 2nd October 2025).

11. Britchenko, I., Cherniavska, T., & Machashchik, P. (2018). New trends in development of services in the modern economy. https://core.ac.uk/download/323058294.pdf (accessed on 2nd October 2025).

12. Carpio, F., Dizdarevic, J., Jukan, A., et al. (2019). A survey of communication protocols for internet of things and related challenges of fog and cloud computing integration. https://core.ac.uk/download/187266039.pdf (accessed on 2nd October 2025).

13. Ciabattini, L., Di Felice, M., Esposito, A., Forlesi, M., Ghini, E., Kamali, S., Marzani, A., et al. (2024). An IoT-AI toolchain for structural health monitoring. https://core.ac.uk/download/654521699.pdf (accessed on 2nd October 2025).

14. Di Felice, M., Gigli, L., Marzani, A., et al. (2022). Blockchain and Web of Things for structural health monitoring applications: A proof of concept. https://core.ac.uk/download/564349438.pdf (accessed on 2nd October 2025).

15. Duquennoy, S., Höglund, J., Misra, P., et al. (2013). Supporting cyber-physical systems with wireless sensor networks: An outlook of software and services. https://core.ac.uk/download/17037990.pdf (accessed on 2nd October 2025).

16. Ewart, P. (2019). A comparison of processing techniques for producing prototype injection molding inserts. https://core.ac.uk/download/322841895.pdf (accessed on 2nd October 2025).

17. Liu, Z., Mbachu, J., & Zhao, L. (2019). Development of intelligent prefabs using IoT technology to improve the performance of prefabricated construction projects. https://core.ac.uk/download/231841646.pdf (accessed on 2nd October 2025).

18. Mottola, L., & Picco, G. P. (2011). Middleware for wireless sensor networks: An outlook. https://core.ac.uk/download/11435219.pdf (accessed on 2nd October 2025).

19. Perera, C., Wang, M., Zhang, M., Strazdins, P., Jayaraman, P. P., Shyamsundar, R. K., & Ranjan, R. (2015). City data fusion: Sensor data fusion in the internet of things. http://arxiv.org/abs/1506.09118 (accessed on 2nd October 2025).

20. Sharma, M., Singh, G., & Singh, R. (2018). An advanced conceptual diagnostic healthcare framework for diabetes and cardiovascular disorders. http://arxiv.org/abs/1901.10530 (accessed on 2nd October 2025).

CHAPTER 10

FUTURE TRENDS AND RESEARCH DIRECTIONS

Contents

INTRODUCTION

An academic essay introduction serves a crucial role in establishing the context for the ensuing discussion, particularly when analyzing the core subject of Future Trends and Research Directions. Under such circumstances, the introduction must not only establish the relevance of the issue at hand, but also the problems and implications with regard to future directions of research. By offering a general overview of significant concerns, the introduction establishes a foundation for subsequent discussions, guiding readers along the ground of new questions. This model of analysis allows for a discussion of innovative methodologies, interdisciplinarity, and the value of timely research and possible effects on society. Given the continually changing academic environment, these key aspects being defined give scholars and practitioners the context they require to address changes that continue unfolding in their discipline. The introduction should also include analysis of related scholarly literature and evidence, foregrounding prior studies while advancing underlying knowledge gaps that follow-up research can address. For instance, the invasive character of the lionfish and its impact on marine ecosystems provides a model case study to contemplate paths of future research. As lionfish expand around the Southeast U.S. and the Caribbean, their impacts on reef fish assemblages and broader ecological processes prompt urgent questions regarding management priorities. Speaking of such key points in the introduction has the effect of calling the attention to the multifaceted nature of future work, calling for proactive intervention against invasive species. The lionfish provides a tangible point of reference for inquiry into larger patterns in ecological and environmental studies, showing how scientific fields of study are interrelated. Lastly, the introduction provides an opening to discuss the need for integrated research approaches in the future that consider today's issues as well as long-term implications. In this context, the integrated assessment of lionfish offers an insight into genetic, ecological, and management factors that lead us to know the need for equilibrated studies. By placing focus on the immediate need for research on invasive species and their wide-ranging implications—ranging from ecological balance, human health, to economic effect—the introduction lays the groundwork for a more in-depth examination of pressing global concerns. Successful integration of these factors ensures that the essay critically addresses its theme, showing how fundamental findings can shape later areas of study that are both new and essential in relation to sustainable practices across disciplines.

Definition of Future Trends

Identifying the definition of future trends requires a look at the evolving dynamics within various fields, including technology, health, and environmental studies. Emerging trends are typically characterized by their ability to reflect paradigm shifts, society's demands, and innovative techniques in research design. This requires an interdisciplinary examination that can accommodate integrating knowledge from different realms in a manner that creates a deeper insight into how trends change over time. For example, human-environment relationships are on the rise in current research, showing that trends in the future would be more focused on sustainability and health. Investigating these intersections helps researchers develop models that touch on the intricacies of human-nature relationships to provide a more refined understanding of environmental-influenced health outcomes (Adger et al., 2016). Moreover, the development of technologies such as brain-computer interfaces (BCIs) demonstrates the transformative power of future trends in expanding human interaction with technology. BCIs, which are slowly making their entry into everyday applications, demonstrate how technological advancement can transform the user experience and the functionality. As this technology evolves further, it also raises significant questions over implementation and population diversity, necessitating the consideration of multiple population support systems, including the disabled. Researchers must consider pragmatic factors and usability in actual settings, underlining the need for ongoing scrutiny of trends and performance assessments (Allison et al., 2012).

Figure 10.1. The future trends requires an examination of the evolving dynamics within various fields.

Source: https://www.schaeffler.com/remotemedien/media/_shared_media_rwd/01_company_1/strategy/roadmap_2025/16_9-schaeffler-roadmap-2025-future-trend-new-mobility-electrified-powertrain_rwd_600.jpg.

Ultimately, the course of BCIs is a broader movement toward more natural and connected technology solutions anticipating user needs and capabilities, illustrative of how future trends will likely unfold. In synthesizing these developments, it is clear that the explanation of future trends includes not only their identification but also the critical exploration of their implications in different fields. The intersection of converging technologies and the evolving definition of health and well-being represents a shift toward the application of an integrated model in research designs. As the discussion of human-nature relations continues to shift, researchers have the responsibility to develop conceptual structures that reflect these complexities and avenues of future study. This interdisciplinary style will provide more insight into how trends in the future can be harnessed to the benefit of humanity, ultimately solving issues of pressing international concern and improving overall health outcomes (Adger et al., 2016). Thus, the vision of future trends is anchored in a vision of interdisciplinary collaboration and innovation and sets the tone for transformative research agendas in the coming couple of years (Figure 10.1).

Importance of Research Directions

Research directions are the principal guiding protocols in areas of study, setting the direction for knowledge and innovation. As the face of research evolves, it becomes ever more necessary to recognize and explore new trends to foster development across fields. For instance, the growing intersection of technology and education necessitates focused research into pedagogy that utilizes digital technology most effectively. Not only do such questions benefit teachers and students, but they also align with industry need, rendering efforts at education sound and effective. As highlighted by logistics education literature, academia-industry partnerships can potentially influence curriculum developments, which in turn reinforce the value of well-delineated research directions in facing contemporary problems (Farris et al., 2008). Moreover, clear research directions can also potentially make academic scholarship more relevant through bridging theoretical-practice gaps. In places characterized by rapid technological and societal changes, like knowledge economies, there is always a need for constant review and adjustment of research themes based on their applicability to the actual world. Through multi-linguistic arguments that incorporate perspectives from fields like happiness studies, researchers can gain insight into how educational policy and practice can be made to achieve greater social purposes. The convergence of such areas of research underscores the imperative for an adaptive orientation of research that addresses altering needs and expectations of society (Engelbrecht et al., 2008). Finally, the importance of research directions resonates with the persistent search for innovation and sustainable growth. As new challenges of climate change and economic disparity necessitate action, cross-disciplinary research domains become necessary in order to create balanced solutions. By identifying priority areas and facilitating interdisciplinary collaboration, academics can usefully mobilize resources to address major global problems. Such collaborative effort not only gives amplified effect to individual research work, but it also lays the ground for future practice and knowledge developments. Finally, it is through the establishment of a culture that prizes the sort of reflective guidance of research that will an evolving and proactive research community be able to address the issues of tomorrow (Engelbrecht et al., 2008; Farris et al., 2008).

10.1. TECHNOLOGICAL ADVANCEMENTS

Technological advancements have significantly changed most fields, particularly the Architecture/Engineering/Construction (AEC) industry. While corporations become more aware of the strategic relevance of Information and Communication Technologies (ICT), enhanced collaboration and efficiency loom as a possibility. Key areas of focus in Construction Informatics, according to current research, are interoperability and intelligent site management that are crucial in guaranteeing effective communication among stakeholders. The use of these technologies is capable of considerably reducing project delivery time and cost, showing how ICT innovation can act as an agent of change. This shift in emphasis puts greater focus on growing awareness of ICT as a necessary resource that impacts business processes but also assists in gaining a competitive edge in the ever-more digital environment, setting the AEC sector in good stead for challenges in the years to come (Acikalin et al., 2009). Trends in Construction Informatics reflect a clear need for academia and industry to collaborate more effectively on ICT strategy implementation. Traditional practices in the AEC industry have typically overlooked the strategic importance of technology investment, favoring short-termism. However, a recent research outlines the shift in regard to ICT as an additional-value resource, highlighting the need for organizations to establish a long-term vision and marry technological advancements with their key strategies.

Figure 10.2. The evolution of technological advancements.

Source: https://aosgroup.ca/cdn/shop/articles/blog_DT_1200x.jpg?v=1670941308.

This shift calls for increased ICT training on a mass scale, a subject priority recognized from an academic point of view. By advancing training programs and developing a pool of personnel that is able to adapt to new technology, the industry can generate innovation and sustainability, which will end up producing improved project outcomes and business viability (Acikalin et al., 2009). In the future, the meeting point of technological progress has the potential to redefine paradigms in the AEC sector. Emerging technologies such as artificial intelligence, machine learning, and the Internet of Things (IoT) are gradually gaining traction and demonstrating their value to revolutionize project management and execution. Besides speeding up decision-making, these technologies also have intelligent, connected spaces that simplify resource optimization and project monitoring. As the practice evolves smart and cooperative practice models, it becomes important to develop research with the aim of comprehending the implication of such technologies. Building research partnerships across disciplines can accelerate knowledge transfer from academy to industry and ensure technological innovation is effectively utilized in addressing future challenges while promoting sustainable construction and engineering practices (Acikalin et al., 2009) (Figure 10.2).

10.1.1. Artificial Intelligence (AI) and Machine Learning (ML)

The evolution of AI and ML is fundamentally reshaping various industries with new ways of data analysis and decision-making. AI and ML have emerged as powerful tools in the field of data mining, pushing the traditional ways to their complexity and amount of data. ML and AI integration offers a mechanism of process automation, particularly by way of Auto-ML's simplification of machine learning processes. Federated Learning and XAI are also critical trends that seek to safeguard privacy and promote model explainability, concerns of a paramount nature in the current data-centric era. Not only do these advancements make data mining more alive but also impinge upon the ethical considerations of AI use. With technology being used more and more in different industries, the most critical factor for future research is scalability, efficiency, and law abidance in AI-driven systems that create opportunities for creative solutions across industries (Dr. Kumar et al., 2025). In tribology, applications of AI and ML have become more popular in recent years due to the fact that it can improve predictive capabilities of tribological properties such as friction coefficients and film thickness of oil. Application of artificial neural networks to tribology is a perfect demonstration of what AI can bring to this niche field of study. With data-driven models, scientists are able to make accurate and efficient predictions, enabling improved knowledge on material interaction under varied conditions. Integration of AI and ML with tribology fosters innovation, along with opening up new avenues for the study of design and characterization of lubricated systems. Future research will likely center around the incorporation of AI methods to improve surface topography examination and lubrication technology, therefore having a significant impact on mechanical engineering and materials science. These innovations are part of an even broader shift towards embracing technology to optimize traditional processes (Aragon et al., 2021). In the years to come, continued advancements in AI and ML will inevitably enable new approaches to solving complex issues across different areas. Ongoing convergence of these technologies with other sectors—healthcare, financial services, and manufacturing—will drive a redefinition of industry norms and practices. Experts will focus more on optimizing algorithms to process unstructured data more efficiently, whereby AI systems can provide valuable insights from big datasets. In addition to technological advancements, attention to ethical matters, like reducing bias and protecting information, will remain a factor in forthcoming exploratory directions. The lessons learned from previous deployment will be significant pillars in ensuring responsible AI deployment, thus affirming trust and accountability in AI-informed decision-making. Thus, the course of AI and ML research will chart an educated future with technology complementing decision-making processes while maintaining ethical practices (Aragon et al., 2021; Dr. Kumar et al., 2025).

10.1.2. Internet of Things (IoT)

The IoT is constantly evolving, providing new opportunities and challenges that will shape its future research streams. As IoT devices expand to sectors such as manufacturing, healthcare, and home automation, the integration of artificial intelligence (AI) becomes increasingly vital. AI enhances IoT by enabling data driven decisions and process automation, thereby making processes more efficient and precise. The projected growth rates within both sectors highlight the need for research directed towards examining cutting-edge algorithms and machine learning models tailored specifically for IoT usage. For instance, AI-integrated civil infrastructure health monitoring is one example of how intelligent analytics can transform traditional practices. The convergence of AI and IoT introduces more sophisticated systems with predictive maintenance and real-time monitoring capabilities, thus revolutionizing infrastructure management practices and making proactive rather than reactive maintenance decision-making possible (Chen et al., 2022). Also, the cloud computing infrastructure is changing at a fast pace and will deeply impact IoT in the future. While more devices become connected, data storage capacity and processing capacity are needed more, but the usage of centralized data centers is decreasing. Decentralized cloud facilities are a trend that is pivotal for future research because it has the capability of enhancing data accessibility, security, and processing speed. This is particularly the case for IoT deployments that generate dense data to be processed in real-time. The emerging cloud systems are going to deliver heterogeneous computing frameworks to support various forms of IoT capabilities, ranging from emergency response low-latency communications to complex data-intensive computations. Therefore, understanding how these new architectures can be effectively developed and deployed will be the key to further IoT technology advances (Buyya et al., 2017). Future IoT research trends should also encompass the necessity for security and privacy in an urgent manner. With more devices placed on the network, the opportunities for cyber-attacks that will compromise data integrity and user confidentiality increase. Innovative research work must be conducted to develop robust security architectures that ensure secure storage and transmission of sensitive information within IoT environments. This includes not only supporting more robust encryption techniques but also exploring decentralized authentication mechanisms that minimize the threat of single points of failure. With the increase in IoT applications, it will become important to have open regulatory policies to guide best practices in data handling and user consent. Researchers in the field must therefore place greater emphasis on multidisciplinary collaboration, drawing on cybersecurity, law, and ethics to create end-to-end solutions that future-proof IoT ecosystems and safeguard user interests (Figure 10.3).

Figure 10.3. The Internet of Things (IoT) continues to evolve.

Source: https://www.mroelectric.com/blog/wp-content/uploads/internet_of_things_iot.jpg.

10.1.3. Quantum Computing

The trajectory of quantum computing is a quantum leap in computing power, with the potential to transform fields from cryptography to the simulation of complex systems. Whereas traditional computers labor with increasing amounts of data and computational needs, quantum computing comes with a paradigm using quantum bits or qubits that can exist in numerous states simultaneously. This is the fundamental nature of quantum computers to compute what would be beyond the capacity of classical machines, potentially uncovering solutions for problems of intractability, e.g., molecular interaction simulation and large-scale optimization. The marriage of quantum technologies, in particular cryptography, is particularly poignant because it upends data security standards and requires current systems to adapt in response to the unprecedented power of quantum mechanics (Heng, 2025).

Figure 10.4. The trajectory of quantum computing.

Source: https://www.science.org/do/10.1126/science.adh2923/abs/_20230224_nid_quantumcomputing_sq.jpg.

Quantum computing pursuit also entails building associated fields such as quantum cryptography and the development of quantum algorithms. In recent years, much focus has been accorded to the creation of quantum-mechanical cryptographic protocols that offer increased security, a major concern in the postmodern age of heightened cybersecurity threats. Further, with new quantum algorithms being developed by theorists, the efficiency of quantum computing systems is anticipated to grow, hence increasing their real-world applicability. These advances are set to drive a new generation of innovation spanning various fields like finance, medicine, and artificial intelligence. The joint research efforts highlight the mutual dependence of quantum theoretical ideas and computational uses, promising to usher in breakthroughs beyond basic computational speedup (Kechris et al., 2001). Future quantum computing research will aim to break existing boundaries with regard to error rates and qubit coherence times. With the technology advancing, qubit design optimization and the building of fault-tolerant quantum computing systems present core issues that researchers must tackle. Improvements in quantum error correction codes and materials science to enhance qubit stability will be critical for realizing scalable quantum systems. The three-way meeting point of proof theory, model theory, and recursion theories of quantum computation necessitates interdisciplinary work between mathematicians, physicists, and computer scientists. By interdisciplinary collaboration, the field of quantum computing can navigate its complexity effectively and thus open up far-reaching breakthroughs that revolutionize our understanding of computation as well as its application in the 21st century Heng (2025) and Kechris et al., 2001) (Figure 10.4).

10.2. ENVIRONMENTAL SUSTAINABILITY

Environmental sustainability is emerging as a critical field of study for future research and policy making with specific mention of agricultural practice and food systems. The trends of the times imply that there is a need for the re-make of policies like the Common Agricultural Policy (CAP), which impact farming practice in the European Union. As documented in recent literature, the CAP is being placed under scrutiny for its ability to address environmental concerns, improve nutritional performance, and support livelihoods in the rural sector. Literature highlights the intrinsic problems associated with the intertwining of these fields, necessitating a comprehensive reaction. Subsequent research must prioritize the need for interdisciplinary research strategies, easing the consolidation of environmental sustainability with rural food production programs. This coordination can enhance stakeholder cooperation and ultimately more effective policy guidelines that are specifically geared towards offering sustainable agricultural methods responsive to the demands of society and the objectives of the environment in an environmentally aware era (Alexandratos et al., 2019). Supply chain development is also directly associated with environmental sustainability principles. For the last two decades, there has been a growing academic interest in sustainable supply chains, driven by increased awareness of environmental impacts across industries. The literature published in Resources, Conservation and Recycling (RCR) has showcased various dimensions of this field, including economic, social, and environmental pillars that must be harmonized within supply chain operations. This book supports the call for comprehensive integration of sustainability into supply chains since it includes critical review of existing methodologies, top authors, and current trend-changing topics. The evolution of this literature highlights the need for industries to evolve their practices due to sustainability challenges and provides directions for future research that will define more advanced sustainable supply chain models and techniques (Ghadimi et al., 2019). In the future, it is critical that research efforts address leading-edge strategies and technological innovations that promote environmental sustainability across all disciplines. Emerging trends suggest a potential shift toward circular economy models that redefine resource utilization and waste management. Encouragingly, existing research emphasizes the imperative of theoretical and practical breakthroughs that can facilitate this transition. An inter-disciplinary strategy, which incorporates insights from economics, ecology, and social science, will be essential to address complex problems of sustainability. Furthermore, there is increased interest in participatory research approaches, involving communities and stakeholders in decision-making directly. Such participatory efforts not only enhance the usability and usefulness of the outcomes of research but also allow communities to self-organize and be responsible for sustainability initiatives, paving the way for inclusive solutions to environmental challenges in the future (Alexandratos et al., 2019; Ghadimi et al., 2019).

10.2.1. Renewable Energy Sources

As the world keeps attempting to wean itself off fossil fuel dependence, the move towards renewable energy sources is becoming increasingly necessary. Among the big problems of such a shift is the need for new ways of integrating renewable power into existing infrastructures. The challenge of accomplishing this is exemplified in the advanced energy infrastructures of countries like the UK, where heat demand transfer into the electrical grid can lead to extremely high electrical demand. This shift is not simple, as it adds more variability to the supply of energy, necessitating the development of hybrid energy vector and storage systems. These controls are necessary in the shift toward renewables and can provide the guarantee of continuity for electrical systems in periods of fluctuating demand and supply, in the end serving the decarbonization agenda well (Hall et al., 2013). The ASEAN region is the best example of the complexity of the transition to renewable energy resources in the face of varied economic and infrastructural landscapes. Pre-set goals would witness the share of renewable energy increase to between roughly 13% and 23% of the energy mix by 2025. The realization of such an aspiration is, however, hampered by challenges like grid flexibility and variability of supply from renewables. The ability of the electricity grid to adapt to changing demand and deliver a secure energy supply is key to overcoming these hurdles. A method of grid flexibility analysis in the ASEAN nations illustrates significant disparity in preparedness, with members required to enhance capability to enhance greater renewable electricity integration. Regional collaboration will be needed to develop strategies for the resolution of these challenges in an effective manner, thus securing high ambitions in renewable energy while maintaining system costs in general as low as possible (Huang et al., 2019). Future research horizons in renewable energy need to concentrate on advancement in smart grid technologies, energy management systems, and predictive analytics. By leveraging these technologies, energy systems can respond to the dynamic nature of renewable energy output and consumption more easily. This includes the ability to forecast electricity demand and supply with more accuracy, hence enhancing the grid stability and reliability. Additionally, investment in energy storage equipment will be critical in mitigating the intermittence character of the renewable sources. With ongoing advancements of battery technologies and energy management strategies, the horizon of high use of renewable energy becomes very wide. When combined, these innovations have the potential to not only facilitate increased penetration of renewable energy sources but also to make next-generation energy systems more resilient and efficient. This multi-disciplinary approach will play a critical role as countries move through the transitional stage to sustainable energy systems (Figure 10.5).

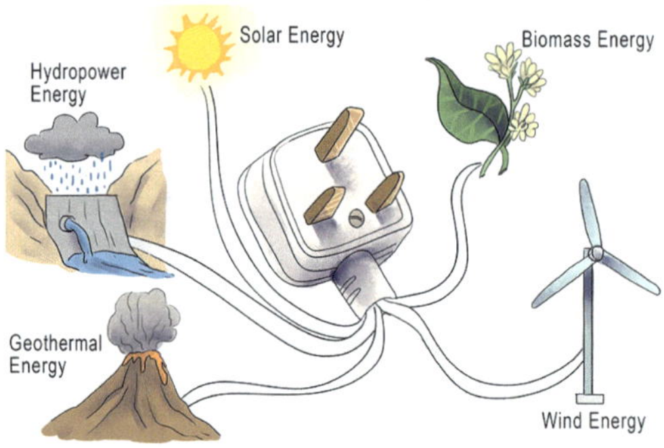

Figure 10.5. Renewable energy sources.

Source: https://mammothmemory.net/images/user/base/geography/energy/remember-renewable-energy-in-energy-geography.243f26f.jpg.

10.2.2. Climate Change Mitigation Strategies

In considering future directions in policies on climate change mitigation, the need for centralized communication cannot be overstated. As climate increasingly affects human health as well as world food security, communication policies take center stage in coordinating grassroot action and boosting participation at the community level. Research has indicated that social scientists are pivotal in bridging the divide between scientific evidence and public awareness, ensuring that adaptation and mitigation procedures are employed equitably and effectively. Short-term concerns and local communities being their concern, researchers can better understand and communicate the urgency of climatic concerns. Subsequent research must focus on geographic representation, namely in under-represented regions such as Africa, the Caribbean, and Latin America, in order to build solutions with which varying populations can identify (Adekunle et al., 2020). An emphasis on seeking out advanced technologies is also central to future climate change prevention strategies. Based on evidence, integrating innovative visualization tools and analytical methods can serve to further amplify our understanding of climate trends as well as facilitate informed decision-making. Initiatives like the Pardee Centers Searchlight project attempt to synthesize anecdotal data from around the world and thus enable authorities to spot trends and devise strategies applicable across various contexts. Not only do such methods highlight local climate issues, but they also initiate an international dialogue regarding successful mitigation strategies. The convergence of these technologies can transform climate information interactions among stakeholders towards possibly increasing levels of public awareness and initiating global action towards sustainable measures (Gopal et al., 2012). In addition, it is crucial to have interdisciplinary work in developing effective climate change mitigation strategies for the future. By generating synergies between scientists, policymakers, and community leaders, more integrated climate vulnerability management strategies can be established. Levying on concepts from various disciplines can enhance the flexibility and effectiveness of mitigation measures, allowing for easier incorporation of indigenous and local knowledge and experience into strategy development. An inclusive approach can lead to greater social equity and cultural suitability of mitigation measures and hence increased public acceptance and active participation. Future research should therefore be aimed at developing networking relationships that empower marginalized groups and synthesizing divergent perspectives to formulate holistic approaches appropriate to their climatic and socio-economic environments (Adekunle et al., 2020).

10.2.3. Sustainable Agriculture Practices

Sustainable agriculture practices have become a hot issue in recent history, a sign that the world is racing against time to tackle environmental issues while maintaining food security. With the rising global populations and climate change hitting the sustainability of agriculture, sustainable agriculture is a vital strategy against ecological loss. Sustainable agriculture is directed at methods that enhance soil fertility, conserve water, and promote biodiversity. The adoption of these methods not only reduces the application of chemical fertilizers and pesticides but also makes the agriculture system more stable to handle changing climatic trends. Prioritizing sustainability becomes essential not only for farmers' sustained productivity but also for the well-being of communities and ecosystems reliant on farm production. The practices reconcile agriculture with environmental stewardship principles, and this may lead to new avenues for future research and agricultural practice (Bayala et al., 2020). For agricultural development, sustainable agriculture then becomes the determining factor in the guarantee of economic viability and social justice. As evident in various reports, there is a pressing call for rebalancing imbalances between rural and urban areas, particularly to guarantee the availability of resources and opportunities for development.

Figure 10.6. Sustainable agriculture practices.

Source: https://saiplatform.org/uploads/indialj.jpg.

By sustainable agriculture, the local development is encouraged such that communities get to benefit from their strengths in harmony with the European Union directives aimed at enhancing social and territorial cohesion (BULATOVIC et al., 2015). The practice encourages investment in rural industry, which can trigger the reawakening of the local economies, create employment opportunities, particularly among the youths, and attract new competencies into farming through the use of innovative agricultural approaches. Therefore, well-implemented sustainable agriculture can be a tool for promoting bigger rural development objectives, balancing economic needs with environmental issues, leading the way to a long-term strategy towards future farm policy. Additionally, the development of sustainable agriculture methods involves a big push in research and technology. Methods such as precision agriculture, agroecology, and integrated pest management are emerging as key factors in making a transition to sustainability. These practices maximize the efficiency of resource use while lowering the environmental charges of standard agriculture practice. Future research agendas ought to be driven by the development and dissemination of such technologies that can enhance productivity without violating the principles of sustainability. Interdisciplinary collaborations between research organizations, governments, and farmers will be critical to this endeavor, facilitating learning and empowering communities to practice optimally. With the might of science research and innovation, agriculture can put itself in a position to meet challenges in the future, construct productive farming systems, and achieve global sustainability goals (Bayala et al., 2020) (Figure 10.6).

10.3. HEALTH AND MEDICINE

In venturing into the intersection of medicine and health, precisely trends of the future, the use of complementary and alternative medicine (CAM) among pregnant women is a new area of investigation. As maternity care practitioners traverse the difficult landscape of patient preference and evidence-based practice, they have the mandate of ensuring that alternative therapies are not only effective but also safe. The popularity of CAM among pregnant women necessitates the thorough analysis of its impact on fetal and maternal well-being. As elucidated by current studies, a need for the reappraisal of studies investigating the consequences of CAM use during pregnancy is imminent, outlining areas that need more research. Thus, proceeding with awareness of how such practice is used in mainstream medical paradigms will lead to optimizing care while honoring patient preference and promoting informed decision-making by clinicians (Adams et al., 2015). Moreover, urbanization is the major impetus in determining trends in health as well as in research in modern society. As an increasing percentage of the world's population resides in urban areas, the implications for public health are significant. The particular issues and opportunities of urban life necessitate a comprehensive grasp of how living in cities affects health outcomes. In the context of densification and diversification of cities, determinants of health are shaped by many municipal, national, and global influences, making the landscape for public health interventions complicated. An integrated approach that considers these complexities can offer targeted interventions and policy design for the improvement of health outcomes among city residents. This model not only includes current evidence but also offers new avenues of research, particularly concerning the social determinants of health in urban spaces (Freudenberg et al., 2005). Looking into the future, the intersection of advanced technologies in health and medicine is another major trend deserving research. Incipient telemedicine, wearable tech, and artificial intelligence (AI) tools in healthcare are going to revolutionize patient care and management forever. With these technologies spreading more and more, understanding their implications on patient outcomes and access becomes crucial. For instance, telemedicine has been shown to be successful in increasing access to healthcare for those in rural or underserved communities. But issues involving data privacy, the digital divide, and the quality of care provided through virtual platforms also need to be examined. Future research must address these concerns with the benefit of technological advancements to create more effective health interventions and promote overall population health, thereby creating a more equitable healthcare system in the future.

10.3.1. Telemedicine and Remote Healthcare

As telemedicine is emerging as the backbone of modern healthcare at an incredible pace, its implications extend beyond mere convenience, signaling a paradigm shift towards the manner in which healthcare is presently being delivered worldwide. With geographical limits lessening, healthcare access becomes greatly enhanced, particularly among patients in rural or disadvantaged areas. Technologies like mobile health applications, specifically those developed by institutions like the Indiana University-Purdue University Fort Wayne, are essential in increasing the extent of care provision. For example, the WoundView app enables nurse practitioners to receive and manage all types of patient concerns remotely, accelerating as well as enhancing the quality of the care. However, implementations of such technologies are faced with issues like needing robust evidence-based research to inform practice and simplifying such applications in a manner that they can be easily integrated into everyday healthcare practices (Reimer et al., 2018). The coming of age of combined telehealth platforms is an important step towards the merging of technology and provision of healthcare services. These locations, as mentioned in recent research articles, particularly in the southeast Mediterranean region, are targeted to provide a package of eHealth tools particularly to the citizens and tourists in rural areas. The emphasis on telecollaboration, teleconsultation, and emergency care is central in enhancing the treatment of risk populations such as the elderly and patients who are coping with chronic conditions. By providing a stable platform that facilitates real-time interaction between healthcare providers and patients, such systems can potentially yield long-term gains in different healthcare environments, hence closing existing loopholes in service provision (Androulidakis et al., 2011). Furthermore, the focus on data interoperability allows various healthcare stakeholders to seamlessly engage with each other, enhancing the bargaining power of telemedicine. In the coming years, it is difficult not to consider the over-riding need for ongoing research into means of facilitating technological innovation in telemedicine and distant care. As exemplified in various rural and urban settings, the viability of such telehealth platforms is based on continuous evaluation and adaptation to regional healthcare needs and constraints of culture. Embedding technology and succeeding will rely on mapping user experience and operational issues for both patients and providers. This focus should likewise be devoted to systemic hindrances to broader application, such as regulatory barriers and technological disparities. It is only through comprehensive and holistic studies that telemedicine can reach its greatest potential, ensuring optimal expansion of access to care while promoting the overall quality and effectiveness of healthcare delivery in the near future (Androulidakis et al., 2011; Reimer et al., 2018) (Figure 10.7).

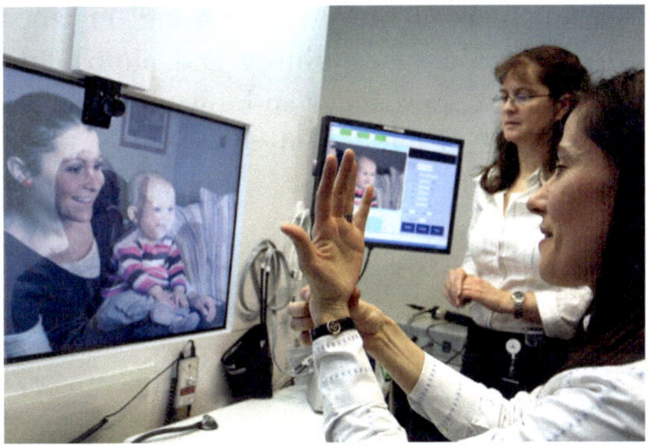

Figure 10.7. Telemedicine and remote healthcare.

Source: https://wp.healthdatamanagement.com/wp-content/uploads/2021/08/tele-health-pic.jpg.

10.3.2. Personalized Medicine

Emerging technology and greater insights into human biology are paving the way for personalized medicine, at its essence transforming healthcare. In contrast to the traditional medical practice in which a one-size-fits-all strategy is applied, personalized medicine focuses on providing treatment specifically designed to meet the unique genetic makeup, living conditions, and lifestyle of each patient. This paradigm shift consolidates the complexities of conditions such as hypertension, which, despite being on multiple effective antihypertensive drugs, remains a challenge to management due to drug resistance and patient compliance. For individual risk factors and genetic determinants, personalized medicine can optimize the efficacy of drug regimens, ultimately translating into improved patient outcomes. With advancing systems biology research and ever more sophisticated high-throughput technologies, the promise of personalized treatment regimens is now no longer a hypothesis but a fast-changing reality in modern healthcare (Borghi et al., 2017). The prospect for personalized medicine is tied directly with ongoing research in various fields of biomedical science, including genomics, proteomics, and metabolomics. With these lines of research further developing, they will have the potential to unlock unparalleled knowledge about disease mechanisms and treatment response. This is a multidisciplinary field that allows for a better understanding of health states at the molecular level, thus allowing the development of targeted treatments that have the potential to minimize unwanted side effects normally associated with conventional treatments. Furthermore, the integration of health data analytics has potential for treatment outcome prediction and enhancement of patient stratification based on genetic susceptibility and biological markers. As the field shifts from reductionist approaches and toward more holistic data-driven systems, the field is poised to redefine disease prevention and management standards (Fortina et al., 2015).

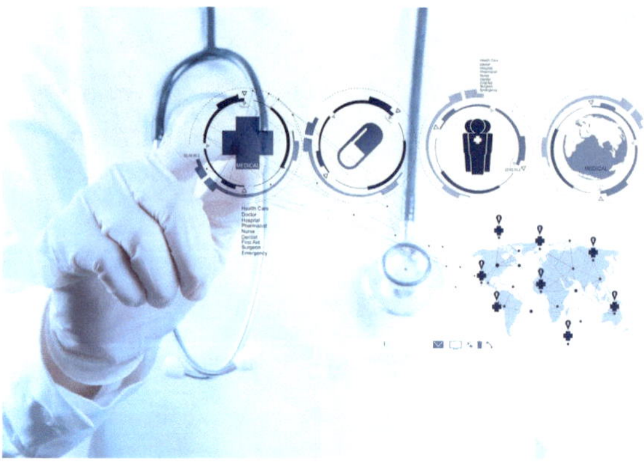

Figure 10.8. Personalized medicine.

Source: https://cdn.prod.website-files.com/686d4fca96e4813b96758e56/686d4fca96e4813b9675a081_6706eaaca5b5f5a3ee319323_66f07832a11a1dc1459eaa86_personalized-medicine-jpeg.webp.

The implications of personalized medicine extend beyond therapeutic efficiency; economic considerations along with access to care are also encompassed. The incorporation of personalized techniques might initially appear costly; however, improved outcomes can mean significant long-term savings through decreased hospitalization and additional treatments. Moreover, with an enhanced understanding of personalized medicine, healthcare systems will need to address disparities of access to such advanced therapies. Ensuring that every patient population benefits from the innovations brought by personalized medicine entails the involvement of researchers, policymakers, and health providers. The real path forward for future investigations is not only to continue seeking enhanced personalized therapeutic modalities but also ensuring that equitable frameworks facilitating shared access are created across heterogeneous populations, enriching the overall healthcare environment (Fortina et al., 2015) (Figure 10.8).

10.3.3. Advances in Biotechnology

Biotechnological progress is increasingly shaping the shape of urban administration, particularly through infusing information and telecommunications technologies. The technological progress that developed towards the end of the 20th century has yielded impressive devices for the improvement of urban living environments. The synergy of these technologies enables a holistic approach to managing urban ecosystems, referred to as Ubiquitous Eco Cities. These municipalities use integrated systems that intersect economic, socio-cultural, and infrastructure elements for the improvement of the quality of life for the inhabitants. Using data-driven decision-making processes, city planners are in a position to utilize advances in biotechnology for promoting sustainability and optimizing the allocation of resources, creating responsive and adaptable urban environments. The implications of such technology convergence are much more than mere efficiency, providing a democratizing space where stakeholders are able to engage with dynamic change in city life, leading ultimately to improved community welfare and environmental control (Yigitcanlar et al., 2009). Biotechnology has also evolved considerably in the fields of health and medicine, notably through mass spectrometry usage. This powerful analytical technique provides information about the molecular make-up of complex biological samples and enables researchers to identify new compounds and quantify known compounds accurately. In the past couple of years, the technology has been enhanced in terms of accuracy and speed, so that the entire world can be scanned for biomolecules such as lipids in blood plasma or serum. Identification of lipid biomarkers of cardiovascular disease is a vivid example of the therapeutic value that mass spectrometry carries for patient management. As scientists seek greater knowledge of their implications, the ability to correlate unique lipid profiles with conditions such as coronary heart disease and potential heart attacks chronicles the contribution that biotechnology makes toward improving patient outcomes and diagnostic technology (Hassan et al., 2015). At the distance, the biotech sector is on the threshold of transforming its life in several sectors with gene editing, synthetic biology, and artificial intelligence being some of the driving technologies. The emergence of technologies like CRISPR-Cas9 transformed the face of biotechnology by offering precise methods for regulating genetic material, thereby enabling scientists to treat complex genetic disorders with unprecedented precision. Parallel to this, synthetic biology drives the development of novel biological components, devices, and systems that can give rise to advances in agriculture, environmental stewardship, and pharmaceuticals. Integrating AI into these fields enables scientists to make sense of enormous datasets, providing knowledge that was inaccessible before. As they advance further, the discipline of biotechnology will play a leading role in addressing issues of world food security, climate change, and public health, demonstrating the magnitude of possible research and applications in the future (Yigitcanlar et al., 2009) (Figure 10.9).

Figure 10.9. Advancements in biotechnology.

Source: https://www.webnx.in/wp-content/uploads/2023/06/Biotechnology-Advancements-and-Genetic-Engineering-Transforming-the-World.jpg.

10.4. EDUCATION AND LEARNING

The learning and education environment is undergoing radical transformation, driven mainly by developments in technology and pedagogy. New models of learning are transforming the boundaries of what previously would have been regarded as a traditional learning environment, emphasizing the importance of adaptability and imagination. As schools are forced to prepare learners more and more for a changing workforce, the use of digital resources and tools has become a common denominator in transforming curricula. This is seen in the growing emphasis placed upon competency-based education, where skill acquisition over mere memorization of content comes to the forefront. Associated with this is the use of a diverse range of learning modalities, including project and experiential learning, in order to foster the critical thinking as well as the resilience of the students. These trends do not just alter the transmission of knowledge but also influence the role of teachers, with a need for continuous professional development to address these new challenges of teaching. In reference to future trends in education, the role of research in informing pedagogy cannot be stressed enough. Reports such as the BERA/TACTYC Review serve to heighten the demand for standard frameworks able to underpin early childhood education and care. The reviews holistic method, one engaging a variety of stakeholders, emphasizes the relationship between policy, research, and practice, and demonstrates how evidence-based outcomes can be employed to enhance education. By covering subjects like professionalism and play pedagogy, this type of research offers significant insight into current best practice and areas to research in the future. Furthermore, the debates are particularly pertinent in the context of the evolving educational reality, defined by a sharp need for models that have the ability to adapt to the complexities of teaching in diverse contexts. Such collaboration among policymakers, practitioners, and researchers is essential to define education for the future. The debate around brain-based learning provides further evidence in favor of effective schooling, particularly adult schooling and human capital formation. As the field grows, it is essential to understand the interaction between educational practice and neuroscience to facilitate rich learning environments. This movement towards using cognitive science principles in instructional design aims to maximize retention and knowledge transfer, which is especially applicable to adult learners. The construction of an integrative theoretical framework, as proposed in recent research, provides a platform for teachers to be creative and adapt their approaches to accommodate the diverse needs of their students. Besides expanding the field of educational research, it stimulates critical examination of pedagogical practices. Hence, such studies not only inform the learning environment but also contribute to arguments regarding the most effective method of learning throughout life in an increasingly demanding world (Chesworth et al., 2017; Jang et al., 2019).

10.4.1. Online Learning Platforms

The rise of online learning platforms has witnessed the most significant transformation occur in education paradigms, thanks to technological advances as well as calls for learning flexibility. As cloud computing continues to evolve, decentralized infrastructures it enables are revolutionizing how learning content is delivered. This revolution enables platforms to deliver scaled personalized learning experiences, significantly enhancing access to diverse groups of learners. As an example, integration of computationally intensive computing enables real-time analysis and adaptive learning pathways that are customized to meet the unique needs of individual learners. As they become sophisticated, they raise questions of data privacy and ethical issues, thus delineating an essential research agenda for the future of educational technology. Addressing these challenges will not only unlock opportunities for improved learning outcomes but also ensure that the potential of such platforms is tapped responsibly (Buyya et al., 2017). The second vital characteristic of web-based learning platforms is their ability to build collaborative areas that are geographically dispersed. The utilization of academic social networking sites like Academia.edu and ResearchGate has been an allied strategy, which has enabled researchers to engage with other researchers and share research products with ease. The platforms are not merely repositories of knowledge but also arenas for the negotiation of professional identities.

Figure 10.10. The emergence of online learning platforms.

Source: https://collegexpad.com/wp-content/uploads/2024/04/online-learning.jpg.

The structured networks built within such SNSs have profound implications for research collaboration, given that they provide for the sharing of information and resources across groups. Enhanced visibility of individual researchers facilitates greater culture of openness in research practice, thus enriching research discourse. To investigate the effectiveness of such interactions within online learning contexts thus presents a promising avenue for future study, as it lies at the interface of technology and teaching practice (Jordan et al., 2016). In the future, the fate of innovative online learning environments appears likely to be constantly reformulated through the implementation of new, revolutionary technologies such as artificial intelligence and machine learning. These innovations can have the potential to deliver intelligent tutoring systems that provide immediate feedback and adaptive instruction, in accordance with individual students' learning styles. Further, transitioning towards more interactive learning experience through virtual and augmented reality, online learning platforms will be forced to retool their approaches in a way that integrates these technologically enhanced interventions effectively. The test will be to measure the impact of such technologies on levels of learner engagement and retention, towards more significant learning experiences. Aligning with these emerging trends will entail inter-disciplinary work from educators, technologists, and researchers to make sure that online learning spaces keep evolving in ways that enhance educational equity and accessibility to all (Buyya et al., 2017) (Figure 10.10).

10.4.2. Gamification in Education

With institutions of learning continually changing in response to technologic advancements, gamification has also emerged as a potential remedy to facilitate engagement and motivation among learners. Through the integration of game design mechanisms into learning spaces, such as leaderboards, badges, and points, faculty can create a participatory learning environment. In particular, an integrated literature review on gamification within higher education in STEM-related fields has revealed that the integration of these game components has a positive effect on students' performance and general attitude toward subjects, especially computer science. In spite of this increased pool of research, there is still an evident gap regarding what particular game components are most effective and how they interplay with various student backgrounds. To realize the full potential of gamification, subsequent research must close these gaps in knowledge to rationalize instructional designs attuned to broader learner groups (Chiluiza et al., 2016). Parallel to the increased application of gamification in traditional classroom settings, the advancement of serious games also indicates a complementary movement toward the exploitation of computer games for purposes beyond learning at a societal level. These games are not just designed for entertainment but to address social issues, empower individuals, and bring about social integration. Serious games, which belong to the videgames ecosystem, are an indication of growing recognition of the potential that game-based interventions possess in addressing problems among marginalized groups. As researchers delve further into this field, it is important to explore the use of gamification in productive means not just for education but even for community engagement and public services. The meeting point of educational initiatives and serious games offers exciting prospects for future research avenues with an emphasis on reaping maximum outcomes of gamification in real contexts (Misuraca et al., 2013). The intersection of serious games and gamification in education raises some very basic questions regarding the future of learning. While the demand for professional professionals in STEM fields is increasing, schools and institutions of learning need to ensure that their students are not only knowledgeable but also possess the skills they require to be successful in a more dynamic job market. Gamification provides a valuable framework for creating engaging learning experiences to be utilized for the improvement of STEM education and for limiting student attrition. However, this model has to be paired with empirical studies defining best practices for implementation and quantifying long-term impact on student achievement. In the process, educational researchers open doors to new pedagogical practices that benefit from the dynamic nature of gamification, thus shaping the future of education and fanning a newfound interest in STEM areas (Chiluiza et al., 2016).

10.4.3. Lifelong Learning Trends

As society advances into an era guided by rapid technological advancements, the concept of lifelong learning has evolved as an integral trend redesigning personal and professional settings. This model understands that learning is not only confined to formal schools or early life but is an ongoing activity vital for change adjustment. Adults, particularly those over 50 years of age, are increasingly making use of various forms of education, driven by personal as well as professional motivations. Adult education programs enable this segment to not only learn new skills but also remain engaged and productive members of society. With an approach to participatory education, institutions and teachers are encouraged to re-examine the salience and availability of curriculum to meet particularly this diversifying population's demands, therefore implying greater innovation and expansion of higher education (N/A, 2008). Moreover, the technological revolution has greatly influenced patterns of lifelong learning as education is more accessible than ever before. Websites and software are propagating, providing the authority to learn on one's own time from one's own home. This shift towards online learning settings demands a more individualized approach, providing the learners with the autonomy to choose what, how, and when to learn. Also, it supports the idea of lifelong learning in that individuals can seek out micro-credentials, certifications, and degrees that enhance their knowledge while juggling other ways of life. As such, institutions need to revamp their outreach and engagement to support the integration of technology in accessing the learning cycle beyond the physical classroom setting (Kidd et al., 1983). In the future, the intersection of new technologies and lifelong learning promises an exciting research and development landscape. The amalgamation of virtual reality, artificial intelligence, and personalized learning algorithms is likely to revolutionize the experience and delivery of learning content by learners. Researchers will need to investigate how best these technologies can be employed to make learning experiences cater to the unique demands of diverse populations, particularly the older generations who may not be as familiar with these technologies. Improved partnerships between educational institutions, technology developers, and community organizations will be essential in driving the trends. By facilitating co-operative research work, the learning environment can learn and evolve based on the life-long learning needs of persons at different stages of their lives so that education is meaningful and productive (N/A, 2008).

10.5. SOCIAL DYNAMICS

The social dynamics theory is important in shaping the future landscape of fields, particularly innovation and entrepreneurship. As societies develop, relations between various actors increase in complexity and demand a complex understanding of such dynamics. The connection between users, entrepreneurs, and innovation policies emphasizes the function of strategic intelligence for enacting systemic innovation. Based on current research, the empowerment of non-traditional actors can enhance substantial value to the effectiveness of innovation strategy. By including these players in the innovation process, we can have an inclusive setting that not only drives technological advancement but also addresses social needs, and therefore aligns with societal goals over the long term. In this, comprehending social dynamics is key to developing frameworks that are favorable to sustainable and inclusive innovation growth. Entrepreneurship can be explained from the social dynamics approach in which emergence is one of the main generators of emerging phenomena. Entrepreneurship, according to recent research, is about generating emergent properties resulting from complex interactions within society. When interacting players engage with the environment around them, emergent processes are formed that generate new products, services, and even emergent patterns of behavior. This dynamic process emphasizes the importance of social structures and contextual knowledge in entrepreneurial processes. By applying principles of social and complexity theory, it is easier for researchers to understand how entrepreneurship yields social change. The challenges and opportunities that these dynamics present highlight the importance of interdisciplinary approaches in studying social interactions and their impact on the emerging entrepreneurial settings since they shape market trends and consumer behavior. The introduction of social dynamics in innovation and entrepreneurship research places serious implications in terms of policy and practice in the forefront. For instance, effective use of strategic intelligence in systemic innovation policies emphasizes coordination among actors. Through applying Constructive Technology Assessment (TA) approaches, innovative new initiatives demonstrate how to operationalize the function of strategic intelligence in generating innovation. Not only do these schemes facilitate dialogue between mainstream and non-mainstream actors, but they also promote mutual understanding of the system of innovation. Second, as more research continues to explore the intersections of social change and entrepreneurship, the focus on the emergent features of social dynamics can reveal new paths to development and resilience. Finally, struggling with these dynamics is essential in framing policies that are dynamic and can accommodate an innovative environment that aligns future trends with the evolving needs of society (Figure 10.11).

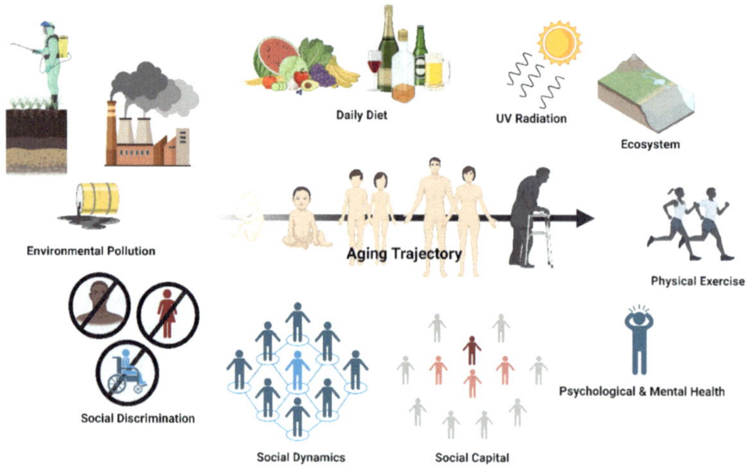

Figure 10.11. The concept of social dynamics.

Source: https://pub.mdpi-res.com/ijerph/ijerph-21-00554/article_deploy/html/images/ijerph-21-00554-g001.png?1714467286.

10.5.1. Impact of Social Media on Communication

Social media has significantly transformed the face of communication to make interactions more immediate and expansive than ever. With Facebook, Twitter, and Instagram, for instance, there is live chat that allows one to share his or her thoughts, images, and updates at the same time to a global audience. This transformation has not only changed the face of personal relations but also affected the way organizations communicate with their publics. Social media help companies project their messages straight to customers so that there is a more personal relationship that could not be obtained otherwise. As such, individuals are increasingly utilizing the sites for information and self-expression, ultimately altering the social norms of communication and connectivity. While the benefits of social media are enormous, the problems in this new paradigm of communication are equally enormous. Issues such as misinformation, diminishing attention span, and potential social isolation suggest the complexities of excessive reliance on online interfaces for interactions. Furthermore, the gravity of traditional communication has lost its hold, and colloquial language and abbreviations are at the forefront in online discourse. This transformation raises questions concerning the destiny of written communication skill for future generations who may find formal writing conventions difficult. As educators and communicators navigate this evolving terrain, it is now crucial to critically assess the influence of social media on individual and collaborative communication capacity. Future studies will need to take into account evolving dynamics of communication in terms of social media. Education institutions will play a critical role within this endeavor, examining how their information literacy programs reflect evolving digital environments. For instance, understanding the communication habits in different demographic groups can provide insight into effective messaging strategies, as well as indicate areas of further research. In addition, as libraries transform their instructional methods to include online interactions, they must evaluate the technology tools and pedagogical methods needed to build effective digital communication skills among users. Such studies would be likely to reveal rich insights for educational practice improvement and public debate, promising trajectories toward thriving communication systems in an increasingly digital future (Chatterjee et al., 2006; Dewan et al., 2013).

10.5.2. Changes in Workforce Dynamics

Workforce dynamics transformation is a multifaceted phenomenon shaped by technological advancements, changes in labor market needs, and changing socio-economic situations. As organizations strive to remain competitive, they are more intent on agility and flexibility among their employees. This shift calls for not only a reworking of the traditional work arrangements but a significant investment in talent management practices customized to a particular organization's environment. Particularly in the small and medium-sized enterprises (SMEs), the demand for particular talent management practices has been escalated, laying bare the complexity of coordinating workforce dynamics in varying economic conditions. Understanding such dynamics requires a delicate touch working on the individualistic needs and aspirations of employees, which is important in creating lasting performance in any organizational setting (Kohont et al., 2019). In recent years, remote work and the gig economy have contributed to shifting workforce patterns. Conventional concepts of a static, long-term employment relationship are being replaced by task-based, flexible engagements. The change compels employers to reconsider their human resource strategies, extending beyond recruitment to incorporate comprehensive employee growth and retention methods. The use of technology in this case is the tipping point; organizations are leveraging data analysis and collaboration software to enhance workforce involvement and productivity while providing greater autonomy to employees. Furthermore, with workers from different generations entering the labor market, there is a need to understand diverse motivations and values when creating good policies for the workplace that will resonate with a wide range of employees (Cornell University ILR School, 2017). In the future, deeper research into the dynamics of the workforce will be essential to the ability of organizations to respond effectively to an ever-changing world. It is the responsibility of researchers and practitioners to focus on how changes will play out for organizational culture, talent recruitment, and workforce planning. Researching the different contexts—such as post-socialist economies—and how they influence talent management and workforce dynamics will offer valuable recommendations for best practices that are relevant to specific environments. This ongoing research needs to center around dynamic relationships between external factors such as economic climates and internal organizational capabilities, proving the necessity of adaptive talent solutions. This extensive exploration not only enables organizations to cope with the issues of the present but equips them to capitalize on future prospects in the ever-changing world of work (Kohont et al., 2019).

10.5.3. Globalization and Cultural Exchange

The multifaceted relationship between globalization and cultural exchange represents a pivotal point of analysis for the evaluation of future trends. As cultural practices and ideologies travel across borders, they are converging and diverging in various societies. This is most clearly illustrated in the context of industrial relations, where globalization facilitates the convergence of labor legislation and organizational structures between various nations. The simultaneous prevalence of convergence in certain domains such as the international labor practice and still the continued existence of local special traditions is an indication of the paradigmatic paradox discovered by theorists. As described in the literature at hand, globalization is the ubiquitous expansion of markets and ideas so that it is increasingly difficult for national governments to impose regulatory frameworks on firms within their respective countries. These regulatory problems coupled with economic interdependence create a dynamic environment in which cultural exchanges offer the promise of enrichment and risk of homogenization of cultures (Turner et al., 1998). As there are cultural elements flowing through globalization, territories worldwide can leverage the exchanges to innovate and change their tradition. For instance, in Africa, emerging trends focus on the uptake of globalization without losing indigenous cultural identities. Examples such as the Africa 2060 conference prove the value of communication among scholars, policymakers, and the general public on how cultural exchange can guide economic and governance policy within the continent. Not only does this platform provide an opportunity for experts to analyze Africa's direction in light of the world, but it also enables an open discourse to give voice to local voices on the global stage. Examining these dimensions illustrates how globalization can serve as a stimulus for the renewal of culture, providing room for new and old ideas and values to meet, resulting in a rich diversity of culture that not only encourages change but also continuity (Clapp et al., 2010). In the future, it is essential to value the dual nature of globalization as facilitator of cultural interaction and simultaneously as a source of potential cultural dislocation. While globalization continues to unfold, scholars and practitioners must navigate the dynamics of such a process to forecast future trends effectively.

Figure 10.12. Globalization and cultural exchange.

Source: https://media.licdn.com/dms/image/v2/D5612AQEYMxMefEVB1Q/article-cover_image-shrink_600_2000/article-cover_image-shrink_600_2000/0/1709532002484?e=2147483647&v=beta&t=UfzgBX6depJExjVIos-iC1GqTFJO-SoZeYtCAOhLQqGY.

This requires a critical examination of the ways in which globalization is transforming cultural identities, especially in multicultural countries. Where global and local practices intersect, it may lead to the hybridization of cultures, whereby particular local customs become intermingled with foreign inputs in order to forge new forms of cultural expression. But the dangers of cultural loss must be tightly monitored, or else dominating world narratives, in their zeal to enclose more heritage, risk overshadowing local traditions. The challenge is to use globalizations strength to promote sustainable cultural exchange while pushing for the preservation of cultural diversity, so that the future is a rich brocade woven from an assortment of traditions (Turner et al., 1998) (Figure 10.12).

10.6. ECONOMIC TRENDS

With the changing global economic scene, some trends have been witnessed that require very careful observation and analysis. The growth in automation and artificial intelligence deeply affects labor markets with opportunities and challenges that decide the growth and stability of the economy. Most industries are going through optimization with the help of technology, but this also brings along with it the risk of job displacement and the lack of skills. With these trends, there is a need for research on the implications of automation in influencing labor forces and income inequities.

The economic theories must evolve to help cover the realities of a technology-driven era in a bid to introduce lasting progress for all stakeholders in economic activities and labor markets. The changing patterns of global economic power among countries also have further implications for research agendas to come. Economies in the Global South are increasingly making their mark in global economic matters, disrupting entrenched economic models that have dominated international trade and finance discussion for so long. This transformation results from varied drivers like demographics, urbanization, and investment in innovation. Scholars must take special notice of where economic development overlaps with social factors, employing social and economic data to make holistic estimates of regional developments. As new players in the economy appear, knowing what they do, where they are struggling, and what they contribute will be essential in the development of fair policies that promote inclusive growth and hence reaffirm the significance of integrated analyses amid shifting trends (N/A, 2013).

Lastly, the relationship between economic trends and climate change cannot be overemphasized as environmental issues increasingly influence economic policy and practice. The impacts of climate change on sectors such as fisheries, agriculture, and tourism demand serious and specialized research that determines the socioeconomic impacts on vulnerable populations. The need for strategies in favor of sustainable activities, for instance, is already beginning to be felt because businesses are being compelled to use more sustainable processes by virtue of consumer pressure as well as governmental pressure. Scholarship must be focused on the evaluation of the effectiveness of climate adaptation measures as well as the economic effects of unsuccessful interventions. Research on such issues will set the stage for identifying appropriate solutions that balance economic development and environmental resilience to set the stage for future discussion regarding the interrelated nature of such conflicting challenges (Kittel et al., 2025).

10.6.1. Cryptocurrency and Blockchain Technology

With the demand for cryptocurrencies continually on the rise, the underlying blockchain technology is rapidly becoming famous for its disruptive features across most sectors. Initially developed to support Bitcoin, blockchain serves as a digital record book that is decentralized with each transaction being recorded publicly and irretrievably, thus establishing trust within the users. Existing research shows that blockchain usage is capable of revolutionizing industries from finance to supply chain management, where it can automate and reduce fraud. In addition, the scalability of blockchain models allows for various features, such as identity management and transaction analytics, to be integrated, which can provide further insights into user behavior in cryptocurrency systems, hence further stimulating innovation and utilization in a broad spectrum of applications (Marchesi et al., 2017). The foundations of blockchain technology—like peer-to-peer networks, distributed databases, and cryptographic security—are primarily responsible for creating a secure and optimized transactional environment. The peer-to-peer form allows decentralized management with active participation of various nodes and strong consensus mechanisms to ensure this. This, in the process, not only raises security but also makes it immune to centralized failures. Secondly, use of distributed databases operates towards increasing data integrity and immutability, which are central facets in preventing unauthorized modifications. Therefore, integration of these elements allows organizations to seek new business models and operational efficiencies and decrease risk from traditional transaction methods (LeBlanc et al., 2020). In the future, blockchain technology and cryptocurrency hold great promise, but one fraught with challenges that necessitate strict study and inquiry. These major concerns are regulatory systems, technology scalability, and energy efficiency that must be worked on in order to usher in the widescale adoption of these technologies. As researchers seek to address these challenges, the overlap of blockchain technologies with other emerging technologies, such as artificial intelligence and the Internet of Things, can create new opportunities unprecedented in magnitude. Through interdisciplinary approaches and fostering cross-disciplinary collaborations between academia and business, innovation in how we think about, utilize, and benefit from blockchain and cryptocurrencies will be accelerated enormously. This comprehensive focus on future trends will shape a stronger and more creative environment in which cryptocurrencies will be able to thrive successfully (Figure 10.13).

Figure 10.13. The cryptocurrency and blockchain technology.

Source: https://media.licdn.com/dms/image/v2/C4D12AQFcqolVLTbAhg/article-cover_image-shrink_720_1280/article-cover_image-shrink_720_1280/0/1643947244541?e=2147483647&v=beta&t=EnnR5QzmgwwcGrI5uH4p-upjj0nEmLQs-bAPVT_IISNY.

10.6.2. Gig Economy and Freelancing

The gig economy is a radical shift in the way individuals engage with work, particularly as digital platforms proliferate that provide opportunities of flexible work. With the rigidity of traditional career routes weakening, particularly in developing nations like Bangladesh, the majority of graduates are now turning to gig work as an alternative source of income. Research indicates that 56% of the graduates in Bangladesh participate in gig economy work, with many others using such employment as a stepping stone to acquiring jobs. This movement is one of a wider trend of individuals searching for alternate sources of income, and thus it is vital to see professional courses from another angle in the light of changing economic paradigms. Besides, gig economy is not merely a part-time job for the majority; it is an essential means of gaining experiential skills, as 61% of the sample think it is helpful for future work activities (Mani et al., 2024). In the umbrella of IT freelancing, gig economy economics is more evident. Collaborative yet competitive online freelance work demands high competence because technology and market requirements keep changing. In a study on IT freelancers, it becomes evident how career path for IT freelancers is affected by progression and degradation mechanisms of digital labor platforms. The exploratory research highlights that the career paths of IT freelancers are not stable; they involve the management of various options for leaving as well as establishing networks to facilitate skill sharing and project leading. These findings highlight the importance of researching not only how individuals find work but also how labor on these platforms influences their career paths and professional identities (Gussek et al., 2022). With the growth of the gig economy, there has also come salient issues about the rights of workers, social security, and institutional recognition, particularly for those who find themselves caught up in the new labor arrangement. Despite all of its promise, the uncertainty of gig work carries with it several drawbacks, including limited social protections that are able to leave workers vulnerable during economic downturns. This vulnerability is particularly relevant to gig workers in developing economies where safety nets may be shallow to start. Addressing these challenges is key to forging a more resilient gig economy. Subsequent research needs to think through how to make gig workers a constitutive part of existing labor arrangements and think through policy interventions that could work to support this increasingly significant segment of the labor force. Such work will not only add to the academic discourse but also be critical to shaping the future of work (Mani et al., 2024).

10.6.3. Automation and Job Displacement

The unprecedented advancement in automation installs a challenging dynamic between technological growth and the stability of the labor market, bringing significant concerns about the displacement of jobs in various sectors. With more industries embracing advanced technologies, tasks previously performed by humans are being automated, and this is likely to create job losses across various occupations. The trend will likely widen economic disparities, especially among less skilled workers who are likely to be displaced. This job polarization phenomenon is being observed as higher-skilled jobs that utilize advanced technologies are thriving, but middle-skilled jobs are on the decline. Therefore, understanding the automation dynamics and their impact on employment trends is critical in crafting useful workforce strategies that can counteract job displacement negative impacts. Demographic dynamics such as population aging complicate further the automation-job displacement dynamics. As highlighted by Kaniovski et al. (2019), the following decades are likely to witness drastic changes in the magnitude and nature of labor due to demographic factors. The working age will alter as current older workers retire and a perceived shortage of younger skilled workers join the workforce. This population problem may incentivize firms to further enhance their reliance on automation to maintain the level of labor productivity, further exacerbating job replacement problems. With businesses changing in concert with the evolving population landscape, policymakers are needed to respond to workforce re-skilling and training initiatives strategically. Programmes crafted strategically can help displaced workers make room into new occupations created by automation, with a focus towards labor market resilience. In responding to the scenario posed by automation concerning job loss, attention must also be devoted to long-term economic impacts of the trend. It is crucial for policymakers and economists to have a well-rounded appreciation of directed technological change, as can be seen in Kaniovski et al. (2019). Aging populations can lead to a redirection of investments in automation technology, restructuring labor demand across industries. Beyond this, there is a need to appreciate how this changing technological environment interacts with socioeconomic realities like accessibility and inclusion in the labor market. Programs designed to fill the skills gap and safeguard vulnerable populations can bring about a more balanced labor market amidst the disruption unleashed by automation. Emerging research needs to examine new paradigms of policy intervention so that society can address the challenges of workforce displacement while maximizing the prospects unleashed by technology transformation (Figure 10.14).

Figure 10.14. Automation and job displacement.

Source: https://www.diyesinternational.edu.in/wp-content/uploads/2024/08/unnamed-10.webp.

SUMMARY

In extracting significant Summary from the current research, it is very important in informing future directions and trends in research. A systematic mapping study by Kuhrmann et al. maps the large Software Process Improvement (SPI) literature base, illustrating many of the current frameworks and practices available. However, the research also indicates some significant gaps, such as a deficiency in theory models and empirical studies specific to various organizational contexts, particularly for small and medium-sized businesses. The implication here is that researchers need to develop flexible frameworks that are responsive to various business environments in order to ensure SPI methodologies continue to be effective and applicable. The ongoing pursuit of frameworks and their application will not only enhance organizational performance but also shape academic theory on best practice in software development. In the context of medical research, specifically tongue squamous cell carcinoma (TSCC), recent bibliometric analyses unveil valuable information with regard to trends and future research directions. The intricate breakdown of TSCC articles from 2001–2021, which revealed a staggering surge in publication rates, reveals that although traditional subject matter such as e-cadherin and metastasis have been the stalwart, new themes such as biomarkers and depth of invasion are gaining popularity. The seeming shift in orientation creates an opening for researchers to redirect their research into new areas that could potentially have a major effect on treatment algorithms and prognosis of patients. While clinical inquiry still dominates, a collaborative approach towards an investigation of such emergent areas will be the driving force to extend the cognition and care of TSCC. Finally, integration across study areas highlights the importance of synergy between disciplines in shaping agendas for the future. Sharing both SPI environment and emergent themes regarding TSCC demonstrates how knowledge areas with high degrees of interdependence can reveal new directions of investigation. Focusing on the discrepancies observed in research trend and newly arising problems makes a complete knowledge imperative that is vital for practitioners and researchers alike. By discovering still unknown areas, such as agile approaches utilization in SPI or effects of invasion depth in TSCC, upcoming research can be predictive, and hence it may change practices in software development and clinical care both. With careful and informed efforts aimed at plugging these holes, both the fields of oncology and software engineering can benefit significantly from more knowledge and innovative solutions.

Research into emerging trends in supply chain management has become increasingly significant, as one would notice from the discussions at the Fourth Annual World Class Supply Chain Summit. The conference put into context important issues about the ever-evolving climate that supply chain professionals must operate within. The summit underscored how firms should retain old school supply chain fundamentals such as having solid partners and robust IT systems. But the imperative of having agility in response to accelerating market change from customers and competitors was underscored as the major addition to these fundamentals. With more advanced supply chains, the disorientation expressed by participants in the industry is an indication of a need to adapt to a changing environment with high-tech developments and global uncertainties (Haughton et al., 2019). Another important topic of debate at the summit was how supply chain strategies were required to react to existing threats, including environmental challenges and geopolitics. These are the types of factors that make it a situation where existing assumptions require to be updated. For instance, the role of supply chain analytics has grown, giving information that can support anticipatory decision-making in times of uncertainty. The presence of experienced industry leaders, economists, and academic thinkers at the summit allowed for an all-embracing discourse, facilitating diverse opinions to be shared. This cross-disciplinary endeavor not just explained the complexities supply chains are facing today but also projected that an overall model within stakeholders will be crucial in resolving future challenges (Haughton et al., 2019). Moreover, inequality and poverty analysis in the past decades highlighted substantive new directions for research, setting them at the forefront for future research. Focus on these socio-economic issues has expanded to include multi-dimensional approaches, adding depth to the richness of poverty beyond income-only measures. Longitudinal research has become a fashionable option, providing richer data on inequality dynamics and informing policy interventions. Further, increased data availability as well as improved analysis methods have paved the way for rich modeling of these phenomena. Future studies need to continue investigating these trends while also seeking out new approaches for addressing structural inequities in divergent settings, underscoring the value of empirical research in guiding wise policy interventions.

REFERENCES

1. Acikalin, U., Goulding, J. S., Isikdag, U., et al. (2009). Construction informatics in Turkey: Strategic role of ICT and future research directions. https://core.ac.uk/download/1666747.pdf (accessed on 2nd October 2025).
2. Adams, J., Broom, A., Sibbritt, D., et al. (2015). The outcomes of complementary and alternative medicine use among pregnant and birthing women: Current trends and future directions. https://opus.lib.uts.edu.au/bitstream/10453/66265/1/The%20outcomes%20of%20complementary%20and%20alternative%20medicine%20use%20among%20pregnant%20and%20birthing%20women%20current%20trends%20and%20future%20directions...pdf (accessed on 2nd October 2025).
3. Adekunle, T., Eise, J., Eise, L., et al. (2020). More inclusive, more practical: Climate change communication research to serve the future. https://core.ac.uk/download/361555893.pdf (accessed on 2nd October 2025).
4. Adger, W. N., Ashford, N. A., Badyaev, A. V., Bahn, V., Baird, R. A., Barton, J. H., Baum, J. K., et al. (2016). The human–nature relationship and its impact on health: A critical review. https://core.ac.uk/download/163076696.pdf (accessed on 2nd October 2025).
5. Alexandratos, N., An, R., Anderson, M., Antonelli, M., Ashfield-Watt, P., Barling, D., Barnes, A., et al. (2019). Assessing the role of CAP for more sustainable and healthier food systems in Europe: A literature review. https://core.ac.uk/download/228156155.pdf (accessed on 2nd October 2025).
6. Allison, B. Z., Dunne, S., Leeb, R., et al. (2012). Recent and upcoming BCI progress: Overview, analysis, and recommendations. https://core.ac.uk/download/pdf/11484486.pdf (accessed on 2nd October 2025).
7. Androulidakis, A., Bracale, U., Konnis, G., et al. (2011). A feasibility study for the provision of electronic healthcare tools and services in areas of Greece, Cyprus and Italy. https://core.ac.uk/download/16664902.pdf (accessed on 2nd October 2025).
8. Aragon, N., Marian, M., Profito, F. J., et al. (2021). The use of artificial intelligence in tribology—A perspective. https://core.ac.uk/download/618224771.pdf (accessed on 2nd October 2025).
9. Bayala, J., Chabi, A., & Dembele, C. (2020). *Partnerships for Scaling Climate-Smart Agriculture (P4S) Phase II: 2019 Annual Report*. https://core.ac.uk/download/288633440.pdf (accessed on 2nd October 2025).

10. Borghi, C., Grassi, G., Rosei, E. A., et al. (2017). Personalized medicine—A modern approach for the diagnosis and management of hypertension. https://core.ac.uk/download/96884785.pdf (accessed on 2nd October 2025).
11. Bulatovic, J., & Rajovic, G. (2015). Some geographical aspects of rural development with view of Montenegro: A review. https://core.ac.uk/download/29426707.pdf (accessed on 2nd October 2025).
12. Buyya, R., & Varghese, B. (2017). Next generation cloud computing: New trends and research directions. https://core.ac.uk/download/96664300.pdf (accessed on 2nd October 2025).
13. Chatterjee, K. (2006). An evaluation study of the core courses of the master's degree in training and development with alignment of globalization. https://core.ac.uk/download/5066837.pdf (accessed on 2nd October 2025).
14. Chen, G., & Mondal, T. G. (2022). Artificial intelligence in civil infrastructure health monitoring—Historical perspectives, current trends, and future visions. https://core.ac.uk/download/555320460.pdf (accessed on 2nd October 2025).
15. Chesworth, L., Davis, G., Georgeson, J., et al. (2017). *BERA-TACTYC Early Childhood Research Review 2003–2017*. https://core.ac.uk/download/84342373.pdf (accessed on 2nd October 2025).
16. Chiluiza, K., Ortiz Rojas, M. E., & Valcke, M. (2016). Gamification in higher education and STEM: A systematic review of literature. https://core.ac.uk/download/147027732.pdf (accessed on 2nd October 2025).
17. Clapp, R., Dawit, S., Eagle, N., et al. (2010). Africa 2060: Good news from Africa, April 16, 2010. https://open.bu.edu/bitstream/2144/22891/1/Conf_GoodNewsforAfrica.pdf (accessed on 2nd October 2025).
18. Dewan, P., & Steeleworthy, M. (2013). Incorporating online instruction in academic libraries: Getting ahead of the curve. https://core.ac.uk/download/143689618.pdf (accessed on 2nd October 2025).
19. Diebold, P., Kuhrmann, M., & Münch, J. (2016). Software process improvement: A systematic mapping study on the state of the art. https://core.ac.uk/download/157586901.pdf (accessed on 2nd October 2025).
20. Ding, Y., Huang, Y., Liang, T., et al. (2024). A bibliometric and visual analysis of tongue squamous cell carcinoma: Current knowledge structure and future research trends. https://core.ac.uk/download/642936162.pdf (accessed on 2nd October 2025).
21. Engelbrecht, H. J. (2008). The (un)happiness of knowledge and the

knowledge of (un)happiness: Happiness research and policies for knowledge-based economies. https://core.ac.uk/download/148642160.pdf (accessed on 2nd October 2025).

22. Farris, M. T., & Gravier, M. (2008). An analysis of logistics pedagogical literature: Past and future trends in curriculum, content, and pedagogy. https://core.ac.uk/download/56737709.pdf (accessed on 2nd October 2025).
23. Fortina, P., Kricka, L. J., Park, J. Y., & Polsky, T. G. (2015). The future of laboratory medicine–A 2014 perspective. https://core.ac.uk/download/54509579.pdf (accessed on 2nd October 2025).
24. Freudenberg, N., Galea, S., & Vlahov, D. (2005). Cities and population health. https://core.ac.uk/download/323072008.pdf (accessed on 2nd October 2025).
25. Fuller, T., Warren, L., & Welter, F. (2008). An emergence perspective on entrepreneurship: Processes, structure and methodology. https://core.ac.uk/download/54047.pdf (accessed on 2nd October 2025).
26. Gaston, M. (1994). Barriers to the employment and work-place advancement of Latinos. https://core.ac.uk/download/5124663.pdf (accessed on 2nd October 2025).
27. Ghadimi, P., Lim, M., & Wang, C. (2019). Sustainable supply chain modeling and analysis: Past debate, present problems and future challenges. https://core.ac.uk/download/228155519.pdf (accessed on 2nd October 2025).
28. Gopal, S., & Najam, A. (2012). Connecting the dots: Information visualization and text analysis of the Searchlight Project newsletters. https://open.bu.edu/bitstream/2144/22895/1/Searchlight-Pardee-Report.pdf (accessed on 2nd October 2025).
29. Gussek, L., & Wiesche, M. (2022). Understanding the careers of IT freelancers on digital labor platforms. https://core.ac.uk/download/511301894.pdf (accessed on 2nd October 2025).
30. Hall, P. J., Kelly, N., Rennie, A. J. R., et al. (2013). A daily representation of Great Britain's energy vectors: Natural gas, electricity and transport fuels. https://core.ac.uk/download/20444143.pdf (accessed on 2nd October 2025).
31. Hassan, M., Jacob, S. S., & Yacoub, M. H. (2015). Utility of mass spectrometry for the diagnosis of the unstable coronary plaque. https://core.ac.uk/download/77008832.pdf (accessed on 2nd October 2025).
32. Haughton, M. (2019). *World Class Supply Chain 2019: Next Generation*

Ideas. https://core.ac.uk/download/225556480.pdf (accessed on 2nd October 2025).

33. Heng, S. (2025). Rising stars in information and communication technology. https://core.ac.uk/download/pdf/9309277.pdf (accessed on 2nd October 2025).
34. Huang, Y. W., Kammen, D. M., & Kittner, N. (2019). ASEAN grid flexibility: Preparedness for grid integration of renewable energy. https://core.ac.uk/download/323310932.pdf (accessed on 2nd October 2025).
35. Jang, C. S., Lim, D. H., Yim, J., et al. (2019). An analysis of research trends in brain-based learning in adult education and HRD fields: The content analysis and network text analysis. https://core.ac.uk/download/267197322.pdf (accessed on 2nd October 2025).
36. Jenkins, S., & Micklewright, J. (2007). New directions in the analysis of inequality and poverty.
37. Jordan, K. (2016). Academics' online connections: Characterizing the structure of personal networks on academic social networking sites and Twitter. https://core.ac.uk/download/42537389.pdf (accessed on 2nd October 2025).
38. Kaniovski, S., & Url, T. (2019). *Macroeconomic Consequences of Ageing and Directed Technological Change. Bertelsmann Stiftung Studies*. https://core.ac.uk/download/322353888.pdf (accessed on 2nd October 2025).
39. Kechris, A. S., Pillay, A., Richard, A., Shore, R. A., & Buss, S. R. (2001). The prospects for mathematical logic in the twenty-first century. https://core.ac.uk/download/12817608.pdf (accessed on 2nd October 2025).
40. Kidd, J. R. (1983). *Learning and Libraries: Competencies for Full Participation*. https://core.ac.uk/download/4816616.pdf (accessed on 2nd October 2025).
41. Kittel, B. (2025). EMU, EU enlargement, and the European social model: Trends, challenges, and questions. https://core.ac.uk/download/pdf/6543628.pdf (accessed on 2nd October 2025).
42. Kohont, A., Skoumpopoulou, D., & Stalker, B. (2019). Talent management in European SMEs: Case analysis between Slovenia and Poland. https://core.ac.uk/download/288430588.pdf (accessed on 2nd October 2025).
43. LeBlanc, N. (2020). An analysis and enumeration of the blockchain and future implications. https://core.ac.uk/download/322850883.pdf (accessed on 2nd October 2025).
44. Mani, L., & Mimi, A. (2024). Gravitating the gig economy for reshaping the careers using technological platform in the digital age in an emerging

economy. https://core.ac.uk/download/655274738.pdf (accessed on 2nd October 2025).

45. Marchesi, M., Orrú, M., Pinna, A., et al. (2017). A Petri nets model for blockchain analysis. http://arxiv.org/abs/1709.07790 (accessed on 2nd October 2025).
46. Misuraca, G., & Stewart, J. (2013). The industry and policy context for digital games for empowerment and inclusion: Market analysis, future prospects and key challenges in videogames, serious games and gamification. https://core.ac.uk/download/28977794.pdf (accessed on 2nd October 2025).
47. Morris, J. A., & Whitfield, P. E. (2009). Biology, ecology, control and management of the invasive *Indo-Pacific Lionfish*: An updated integrated assessment. https://core.ac.uk/download/11018553.pdf (accessed on 2nd October 2025).
48. N/A. (2007). *CAHRS hrSpectrum (May–June 2008).* https://core.ac.uk/download/5128583.pdf (accessed on 2nd October 2025).
49. N/A. (2008). *Mapping New Directions: Higher Education for Older Adults.* https://core.ac.uk/download/71352944.pdf (accessed on 2nd October 2025).
50. N/A. (2013). *PICES Press, Vol. 21, No. 2, Summer 2013.* https://core.ac.uk/download/17201005.pdf (accessed on 2nd October 2025).
51. N/A. (2017). *ILR Faculty Research in Progress, 2016–2017.* https://core.ac.uk/download/127599621.pdf (accessed on 2nd October 2025).
52. Reimer, N., & Shelton, A. (2018). Telehealth wound applications: Barriers, solutions, and future use by nurse practitioners. https://core.ac.uk/download/228066223.pdf (accessed on 2nd October 2025).
53. Sarewitz, D., Guston, D. H., van Merkerk, R., & Smits, R. (2025). The role of TA in systemic innovation policy. https://core.ac.uk/download/pdf/7372173.pdf (accessed on 2nd October 2025).
54. Suresh Kumar, B., & Priyanka, R. (2025). Empirical evaluation in data mining in association with artificial intelligence and machine learning. https://core.ac.uk/download/661429514.pdf (accessed on 2nd October 2025).
55. Turner, L., & Windmuller, J. P. (1998). *Convergence and Diversity in International and Comparative Industrial Relations.* https://core.ac.uk/download/19958957.pdf (accessed on 2nd October 2025).

INDEX

E

F

G

H

I

J

L

M

N

O

P

T

V

W